Landfilling of Waste:
Leachate

Landfilling of Waste: Leachate

Edited by

T. H. CHRISTENSEN
Department of Environmental Engineering
Technical University of Denmark
Lyngby, Denmark

R. COSSU
Institute of Hydraulics
University of Cagliari
Cagliari, Sardinia, Italy

R. STEGMANN
Institute of Waste Management
Technical University of Hamburg-Harburg
Hamburg, Germany

CRC Press
Taylor & Francis Group
Boca Raton London New York

CRC Press is an imprint of the
Taylor & Francis Group, an **informa** business

A TAYLOR & FRANCIS BOOK

CRC Press
Taylor & Francis Group
6000 Broken Sound Parkway NW, Suite 300
Boca Raton, FL 33487-2742

First issued in paperback 2019

© 1992 by Taylor & Francis Group, LLC
CRC Press is an imprint of Taylor & Francis Group, an Informa business

ISBN-13: 978-0-419-16140-0 (hbk)
ISBN-13: 978-0-367-86405-7 (pbk)

A Catalogue record for this book is available from the British Library

Library of Congress Cataloging-in-Publication Data 91-35776 CIP

Visit the Taylor & Francis Web site at
http://www.taylorandfrancis.com

and the CRC Press Web site at
http://www.crcpress.com

Preface

During the last couple of decades landfilling of waste has developed dramatically and today in some countries involves fully engineered facilities subject to extensive environmental regulations. Although much information and experience in landfill design and operation has been obtained in recent years, only a few landfills meet the current standards for an environmentally acceptable landfill. In view of the increasing public awareness and new scientific understanding of waste disposal problems, the standards for landfilling of waste may improve even further within the years to come.

In view of this great demand for information on landfills we decided to establish, as Editors, a series of international reference books on landfilling of waste. No worldwide, long tradition of publishing information on landfilling exists and this book series is seen as an attempt to establish a common platform representing the state-of-the-art, useful for implementing improvements at actual landfill sites and for identification of new directions in landfill research.

This first book deals with leachate, the strongly contaminated wastewater developing in landfills. Aspects such as landfill hydrology, leachate characterization and composition, factors controlling leachate quality, principles of leachate treatment, and effects of leachate in groundwater are discussed in great detail. The introductory chapter presents an overview of the main aspects covered in the book. Some of the contributions in the book may express different opinions and strategies, allowing the reader to develop his own balanced views. Leachate collection, including liner and drainage systems, is not described in this book, but is the topic of a following book.

This book on leachate consists of edited, selected contributions to the International Symposia on Sanitary Landfills held in Sardinia (Italy) every second year and of chapters specially written for this book. The responsibility for the technical content of the book primarily rests with the individual authors and we cannot take any credit for their work. Therefore reference should be made directly to the authors of chapters. Our role as Editors has been one of reviewing and homogenizing the chapters and of making constructive suggestions during the preparation of the final manuscripts.

We would like to thank all our contributors for allowing us to edit their manuscripts and ask for everybody's forbearance with our unintended mistreatment of the English language. For this reason we would like to give credit to Elsevier Science Publishers for correcting the English where necessary and for giving shape to the book.

Finally we would like to thank Ms Anne Farmer at CISA, Cagliari, for her patient indispensable work on the many drafts of the book.

Thomas H. Christensen
Raffaello Cossu
Rainer Stegmann

Contents

3   LEACHATE TREATMENT

Contents

4 ENVIRONMENTAL ASPECTS OF LEACHATE

4 ENVIRONMENTAL ASPECTS OF LEACHATE

List of Contributors

L. Ahnert
Universität Wuppertal, Abfall- und Siedlungswasserwirtschaft, Paulus-kirchstrasse 7, D-5600 Wuppertal, Germany

H. Albers
Ingenieurbüro Dr Born – Dr Ermel, Finienweg 7, D-2807 Achim-Baden, Germany

G. Andreottola
CISA, Environmental Sanitary Engineering Centre, Via Marengo 34, I-09123 Cagliari, Italy

F. Avezzù
Department of Environmental Sciences, University of Venice, Calle Larga Santa Marta, Venice, Italy

P. Baccini
Swiss Federal Institute of Water Resources and Water Pollution Control, CH-8600 Dübendorf, Switzerland

J. M. Ball
University of the Witwatersrand, P O Wits, 2050 I, Jan Smuts Avenue, Johannesburg, South Africa

C. Barber
CSIRO, Groundwater Research, Wembley, WA 6014, Australia

H. Belevi
*Swiss Federal Institute of Water Resources and Water Pollution Control,
CH-8600 Dübendorf, Switzerland*

G. Bissolotti
SIAD SpA, Via San Bernardino 92, I-24100 Bergamo, Italy

N. C. Blakey
*Water Research Centre plc, Medmenham, PO Box 16, Marlow, Bucks,
UK, SL7 2HD*

G. E. Blight
*University of the Witwatersrand, PO Wits, 2050, 1 Jan Smuts Avenue,
Johannesburg, South Africa*

P. Cannas
*CISA, Environmental Sanitary Engineering Centre, Via Marengo 34,
I-09123 Cagliari, Italy*

T. H. Christensen
*Department of Environmental Engineering, Technical University of
Denmark, Building 115, DK-2800 Lyngby, Denmark*

C. Collivignarelli
*Department of Civil Engineering, University of Brescia, I-25060 Brescia,
Italy*

R. Cossu
*Institute of Hydraulics, University of Cagliari, Piazza d'Armi, I-09123
Cagliari, Italy*

P. Courant
*France Dechets, F.D. Conseil, 71 rue Henri Brettonet, F-78970
Mezieres-sur-Seine, France*

H. Doedens
*Institut für Siedlungswasserwirtschaft und Abfalltechnik, Universität
Hannover, D-3000 Hannover, Germany*

H.-J. Ehrig
*Universität Wuppertal, Abfall- und Siedlungswasserwirtschaft, Paulus-
kirchstrasse 7, D-5600 Wuppertal, Germany*

M. Ettala
*Department of Environmental Sciences, University of Kuopio, PO Box
1627, SF-70211, Kuopio, Finland*

A. Foverskov
Department of Environmental Engineering, Technical University of Denmark, Building 115, DK-2800 Lyngby, Denmark

R. Gillham
Waterloo Centre for Groundwater Research, University of Waterloo, Waterloo, Ontario, Canada N2L 3G1

K. Hasselgren
Department of Civil Works, S-268 80 Svalöv, Sweden

D. J. Hojem
University of the Witwatersrand, PO Wits, 2050, 1 Jan Smuts Avenue, Johannesburg, South Africa

F. Holz
Haase Energietechnik GmbH, Gadelander Strasse 172, D-2350 Neumünster, Germany

A. Jaap
Euroconsult, PO Box 441, NL-6800 AK Arnhem, The Netherlands

J. M. Jans
Euroconsult, PO Box 441, NL-6800 AK Arnhem, The Netherlands

P. Kjeldsen
Department of Environmental Engineering, Technical University of Denmark, Building 115, DK-2800 Lyngby, Denmark

K. Knox
Knox Associates, 21 Ravensdale Drive, Wollaton, Nottingham, UK, NG8 2SL

P. Kristensen
Water Quality Institute, 11 Agern Allé, DK-2970 Hørsholm, Denmark

G. Krückeberg
Abfallentsorgungsbetrieb des Kreises Minden–Lubbecke, Pohlsche Heide, D-4955 Hille 1, Germany

J. La Cour Jansen
Water Quality Institute, Agern Allé 11, DK-2970 Hørsholm, Denmark

T. Larsen
Department of Environmental Engineering, Technical University of Denmark, Building 115, DK-2800 Lyngby, Denmark

R. G. W. Laughlin
Ortech International, 2395 Speakman Drive, Mississauga, Ontario, Canada L5K 1B3

U. Lund
Water Quality Institute, Agern Allé 11, DK-2970 Hørsholm, Denmark

J. Lyngkilde
Department of Environmental Engineering, Technical University of Denmark, Building 115, DK-2800 Lyngby, Denmark

P. J. Maris
Monitor Environmental Services, 3 Dewpond Close, Stevenage, Herts, UK, SG1 3BL

N. Millot
Labo Services, Route de la Centrale, F-69700, Givors, France

F. E. Mosey
Water Research Centre plc, Medmenham, PO Box 16, Marlow, Bucks, UK, SL7 2HD

A. Muntoni
CISA, Environmental Sanitary Engineering Centre, Via Marengo 34, 09123 Cagliari, Italy

K. Nilsson
VBB VIAK, Division for Water & Environment, Geiersgaten 8, S-216 18 Malmö, Sweden

S. O'Hannesin
Waterloo Centre for Groundwater Research, University of Waterloo, Waterloo, Ontario, Canada N2L 3G1

P. Østfeldt
Water Quality Institute, Agern Allé 11, DK-2970 Hørsholm, Denmark

F. G. Pohland
Department of Civil Engineering, 949 Benedum Hall, University of Pittsburgh, Pittsburgh, PA 15261, USA

L. Rasmussen
Water Quality Institute, Agern Allé 11, DK-2970 Hørsholm, Denmark

H. Robinson
Aspinwall & Company Ltd, Walford Manor, Baschurch, Shrewsbury, UK, SY4 2HH

H. Segato
Water Quality Institute, Agern Allé 11, DK-2970 Hørsholm, Denmark

R. Serra
CISA, Environmental Sanitary Engineering Centre, Via Marengo 34, 09123 Cagliari, Italy

B. Skov
Department of Environmental Engineering, Technical University of Denmark, Building 115, DK-2800 Lyngby, Denmark

R. Stegmann
Institute of Waste Management, Technical University of Hamburg-Harburg, Harburger Schlossstrasse 37, D-2100 Hamburg 90, Germany

U. Theilen
Institut für Siedlungswasserwirtschaft und Abfalltechnik, Universität Hannover, D-3000 Hannover, Germany

P. J. Top
Ortech International, 2395 Speakman Drive, Mississauga, Ontario, Canada L5K 1B3

A. Van der Schroeff
Euroconsult, PO Box 441, NL-6800 AK Arnhem, The Netherlands

V. Vanek
VBB VIAK, Division for Water & Environment, Geiersgaten 8, S-216 18 Malmö, Sweden

G. P. Vicevic
Ortech International, 2395 Speakman Drive, Mississauga, Ontario, Canada L5K 1B3

B. Weber
Haase Energietechnik GmbH, Gadelander Strasse 172, D-2350 Neumünster, Germany

1. LANDFILL LEACHATE: AN INTRODUCTION

1 Landfill Leachate: An Introduction

THOMAS H. CHRISTENSEN

Department of Environmental Engineering, Technical University of Denmark, Building 115, DK-2800 Lyngby, Denmark

RAFFAELLO COSSU

Institute of Hydraulics, University of Cagliari, Piazza d'Armi, I-09123 Cagliari, Italy

&

RAINER STEGMANN

Institute of Waste Management, Technical University of Hamburg-Harburg, Harburger Schlossstrasse 37, D-2100 Hamburg 90, Germany

INTRODUCTION

The sanitary landfill plays a most important role in the framework of solid waste disposal and will remain an integral part of the new strategies based on integrated solid waste management.

The quality of landfill design, according to technical, social and economic development, has improved dramatically in recent years. Design concepts are mainly devoted towards ensuring minimal environmental impact in accordance with observations made concerning the operation of old landfills.

The major environmental concern associated with landfills is related to discharge of leachate into the environment and the current landfill technology is primarily determined by the need to prevent and control leachate problems.

LEACHATE PROBLEMS IN LANDFILLS

The most typical detrimental effect of leachate discharge into the environment is that of groundwater pollution. To prevent this, the first step in landfill design development was to site the landfill far from the groundwater table and/or far from groundwater abstraction wells. Thus more attention was focused on studying the hydrogeology of the area in order to identify the best siting of the landfill.

A further step in landfill technology was to site the landfill in low permeability soil or to engineer impermeable liners to contain wastes and leachate. Containment, however, poses the problem of leachate treatment. Nowadays the leachate control strategies involve not only landfill engineering but the concept of waste management itself.

Leachate pollution is the result of a mass transfer process. Waste entering the landfill reactor undergoes biological, chemical and physical transformations which are controlled, among other influencing factors, by water input fluxes. In the reactor three physical phases are present: the solid phase (waste), the liquid phase (leachate) and the gas phase. The liquid phase is enriched by solubilized or suspended organic matter and inorganic ions from the solid phase. In the gas phase mainly carbon (prevalently in the form of CO_2 and CH_4) is present.

Discharge of leachate into the environment is nowadays considered under more restrictive views. The reasons for this are:

• Many severe cases of groundwater pollution at landfills.
• The greater hazard posed by the size of landfill which is larger than in the past.
• The need to comply with more and more restrictive legislation regarding quality standards for wastewater discharges.
• With integrated waste management strategy the volume of refuse will be reduced but more hazardous waste may need to be landfilled, e.g. combustion residues, hazardous components consequent to separate collection, etc.
• More and more often landfills are located on the ground or on a slope and in both cases accumulation of leachate may be a negative factor with respect to geotechnical stability.

The leachate problem accompanies landfill from its beginning to many decades after closure. This means that leachate management facilities should also last and their effectiveness be ensured over a long period of time—so far, this still remains to be proven.

LEACHATE CONTROL STRATEGIES

In view of all these reasons leachate control strategies involve the input (waste and water), the reactor (landfill) and the output (leachate and gas).

Control of Waste Input

The first step in the waste input control strategy should be that of reducing to a minimum the amount of waste to be landfilled. This could be obtained by separate collection activities, recycling centres, incineration and composting. Separation of the hazardous fraction of municipal waste such as batteries, expired medicines, paint, mercury lamps, pesticides, etc., would reduce leachate concentrations of heavy metals, halogenated hydrocarbons and other toxic compounds.

Another step is that of reducing waste to a non-leachability level. This could be achieved for MSW by incineration followed by fixation of the solid residues.

Pretreatment could also aim to reduce the biodegradability of waste to be landfilled. This would reduce or even eliminate the need for process water in the biostabilization. One way of reaching this aim is to pretreat waste by mechanically sorting organic matter and paper. This material could then be either composted or anaerobically digested.

Control of Water Input

The strategy for water input control is strictly related to the quality of waste to be landfilled. In the case of non-biodegradable waste, according to its hazardous potential for the environment, prevention of water infiltration can be adopted as the main option (normally by means of top sealing). On the contrary, in the case of biodegradable waste, a water input must be assured until a reasonable degree of biostabilization is achieved. In this case the water input should be limited to the strictly necessary amount and minimization techniques should be applied. The most important parameters in this regard are:

- siting of landfill in low precipitation areas, if possible;
- usage of cover and topsoil systems suitable for vital vegetation and biomass production;

- vegetation of the topsoil with species which optimize the evapotranspiration effect;
- surface lining in critical hydrological conditions;
- limitation on sludge disposal;
- surface water drainage and diversion;
- high compaction of the refuse in place;
- measures to prevent risks of cracking owing to differential settlement.

Furthermore, utilization of mobile roofs on the front of the waste deposit could represent a further useful minimization technique.

Control of Landfill Reactor

The main option in controlling leachate quality through controlling the landfill reactor is the enhancement of the biochemical processes (when biodegradable wastes are deposited).

One of the main objectives is to convert and transport as much carbon as possible from the solid phase into the gas phase rather than into the liquid phase. This is achieved by accelerating the methane generation step.

Control of Leachate Discharge into the Environment

As mentioned earlier, this is the parameter traditionally controlled and nowadays the regulations are more restrictive. The following tools are adopted.

Lining. The lining system should be based on the multi-barrier effect (double or triple liners). Quality of material and construction methods should be improved to ensure higher safety and durability.

Drainage and collection systems. A rational drainage and collection system is important to avoid emission or accumulation of leachate inside the landfill. Unfortunately the level of engineering and operation of drainage systems appears to be very poor and represents one of the shortcomings in current landfill design. The main problems are proper choice of material, clogging, durability and maintenance. No drainage system currently in use appears to be safe enough or long lasting.

Treatment. Leachate has always been considered a problematic wastewater from the point of view of treatment as it is highly polluted and the quality and quantity are modified with time in the same landfill. Nowadays according to the increasingly restrictive limits for wastewater discharge, complicated and costly treatment facilities are imposed. Normally a combination of different processes is required.

Environmental monitoring. This aspect is of extreme importance for the evaluation of landfill operational efficiency and for the observation of the environmental effects on a long term basis.

ISSUES IN THIS BOOK

The aspects of leachate production and composition (including water balances, characterization methods and factors controlling composition), leachate treatment (individual processes, combination of processes, and recirculation) and environmental issues of leachate (fate of pollutants, monitoring) are described in this book following the introduction given in the paragraphs below, whilst lining and leachate collection are described in a following book.

Leachate Production

In most climates rain and snowfall will cause infiltration of water into the landfilled waste and, after saturation of the waste, generation of leachate. Determination of the amount of water infiltrating a covered landfill cell can be made from the hydrological balance of the top cover, paying attention to precipitation, surface run-off, evapotranspiration and changes in moisture content of the soil cover. However, the balancing must be timewise discretized in order to identify the periods of excessive water leading to infiltration into the waste below.

The water content of the waste being landfilled is usually below saturation (actually field capacity) and will result in absorption of infiltrating water before drainage in terms of leachate is generated. The water absorption capacity of the landfilled waste and its water retention characteristics are very difficult to specify due to the heterogeneity of the waste. Furthermore these characteristics may change over time as the waste density is increasing and the organic fraction which dominates the

water retention is degraded in the landfill. It has often 'been observed that
new landfills produce leachate soon after the waste has been filled in,
although the total water absorption capacity should not be depleted that
quickly. Accounting for these aspects is still very rudimentary and
definitely the less developed component of the hydrological cycle of the
landfill.

Mathematical models for landfill hydrology are currently being
developed (e.g. Chapter 2.1) and may prove to be efficient tools in
designing leachate-minimizing top covers and determining expected
leachate flow rates to be treated before discharge. It should be
emphasized also that landfills in areas with an annual deficit of rainfall
may produce leachate in the wet season as shown in Chapter 2.2.

Besides improvement of top covers to store moisture for the dry
season, establishing short-rotation tree plantations on landfill sections
and irrigation with leachate may also prove—even in relatively cold
climates—an effective means of reducing the leachate generation rate.
Chapter 2.3 describes some extensive Finnish experiments where willow
plantations, even in shallow soil covers, increased the evapotranspiration
by 400 mm per year, supposedly sufficient to reduce leachate generation
substantially in many countries.

Leachate Composition

The characteristics of landfill leachate, as described in Chapter 2.4, are
relatively well known, at least for the first 20 years' life of the landfill, the
period from which actual data are available. On the other hand the
leachate composition of later phases of the landfill is hardly known and
the basis for making good estimates is rather weak. However, there is no
doubt that for some components the leaching will continue for more than
a century.

Leachate from landfills contains a vast number of specific compounds,
in particular specific organic compounds in micro-amounts which may
make it an impossible task analytically to determine all relevant
compounds. As a consequence of this, combined with the uncertainty of
the biological effects of mixtures containing many chemicals, biological
characterization methods are now beginning to be applied to landfill
leachate samples. Ecotoxicological tests (e.g. Micro-tox test and *Daphnia*
test) and mutagenic tests (e.g. Ames test) have been applied as described
in Chapter 2.5. The results are still too few to make general conclusions

on the applicability and reliability of biological characterization methods, but the approach is interesting and further data are expected in the years to come.

Before selection of proper treatment processes for landfill leachate, data on the composition of the leachate in question must be available. Usually chemical data are available making up the basis for treatment design. However, in France a specific procedure, as described in Chapter 2.6, has been applied for characterizing leachate treatability. In addition to traditional chemical analysis this procedure involves gel permeation chromatography. In France this procedure has been applied to the design of several full-scale leachate treatment plants.

Most of our current knowledge on leachate composition originates from landfills that are less engineered and managed than the modern sanitary landfills. New developments identifying the influence of various management procedures on landfill stabilization and hence on leachate composition, as reviewed in Chapter 2.7, show that specific procedures for the landfill operation, and maybe controlled leachate recirculation, may lead to a rapid stabilization of the landfill and to a less polluted leachate. These procedures are expected to gain increasing practical attention in the future.

Co-disposal of hazardous waste in landfills is seen in some countries, based on experiences from old-fashioned landfills and dumps, as malpractice while other countries see co-disposal as an efficient and cheap disposal method for hazardous waste if properly engineered. This means that the type and amount of industrial waste disposed at the landfill must not significantly disturb the degradation processes and negatively influence the quality of gas and leachate. Chapter 2.8 provides a comprehensive report on how the effects of co-disposal can be investigated showing data for both organic and metallic wastes. Our knowledge about engineered co-disposal systems is mostly related to biologically active landfill environments. For compounds that are not degraded but attenuated in the landfilled waste by ion exchange, sorption and precipitation, our knowledge about the long-term effects is still rather scarce.

Characterization of landfill leachate by chemical-analytical methods has during the last 20 years to a large extent developed from standard methods for wastewater characterization. However, leachate differs analytically in many ways from wastewater and the chemical-analytical methods must be adjusted to account for this. No standard methods for leachate analysis exist and much reported data may be based on 'local' adjustments of available analytical procedures. Unfortunately, the methods employed are rarely reported and the transfer of experience is very limited. Chapter 2.9 gives an introduction to various analytical characterization methods emphasizing the special aspects of leachate analysis as compared to wastewater analysis.

Leachate Treatment

Since more and more landfills all over the world will be sealed at their bases by means of mineral and/or artificial liners and leachate will be collected, there is a great need for appropriate leachate treatment facilities. The high concentrations of organic and inorganic constituents in leachate have to be treated due to the requirements in different countries. The tendency is for relatively low effluent concentrations in organics, ammonia, halogenated hydrocarbons, heavy metals, and fish toxicity (see Chapter 3.15). There is still discussion as to whether it makes sense to treat leachate down to low COD-values where at concentrations $\leqq 1000$ mg/litre probably mainly humic- and fulvic-like acids are removed. The potential effects of these components in rivers, lakes, etc., cannot be predicted. Investigations are underway where the organic concentrations of the biologically treated leachate are compared with humic sunstances produced in the natural biological cycle (e.g. humic substances in the forest soils). Hopefully there will be answers in the future.

The removal of nitrogen especially from leachate from the methanogenic phase (low BOD_5- and high COD-concentrations) is still a problem. Nitrification can be achieved when the treatment plants are designed for this purpose. Denitrification can be practised with 'young' leachates; 'old' leachates do not have enough degradable organics to supply denitrifying bacteria with the carbon needed for the reduction of nitrate.

The trace organics have to be looked at in more detail. Also in MSW-landfills halogenated hydrocarbons and other organics like aromatic compounds are landfilled with the daily life products. These components can be found in leachate also and have to be removed by means of special treatment.

Due to the kind of treatment there will still remain a variety of components in the leachate; these will be mainly salts, organic trace components, heavy metals and the above-mentioned humic- and fulvic-like acids.

Leachate treatment plants will consist in the future of more than one step. It may start in the landfill by practising the enhanced biological degradation which results from the early stage of landfilling in low BOD_5-concentrations. The removal of organics and nitrification will mainly take place by means of biological processes while further treatment requires chemical/physical methods.

Leachate treatment will be very costly in the future especially if the required effluent standards of Germany have to be met. There is not much experience in full-scale regarding this new generation of leachate treatment plants. Costs in the range of 30–70 US$ per m^3 are expected that will increase the costs for landfilling substantially. This is especially true when the total time of leachate treatment after the landfill has been closed (50–100 years or more?) is respected.

Concerning the different approaches in the different countries, it has to be kept in mind that legislation, landfill management, operation, etc., may be different. In addition it should be pointed out that there is not only one solution for the treatment of leachate to obtain the required final concentrations. One conclusion from the papers in Part 3 on leachate treatment is the need for leachate minimization procedures.

Biological leachate treatment. About eight papers in Chapter 3 deal with biological treatment procedures for leachate. In general it can be stated that biological leachate treatment is a most favourable procedure that should be used also in those cases when chemical/physical treatment is required. Biological leachate treatment is a relatively low cost process and organics are degraded mainly to CO_2, water, and biomass. All the substances that have been eliminated using biological degradation do not have to be treated by means of high cost chemical/physical procedures.

As already mentioned, with increasing landfill age or by means of enhancement techniques, leachate treatment will mainly focus on the nitrification of ammonia. Biological denitrification can be achieved when an external organic substrate is added to the leachate.

The design criteria for sewage treatment plants cannot be used for leachate. For the design and operation of biological leachate treatment plants specific points have to be respected:

• foam production during certain periods;
• precipitation of constituents that results in clogging of pipes, etc.;
• low leachate temperatures during biological treatment due to long detention times in the reactors;

- low phosphorus concentrations in the leachate;
- low BOD$_5$ and high ammonia concentrations in old leachate;
- halogenated hydrocarbons.

No investigations have been made so far to determine the stripping effects on volatile organics during aeration.

Co-treatment of leachate and sewage is the treatment process used in most cases. It may operate well if the relation between the amount cf leachate and sewage is acceptable (see Chapter 3.14). The final comparatively high COD-concentrations from the leachate will increase the total COD-effluent concentration of the sewage/leachate mixture according to its initial ratio. The AOX-concentrations (chlorinated hydrocarbons) are in general not degraded and may be stripped, adsorbed to the sludge or only diluted.

Chemical–physical leachate treatment. Chemical–physical treatment mainly results in the separation and concentration of pollutants from the leachate. The concentrate has to be incinerated, landfilled, or further treated. So it is not a 'real' treatment process compared to biological methods. Other chemical processes such as wet-oxidation as well as ozone, UV, and H$_2$O$_2$ treatment may also result in a conversion of organics to mainly CO$_2$ and water. In Germany at present ozone- and UV-oxidation processes are being tested at semi-technical scale.

The oxidation processes are expected to improve the biological treatment of the organics and/or to totally oxidize them. By means of reverse osmosis, organic and inorganic components of the leachate are accumulated in the concentrate due to their chemical/physical characteristics. Evaporation or incineration of leachate also remove organics and inorganics from the liquid phase. Precipitation using organic and/or inorganic flocculants mainly results in a removal of organics and an increase of the salt content when inorganic flocculants are used. Activated carbon mainly removes organics from the water and gas phase.

When considering chemical/physical treatment the whole process including energy requirement, gas treatment, residue-removal, quality of the treated leachate, stability and efficiency of the process as well as costs have to be respected.

Leachate recirculation. No detailed research on the effect of leachate recirculation on the water budget of a landfill has been performed so far. Leachate recirculation has been practised in the past in many cases, with the aim of totally solving the leachate problem. Experiences show that under middle European climatic conditions this is not possible and that on the contrary a build-up of leachate in the landfill has been observed. In some cases water migrated over the edge of the landfill pit.

Leachate recirculation may be practised only with biologically treated leachate in a controlled way. The main aim of leachate recirculation is the maximization of evaporation. So leachate recirculation should be practised dependent upon the actual evaporation rate over the year; since this procedure is somewhat theoretical the rate of leachate recirculated with time may be related to the average evaporation rates and the water retention potential in the upper 10 cm of waste and/or cover-soil. A certain amount of leachate may penetrate into the landfill, which in general will not cause adverse effects.

Environmental Aspects of Leachate

The main environmental aspects of landfill leachate are the impacts on surface water quality and groundwater quality if leachate is discharging into these water bodies. In addition nuisance may arise from leachate handling and treatment, e.g. in terms of malodours and aerosols, but these aspects are only temporary and very local.

The effects of leachate on surface waters, e.g. streams and creeks, and groundwater have been recognized for many years. Many cases of leachate impact on surface water and groundwater quality may be linked to improper or insufficient landfill technology of the past, e.g. with respect to either lining or drainage collection. These cases demonstrate that leachate control is a compulsory element of the modern sanitary landfill. In fact the need to control leachate in order to avoid uncontrolled discharges into the environment is bearing the greatest influence on landfill siting, design, operation, maintenance and costs.

An efficient and well managed modern landfill should not cause any uncontrolled releases of leachate into the environment, but it seems important to understand the potential environmental aspects of leachate in order to further develop landfill technology regarding these aspects, to form a basis for risk and environmental impact assessment at new landfills, and to take remedial action efficiently if a leachate release has occurred accidentally.

The main question raised with respect to landfill leachate is the length of time that leachate may constitute an environmental risk. No definite answer is currently available but it seems that we are speaking of hundreds of years rather than decades for the most persistent pollutants. Chapter 4.1 presents a Swiss approach to assessing the long-term influence of leachate from landfills indicating that dissolved carbon may constitute a problem for more than a thousand years. This is a rather long time frame in view of our current experiences with the reliability of constructions. Such realizations may also lead to new discussions on what kind of waste is suitable for landfilling.

Leachate entering the soil and groundwater environments will be subject to attenuation processes such as dilution, biological degradation and physico-chemical processes. The balance between these processes and the leachate load and composition will determine the extent of the leachate plume and the significance of the environmental risks to surface water ecology and to public health through water supplies based on surface water or groundwater. Chapter 4.2 discusses these attenuation mechanisms. Although our understanding of these mechanisms has improved, much still remains to be learned before an integrated understanding of these complex processes is available and predictive methods are developed.

Specific organic compounds in landfill leachate are supposedly the contaminants of most concern in addressing the impacts on groundwater. These specific organics will be present in nearly all landfills originating from household chemicals, small industries and maybe as co-disposed hazardous waste. Chapter 4.3 discusses our current knowledge about the degradation of these compounds in landfill leachate-polluted groundwater. It appears that the redox sequence supposedly present in the leachate plume, ranging from methanogenic conditions close to the landfill to potentially aerobic conditions in the most diluted part of the plume, may be advantageous to the degradation of a range of specific organic compounds. However, we also need to identify the specific compounds that will persist while passing through this sequence. These compounds may appear to possess the major environmental risk.

In licensing landfills, groundwater control monitoring is often demanded by the authorities in order to demonstrate that all the measures taken to prevent contamination of groundwater meet expectations. In order to constitute reliable control monitoring systems many aspects must be taken into consideration, as shown in Chapter 4.4. It is very simple to establish a control monitoring system that will never show an accidental pollution of the groundwater. If not properly designed and operated the groundwater control monitoring system may be an expensive system not providing the intended control and may be insensitive to a contaminant plume that later may appear to be very costly to remedy.

2. LEACHATE PRODUCTION AND COMPOSITION

2.1 Model Prediction of Landfill Leachate Production

NICK C. BLAKEY

Water Research Centre plc, Medmenham, PO Box 16, Marlow, Bucks, UK, SL7 2HD

INTRODUCTION

Existing and anticipated regulations related to the landfill disposal of wastes and environmental protection indicate that in future an increasing number of landfills will be designed and operated as containment sites. An inevitable consequence will be that greater attention will be paid both to waste management methods which control the volume of leachate produced and to the disposal routes for the accumulated leachate. In order to test the effects of alternative waste disposal schemes on the timing and the volume variations with time of leachate production, a predictive model has been developed which takes account of variables such as changes in the rate of waste input, compaction, cell geometry, surface slopes, influence of intermediate and final covers and changes in liquid inputs. The same data are also a prerequisite to designing rational leachate treatment and disposal options. The volume of leachate and the length of time over which it is produced strongly influence the sizing of treatment plants and, hence, their capital and running costs, as well as potential charges associated with options such as direct disposal to sewer.

At present, the predictive model deals only with leachate volumes. Leachate strengths are estimated empirically, with predictions based on accumulated information on typical leachate compositions (Robinson & Maris, 1979; Department of the Environment, 1986). Nevertheless, a combination of application of the model with knowledge of the characteristics of the various site operational and leachate treatment options allows rational design of leachate control and management systems.

WATER BALANCE

Predictions of the volume of leachate which can be generated from domestic wastes have been variously described (Department of the Environment, 1978; Blakey, 1982; Campbell, 1982; Ehrig, 1983; Holmes, 1984; Lu, 1985; Canziani & Cossu, 1989). Although the approaches employed in these methods vary, they are basically derived from a water balance or budget principle, the principal factors that control leachate production being:

- the monthly balance between the components of liquid input to the site that give rise to infiltration through the surface; and
- changes in the moisture retention and transmission characteristics of the waste as infiltration percolates through successive layers.

For a given area of landfill this can be represented by the following simplified equation employing the terminology of Canziani & Cossu (1989)

$$L = P - R - \Delta U_s - ET - \Delta U_w \tag{1}$$

where

L = leachate production

P = precipitation

R = surface run-off

ΔU_s = change in soil moisture storage

ET = actual evaporative losses from the bare-soil/evapotranspiration losses from a vegetated surface

ΔU_w = change in moisture content of the refuse components.

Infiltration of liquid into the landfill and absorption of percolating water by wastes are discussed in detail below.

INFILTRATION

The amount of water available for infiltration depends on the relationship between precipitation (P), surface water run-off (R), soil moisture storage (U_s) and actual evaporation/evapotranspiration (ET) as shown in eqn (1).

Daily values of rainfall and calculated estimates of potential evapotranspiration can be derived from Meteorological Office data. Surface water run-off on the other hand is estimated less reliably by using a coefficient appropriate for the surface cover characteristics of the site (slope and type of cover material) (Chow, 1964)

$$R = C . P \qquad (2)$$

where

R = surface run-off peak discharge (mm)

C = run-off coefficient

P = uniform rate of rainfall intensity (mm).

The literature contains many reports of standard run-off coefficients for various agricultural and engineering applications. Table 1 lists three separate examples of run-off coefficients based on individual effects of topography, soil type and surface vegetation. The simplistic approach adopted in the generation of data presented in Table 1 implies that the relationship between run-off and rainfall duration (i.e. (1) short showers only result in wetting of soil surface and filling of depression ponds on the surface and (2) the infiltration rate decreases as the cover material becomes wetter) has not been considered along with the effect of varying the cover material compaction. Although research is currently being funded by the UK Department of the Environment (DoE) into aspects of surface run-off and infiltration through clay-capped landfills, few data are presently available to challenge the validity of this alternative data.

Close examination of the data in Table 1 shows that these estimations are not consistent and data from further investigations of landfill site surface run-off factors are keenly awaited if rational estimation methods are to be used with reasonable accuracy. Despite these reservations, the simplicity of this empirical approach has made it a practical tool used widely by hydrologists and landfill site engineers alike.

The monthly change in soil moisture storage and the effect this has on the potential evaporation can be assessed in the UK by using the Meteorological Office's Rainfall and Evaporation Calculation System— MORECS (Meteorological Office, 1981). Holmes (1984) demonstrated the use of this system for a landfill application and showed that an assessment of soil moisture storage is important in any water balance

TABLE 1. Comparison of Run-off Coefficients for Drainage Areas with Different Topography, Soil, and Cover Conditions

Area type	Run-off coefficient C								
	Flat: slope <2%			*'Rolling':* slope 2–10%			*Hilly:* slope >10%		
Bare earth (clay)	0·60	0·60	0·60	0·66	0·70	0·70	0·70	0·82	0·80
(Clay or silt loam)	—	0·50	0·50	—	0·60	0·60	—	0·72	0·70
Meadows and pasture (Clay or silt loam)	0·25	0·30	0·35	0·30	0·36	0·45	0·35	0·42	0·55
Cultivated—impermeable (clay)	0·50	0·40	0·50	0·55	0·55	0·60	0·60	0·60	0·70
Cultivated/permeable (sandy loam)	0·25	0·10	0·20	0·30	0·16	0·30	0·35	0·22	0·40
Reference[a]	1	2	3	1	2	3	1	2	3

[a] 1. Perry, R. H. (1976). *Engineering Manual*, 3rd edn. McGraw-Hill Book Co., New York, 946 pp.
2. Salvato, J. A. *et al.* (1971). Sanitary landfill leaching prevention and control. *JWPCF*, **46**, 2084–2100.
3. Bernard, M. (1982). Discussion of run-off: Rational run-off formulas. *Trans. Am. Soc. Civil Eng.*, **96**.

calculation since this determines the amount of liquid percolation through the cover soil into the refuse layers beneath.

On a restored site, infiltration through the cover soil can be limited further by plant growth. The amount of water retained by plants is only a small fraction of the total absorbed by the roots. By far the greater part is transported to the aerial parts of plants where it evaporates into the surrounding air. Given an abundant supply of moisture in the soil then the loss of water as vapour from plants is controlled by the prevailing climatic conditions and is termed the potential evapotranspiration. There are two states of water in the soil available to plants; the first is gravitational water which temporarily displaces air from large spaces between the soil particles following rain and gradually percolates downwards under the influence of gravity. The second is capillary water which comprises the bulk of the water remaining in the soil after gravitational water has drained away, and is the main source of water to plants. Soil containing the maximum amount of capillary water and no gravitational water is said to be at its field capacity. When the soil dries and is no longer at field capacity then evaporation is less than the potential rate and a soil moisture deficit will develop.

Where identified in the computer model, this deficit is decreased by subsequent monthly rainfall until it becomes negative. At this point excess rainfall passes through the soil cover and is considered to be hydrologically effective. Estimating infiltration in this way can lead to a high degree of variability in effective rainfall from year to year even though the annual average rainfall levels remain similar. In other words a high proportion of rainfall falling in the low potential evapotranspiration months during the winter will produce greater amounts of effective rainfall compared to a year with a more even rainfall distribution.

An illustration of the relationship between rainfall and calculated infiltration is provided by data generated from a number of experimental landfills sited at Edmonton, North London (surface area of each experimental landfill, 90 m^2). The variation in annual rainfall exceeded 50 percent during the five-year period of investigation (Table 2). Of particular interest is that rainfall during year 3 was 15 percent greater than the two preceding years and yet the calculated infiltration was shown to be less. This was due to heavy rainfall events during the summer months where evapotranspiration was maximised.

These infiltration levels are not atypical for the operational phase of a landfill where surface water run-off is minimal. Values in excess of 50% have been recorded with seasonal variations ranging from 20 to 70% in the

TABLE 2. Annual Rainfall and Calculated Infiltration from Experimental Landfills, Edmonton, North London[a]

Year	Rainfall[b] (P) (mm)	Calculated infiltration (P_i)	P_i/P (%)
1	532	219	41
2	594	255	43
3	685	266	39
4	808	403	50
5	531	222	42

[a] Data on the experimental landfills appear in Table 3.
[b] Average data collected from four Meteorological Office stations within 4 km of the site.

summer and winter, respectively (Campbell, 1982). Ehrig (1983) has shown that values in the range 35–80% have also been recorded at operational sites in Germany.

Typical values of infiltration through cover soils at restored landfills have been shown to range between 14 and 34% of annual rainfall (Holmes, 1984). These levels may be reduced further by installing sub-surface drainage systems or by encouraging surface water run-off by surface contouring.

It is very clear from these observations that gross estimation figures quoted as national or regional guidelines should be avoided where more precise estimation of infiltration is required. It is recommended that careful assessment of the interaction of rainfall and site surface conditions be carried out bearing due regard for the differences in climatic conditions that may prevail in the vicinity of the landfill site (e.g. differences between conditions at the bottom of a deep quarry compared to the surface).

ABSORPTION

In order to clarify the terminology used to describe the assumed moisture-related properties of domestic waste a number of definitions are given below:

- Initial moisture content of domestic waste is the moisture that waste contains when first received at the site for disposal.

Although subject to waste type, seasonal trends and treatment after collection (i.e. wet pulverisation, baling, or mixing with other wastes under co-disposal conditions), a typical value of 35 percent of dry weight (equivalent to 26 percent of wet weight) is often quoted.

- Field capacity is often used to quantify the amount of liquid that a given mass of material will absorb before downward percolation occurs due to gravitational forces. This definition is suitable for homogeneous materials such as soils and perhaps pulverised domestic waste. However, with crude domestic wastes, liquid is released before field capacity is reached. For this reason, it is more meaningful to rename this term the 'absorptive capacity' of the domestic waste, indicating the point at which leachate will first be generated.
- Saturation capacity of the waste is reached over a period of years after the absorptive capacity is achieved. This is the amount of liquid that is retained within the void in which the waste is deposited.

These definitions imply two major and well recognised mechanisms for moisture retention within domestic wastes:

- physical absorption of liquid within the landfill which is determined by capillary forces;
- apparent absorption of water within void spaces in the waste giving localised areas of saturation.

Several factors are known to affect the first of these mechanisms, the absorptive capacity:

- Waste density. At waste densities commonly achieved in the UK ($0\cdot7-0\cdot8$ t/m^3) absorptive capacities of $0\cdot16-0\cdot27$ m^3 per dry tonne have been reported (Department of the Environment, 1978; Blakey, 1982). However, work has shown that where emplacement densities approach 1 t/m^3, absorptive capacities may fall to as little as $0\cdot02-0\cdot03$ m^3 per dry tonne (Campbell, 1982).
- Short-circuiting of liquid through waste (preferential pathways) can be established within a body of waste which is not highly compacted, through which infiltrating liquid may pass rapidly without being absorbed.
- Rainfall intensity. The effect of a high rainfall event would be to encourage short-circuiting.

Additionally, research has shown that where landfilled wastes have received prolonged periods of infiltration the additional uptake of liquid, over and above an average value of 35% of dry weight for the initial moisture content of domestic waste, is of the order 0·40–0·65 m³ per dry tonne of waste (Blakey, 1982). Although it can be argued that at least a proportion of this moisture might eventually drain from the wastes under gravity it is an important factor to be taken into account in any landfill water balance model.

In order to predict the quantity and pattern of leachate discharge from domestic waste a realistic estimation of the amount of absorption of infiltrating rainfall has been calculated for the experimental landfills at Edmonton, North London, using a site water budget. Here, calculated infiltration has been balanced against measured leachate output; infiltration has been estimated using actual rainfall, evaporation and evapotranspiration data where appropriate and taking into account soil moisture deficit where calculated during the summer months.

For the first six months of operation, leachate production accounted for only 6–10% of the estimated infiltration in cells 1, 2 and 4 (Table 3). This indicated that an initial uptake of liquid of 0·07–0·08 m³ per dry tonne of waste had occurred (Fig. 1). The comparatively small volumes of leachate produced during this initial period are considered to represent short-circuiting of infiltration through the domestic wastes since the major proportion of leachate up to this time had been collected from drainage bunds along the edges of each cell wall. These isolate leachate collected from the main body of the waste from such short-circuiting.

In comparison, up to 25% of measured infiltration was drained from cell 2 over the first six months. This was thought to be due to the abnormally high levels of putrescible waste which had dramatically increased the moisture content of the wastes held in this cell (Table 3).

After 12 months when the quantity of leachate, as a percentage of infiltration, continued to rise the amount of liquid resident within the cells also increased, particularly in response to heavy rainfall. This decreased only slightly during periods when little infiltration took place (Fig. 1). This increase in apparent liquid absorption was not observed in cell 2 where liquid storage in excess of initial moisture content fluctuated about a mean of 0·07 m³ per dry tonne.

The apparent discrepancy in liquid storage between the cells indicated in Fig. 1 can be clearly attributed to the initial moisture content of the wastes as emplaced, and in particular the influence of the moisture held in the industrial or putrescible wastes. For example, the phenol lime

TABLE 3. Details of Cell Contents and Water Balance for Four Experimental Landfills at Edmonton, North London (July 1979—May 1984)

Cell	1	2	3	4
Contents				
1. Domestic waste (tonnes-dry wt)	126·2	100 (143 from April 81)	98·1	96·3
2. Other waste (tonnes-dry wt)	— (control)	10 (putrescible food waste) (12·1 from April 81)	28·8 (phenol lime mud)	19·1 (spent oxide)
Period monitored (months)	6 12 24 36 48 59	6 12 24 36 48 58	6 12 24 36 48 58	6 12 24 36 48 59
Calculated infiltration (mm)	114 219 474 740 1143 1365	114 219 474 740 1143 1434	114 219 474 740 1143 1365	114 219 474 740 1143 1365
Leachate volume (mm)	7 28 118 271 515 618	29 100 310 632 1046 1205	11 61 211 402 670 497	11 61 211 402 670 774
Leachate production as a percentage of estimated infiltration for each period (%)	6 20 35 58 61 46	25 68 82 121 103 79	7 17 41 46 45 28	10 48 59 72 67 47
'Absorption'[a] of moisture (litres/dry kg)	0·07 0·15 0·25 0·32 0·43 0·51	0·07 0·10 0·09 0·06 0·05 0·09	0·07 0·13 0·24 0·34 0·49 0·60	0·08 0·12 0·20 0·26 0·36 0·45

Rainfall was measured on site as well as at four Meteorological Office stations.

[a] Water 'absorbed' is that volume which cannot be accounted for as leachate or evaporation.

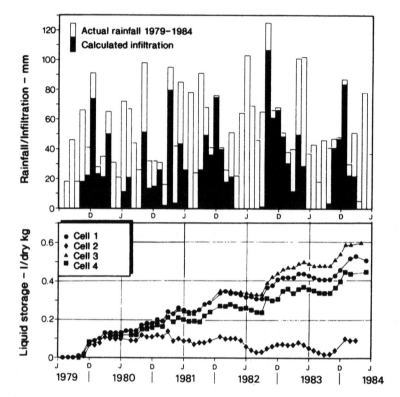

Figure 1. The effect of infiltration on calculated liquid storage of wastes contained in four experimental landfills at Edmonton, N. London. (Cell 1—domestic waste only; Cell 2—domestic/putrescible waste mix; Cell 3—domestic/phenol lime sludge waste mix; Cell 4—domestic/spent oxide waste mix.)

mud contained in cell 3 had a moisture content of 27% of wet weight (Barber *et al.*, 1980) and over the first three and a half years of operation moisture storage in this cell showed no appreciable difference to that of the control cell 1 (Fig. 1). On the other hand, cell 4, containing a spent oxide waste initially holding 49% of wet weight as moisture, showed significant changes in the apparent moisture storage of the wastes. This reduction in the amount of additional uptake of liquid can be compared with cell 2 where the putrescible wastes, containing 80–90% of wet weight as moisture, had a pronounced effect on the apparent moisture storage in the cell (Fig. 1).

These findings show that site operators need to be aware of the potential problems associated with relatively large quantities of waste with high moisture contents. Where these wastes are accepted for disposal, moisture retention within the wastes cannot be exploited during the operational phase of the landfill to delay the production and reduce the volume of leachate.

To illustrate this point further, all the factors so far described have been combined together in the development of a computer model which has been used to predict leachate production from the Edmonton test cells.

COMPUTATIONAL MODEL

The model assumes that once all the refuse in a landfill reaches its saturation capacity an equilibrium is achieved and a uniform leachate generation pattern is obtained. The time taken for leachate to appear is estimated from a moisture movement calculation. The model first calculates infiltration through the cover material or surface of the landfill, and then using these results calculates the rate of vertical movement through the underlying layers of compacted wastes.

The results obtained from the model for the Edmonton test cells are shown in diagrammatic form in Figs 2A and 2B. The widening discrepancy between calculated infiltration and predicted drainage in Fig. 2B represents the steady increase in moisture retention in cells 1, 3 and 4 that was shown in Fig. 1. This effect is absent in cell 2 where the wastes reached saturation capacity relatively rapidly (Fig. 2A and Table 3).

When taking moisture storage fluctuation into account, reasonable agreement between predicted drainage and actual leachate discharge from all four cells has been obtained. For comparison, infiltration and predicted drainage has been estimated using 10-year average data for southern England (Rothamsted, 1960–1969). In this instance predicted drainage bears little relation to actual leachate discharge from the cells. This serves to demonstrate that where realistic estimations of leachate production are to be calculated the importance of using site specific data cannot be overemphasised. Gross estimation figures and national or regional average meteorological data should be avoided wherever possible.

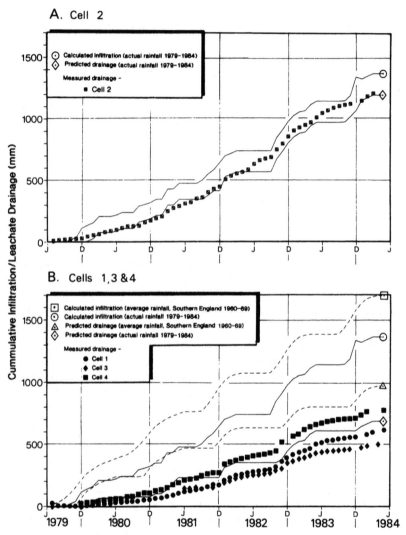

Figure 2. A comparison between the expected pattern of infiltration and leachate drainage with observed flow from four experimental landfills at Edmonton, N. London. (Cell 1—domestic waste only; Cell 2—domestic/putrescible waste mix; Cell 3—domestic/phenol lime sludge mix; Cell 4—domestic/spent oxide waste mix.)

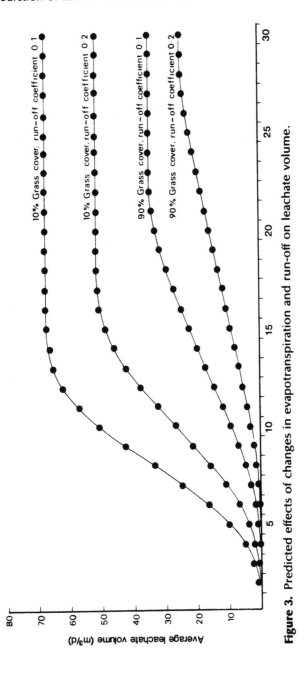

Figure 3. Predicted effects of changes in evapotranspiration and run-off on leachate volume.

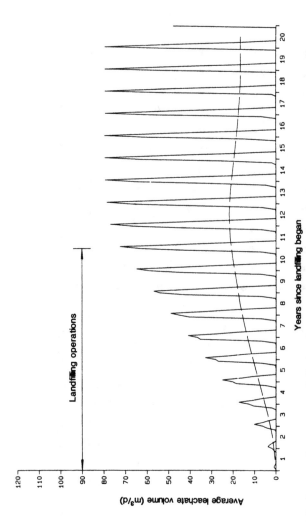

Figure 4. Predicted changes in average daily leachate volume following progressive restoration and grass cover. Monthly basis (full line), annual basis (dotted line). Specific data for landfill: rate of filling = 600 tonnes/week; density of fill = 0·96 tonnes/m³; precipitation = 750 mm/year; run-off coefficient = 0·1.

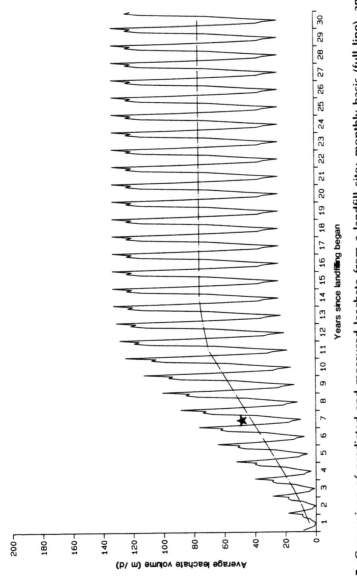

Figure 5. Comparison of predicted and measured leachate from a landfill site: monthly basis (full line), annual basis (dotted line). ★—Actual/measured winter leachate flow, seven years after start of filling. Specific data for landfill: density of fill = 0·8 tonnes/m³; precipitation = 1620 mm/year; run-off coefficient = 0·4.

In this, its simplest form, the model assesses a fixed surface area and depth of waste as would be the case for a completed landfill. However, the model can also assess leachate production at a working landfill where increasing surface area and variable depth of waste is encountered. This enables a more realistic estimation of the landfill water balance and also gives some insight into those features of the landfill operation which may increase or decrease leachate generation (i.e. surface run-off, progressive restoration, etc.).

Figure 3 shows how sensitive the model is to selection of appropriate run-off and evapotranspiration factors for an 8-hectare lined landfill in southern England. Assuming a weekly waste input of 2500 tonnes and an annual rainfall of about 770 mm, the sharp decrease in predicted flows associated with higher run-off values or increased evapotranspiration (10% grass compared with 90% grass cover) is clearly displayed.

In practice, evapotranspiration losses would be expected to increase with time, as a site is progressively restored and vegetation cover becomes fully established. The effect on leachate flow of such a regime is shown in Fig. 4, which is based on a hypothetical landfill covering an area of 5 hectares, with a waste depth ranging between 5 and 8 metres and a surface slope of 2%.

The effect of high rainfall input on leachate production pattern is well illustrated in the case of a natural containment site of about 6 hectares receiving 650 tonnes of waste per week (Fig. 5), where recorded winter discharges of leachate of about 55 m^3/day in 1986/87 correspond well with predicted values from the model.

CONCLUSIONS

A landfill water balance model has been developed to provide a quantitative assessment of landfill leachate generation. This has been used to estimate leachate production from a series of experimental and full-scale landfills. Despite the limitations of the simplistic approach adopted, the model has been found to be reasonably accurate and is now finding application as a planning tool in waste management.

Infiltration has been shown to be site specific and seasonally variable. The experimental landfills demonstrate that under these circumstances it is inappropriate to use gross estimation models using national or regional figures for rainfall and evaporation/evapotranspiration. It is recommended that careful assessment of the interaction of rainfall and site surface conditions be carried out at least on a monthly basis using site specific data.

Substantial reductions in the level of infiltration can be achieved during the operational phase of the landfill, as well as after restoration, by encouraging surface water run-off. This is particularly important during the winter months when evaporation/evapotranspiration is at its lowest.

At emplacement densities of $0\cdot6$–$0\cdot8$ t/m^3, and with an initial moisture content of 35% of dry weight, it is likely that an absorptive capacity of $0\cdot16$–$0\cdot27$ m^3 per dry tonne of waste can be achieved before substantial leachate generation commences. After prolonged periods of infiltration this absorptive capacity can be exceeded to reach a saturation capacity of $0\cdot4$–$0\cdot6$ m^3 per dry tonne of waste.

Liquid retention and migration characteristics in landfilled wastes and its eventual emergence as leachate have been studied using a variety of waste mixes. Experimental evidence shows that where the initial moisture content of waste is substantially greater than 35% of dry weight then the saturation capacity is reached much more rapidly. Under these circumstances the moisture retention characteristics of the waste cannot be exploited in the same way to delay production and reduce volume of leachate during the operational phase of the landfill.

REFERENCES

Barber, C., Maris, P. J. & Johnson, R. G. (1980). Behaviour of wastes in landfill sites: Study of the leaching of selected industrial wastes in large-scale test cells, Edmonton, N. London. WLR Technical Note No. 69, Department of the Environment, UK.

Blakey, N. C. (1982). Infiltration and absorption of water by domestic wastes in landfills—research carried out by WRC. Harwell Landfill Leachate Symposium, Harwell, Oxon, UK, 19 May 1982.

Campbell, D. J. V. (1982). Absorptive capacity of refuse—Harwell research. Harwell Landfill Leachate Symposium, Harwell, Oxon, UK, 19 May 1982.

Canziani, R. & Cossu, R. (1989). Landfill hydrology and leachate production. In: *Sanitary Landfilling: Process, Technology and Environmental Impact*, ed. Christensen, T. H., Cossu, R. & Stegman, R., Academic Press, London, Chapter 9.1.

Chow, V. T. (Ed.) (1964). *Handbook of Applied Hydrology*. McGraw-Hill, New York.

Department of the Environment. (1978). Co-operative programme of research on the behaviour of hazardous wastes in landfill sites. Final report of the Policy Review Committee (Chairman, J. Sumner). HMSO, London, 1978, 169 pp.

Department of the Environment. (1986). Waste Management Paper No. 26, Landfilling wastes. HMSO, London, 206 pp.

Ehrig, H.-J. (1983). Quality and quantity of sanitary landfill leachate. *Waste Management and Research*, **1**, 53–68.

Holmes, R. (1984). Comparison of different methods of estimating infiltration at a landfill site in South Essex with implications for leachate management and control. *Q. J. Eng. Geol. London*, **17**, 9–18.

Lu, J. C. E. (1985). *Leachate from Municipal Landfills. Production and Management*. Pol. Tech. Review, 119, Noyes Publications, Park Ridge, NJ.

Meteorological Office. (1981). Rainfall and evaporation calculation system—MORECS. *Hydrological Memorandum*, **45**, 27 pp.

Robinson, H. D. & Maris, P. J. (1979). Leachate from domestic waste: generation, composition and treatment. A review. WRC technical report TR 108, WRC, UK, 38 pp.

2.2 Production of Landfill Leachate in Water-Deficient Areas

GEOFFREY E. BLIGHT, DAVID J. HOJEM & JARROD M. BALL

*University of the Witwatersrand, PO Wits, 2050,
1 Jan Smuts Avenue, Johannesburg, South Africa*

INTRODUCTION

Water is a scarce commodity in arid and semi-arid areas and pollution of surface and underground water resources can be disastrous to communities and households depending on these sources for domestic supply.

As much of the Earth's land surface consists of water-deficient areas, concern has arisen lest existing and future sanitary landfills are causing, or have the potential to cause, unacceptable water pollution. Such pollution is also most costly and difficult to clear up once it has occurred. If nothing is done to ameliorate the situation, the pollution may persist in the groundwater for decades (Ball & Blight, 1986) even though the source of the pollution has been removed.

A preliminary study (Ball & Blight, 1986; Blight *et al.*, 1987) produced strong evidence that if climatic conditions are such that a perpetual water deficit exists at the site of a landfill, no or very little leachate will be formed or exit from the base of the landfill. Hence, if there is an adequate separation between the lowest level of refuse and the highest level of the regional phreatic surface, no groundwater pollution will occur. By extension, surface water replenished by the groundwater will also remain unpolluted by leachate from the landfill.

There is considerable support for this view in the literature. For example, Keenan (1986) gives figures indicating that landfills receiving more than 750 mm of precipitation per annum will eventually produce leachate, while those in arid regions receiving less than an annual 325 mm are likely never to exude pollution. Saxton (1983) states that for climates where annual precipitation is less than 400 mm, virtually all precipitation is evapotranspired.

Earlier, Fenn *et al.* (1975), Burns & Karpinski (1980) and Holmes (1980) all agreed that if a net annual water deficit exists at the site of a landfill, little if any leachate will exit from its base.

It must also be recognised that good engineering and management of a landfill can be used to maintain a perennial water deficit within the fill even though there may actually be an excess of precipitation over potential evaporation. This can be done by maximising run-off and minimising infiltration into the refuse. A suitably sloping surface and the installation of a carefully designed impervious cover layer can achieve this (e.g. Lundgren & Elander, 1987).

This chapter describes a recent detailed investigation into the movement and retention of water within two sanitary landfills located in water-deficient areas in South Africa. The results largely corroborate the evidence outlined above and have formed the basis for a set of guidelines for the design of sanitary landfills in water-deficient regions, to avoid water pollution.

THE WATER BALANCE FOR A LANDFILL

The water balance for a landfill can be stated as follows

water input = water output + water retained

In this equation, each term represents a rate of accumulation or loss. Water input includes precipitation (P) and the water content of the incoming waste (U_w). U_w, however, only makes a once-off contribution to the annual water balance of a given mass of landfill. Water output includes evapotranspiration (ET), water vapour entrained by gas (G), water lost in leachate (L), and run-off (R). Finally, there is water absorbed and retained by the waste (ΔU_w) and the soil cover (ΔU_s), i.e. for an annual water balance,

$$P + U_w = ET + G + L + R + \Delta U_w + \Delta U_s \tag{1}$$

In the present study G has been ignored, as it is understood to be small in comparison with the other terms. The annual water balance equation,

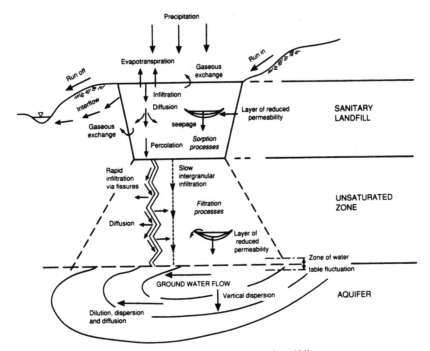

Figure 1. Details of water balance in a sanitary landfill.

as applied to an established landfill, has been simplified to

$$P = ET + L + R + \Delta U_w + \Delta U_s \qquad (2)$$

Figure 1 (after Naylor *et al.*, 1978) illustrates the components of the water balance for a landfill in greater detail. In eqn (2), the only components that can be directly controlled by the engineer are the run-off R, and by limiting infiltration, the terms ΔU_w and ΔU_s. In terms of eqn (2) the leachate production L is given by

$$L = P - ET - R - \Delta U_w - \Delta U_s \qquad (3)$$

Obviously the smaller the precipitation (P) and the larger the evapotranspiration (ET) and run-off (R), the less the potential for the generation of leachate (L). These terms are particularly favourable in water deficient areas.

THE FIELD CAPACITY OF REFUSE AND LANDFILLS

The field capacity is that water content (by mass of dry solids) which the refuse will absorb and store or retain by capillarity. In a first simple concept for an annual water balance, the term ΔU_w will persist and the term L will be zero until the field capacity of the refuse has been reached. At this stage, the refuse will absorb no more moisture, the term ΔU_w will disappear and the term L will appear in its place. Another concept introducing both water absorption capacity and field capacity has been applied by Canziani & Cossu (1989) and Blakey (1990).

It is obvious that as decomposition and compaction of refuse occurs in a landfill, the field capacity will progressively decrease. The literature records values for the field capacity of refuse that vary from 80% for fresh refuse (Campbell, 1983) to between 63% and 74% for refuse more than four years old (Holmes, 1980). These figures obviously depend both on the composition of the refuse and the method of determining the dry mass.

It was considered worthwhile to make a separate study of the field capacity of refuse from one of the sites studied. The results of two independent studies by Roper & Fongoqa (1988) and by the second author are summarised in Fig. 2. The refuse contained on a dry weight basis 54% organics, 23% paper, 9% glass, 8% plastic and 6% metal. Figure 2 shows an even wider variation of field capacity than that recorded above. As a conservative working figure, a field capacity of 60% by dry (50°C) mass of refuse has been assumed in this study.

The theoretical concept that refuse will continue to absorb moisture until the field capacity is reached, and will thereafter release moisture at the same rate as it receives it, is obviously an over-simplification. It can be deduced from Fig. 2 that the field capacity of refuse itself can be reached because of the accumulation of moisture, or because the field capacity is changing as the age and state of compaction and decomposition of the refuse increase, or by a combination of the two processes.

A landfill may start to release leachate long before its overall field capacity has been reached because certain interconnected zones within it have a lower field capacity than others. This is well illustrated by Fig. 3 which shows a record of the precipitation on the surface and the corresponding release of leachate from a 20-m square and 5-m deep cell of refuse at a landfill near Cape Town, South Africa. The cell was underlined with a plastic sheet that trapped all leachate exiting the refuse

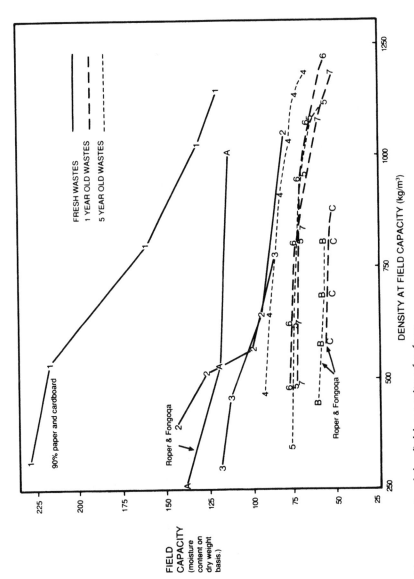

Figure 2. Measurements of the field capacity of refuse.

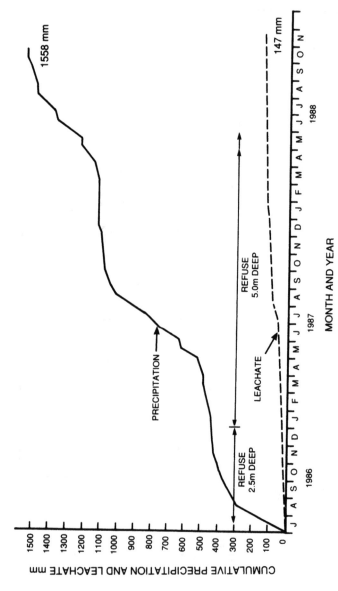

Figure 3. Precipitation and leachate production at a landfill in Cape Town.

and channelled it into a container, which enabled the leachate production to be measured.

As at the end of 1988 the landfill had received over 1500 mm of precipitation, and had released less than 150 mm of leachate over a $2\frac{1}{2}$ year period. The water balance calculation (to be described later) predicted that once the field capacity of the landfill had been reached, the annual output of leachate would be 200 mm per year or 500 mm in $2\frac{1}{2}$ years. It is therefore clear that the overall field capacity has not yet been reached. At the same time, certain zones in the cell are releasing leachate and therefore must have reached their localised field capacity. Leachate is, however, only generated during the cool wet winter months (May to August) when there is a seasonal water surplus.

WATER BALANCE FOR TEST LANDFILLS

The experimental work was performed at a landfill on the Witwatersrand, South Africa, and one near Cape Town. The Witwatersrand landfill is in a summer rainfall area while that at Cape Town is in a winter rainfall area. The climatic parameters for the two sites have been summarised in Fig. 4.

At the Witwatersrand landfill, pan evaporation exceeds precipitation throughout the year, while at Cape Town precipitation exceeds evaporation during the $3\frac{1}{2}$ to 4 winter months. Thirty-year average annual figures for precipitation and rainfall at the two sites are:

Witwatersrand Precipitation = 745 mm
          Pan evaporation = 1550 mm
 Cape Town   Precipitation = 510 mm
          Pan evaporation = 1110 mm

Hence both sites are nominally in water-deficit areas with potential evaporation exceeding precipitation by some 600 to 800 mm.

Figure 5 shows detailed water balances for the two sites calculated by Fenn *et al.* (1975) based on an adaptation of the method of Thornwaite & Mather (1957), and on a weekly calculation interval. Assumptions for both sites were the same, namely 1000 mm of top cover with 100 mm of available storage in the cover. Precipitation was taken as

P = mean rainfall + 1 standard deviation

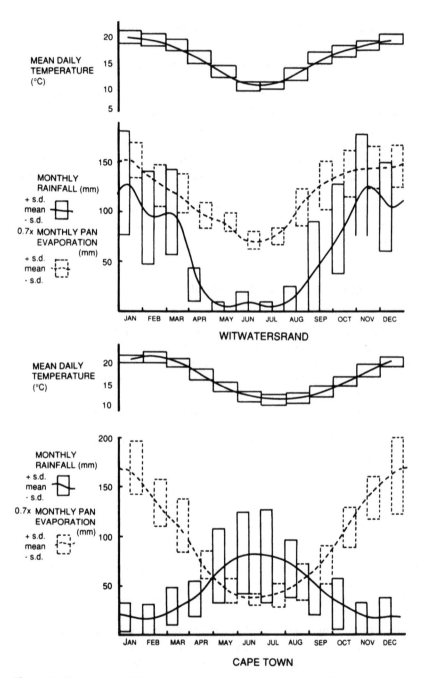

Figure 4. Summary of climatic parameters at two test landfills.

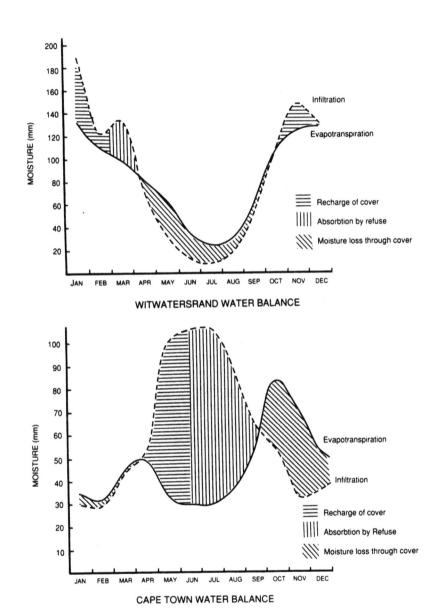

Figure 5. Water balances for Cape Town and Witwatersrand landfills.

and evapotranspiration as

$ET = 0{\cdot}7$ (mean pan evaporation $-$ 1 standard deviation).

Hence some conservatism was built into the calculations.

Studies by Schulz (1974) and others have shown that the evaporation from a soil surface is closely represented by $ET = 0{\cdot}7$ (pan evaporation).

Run-off for the Witwatersrand site, where the cover material is a clayey sand, was taken as

$R = 0{\cdot}11P$

while that at Cape Town, where the cover is a clean fine sand, was taken as

$R = 0{\cdot}08P$

Both of these are probably also conservative figures.

Figure 5 indicates that at Witwatersrand it is only in the two-month mid-February to mid-April period that the refuse will be gaining in moisture, while at Cape Town, moisture is gained for three months from mid-June to mid-September. As mentioned previously, these calculated annual water balances will usually be conservative, and will inevitably vary from year to year. For example, the precipitation at the Witwatersrand site in 1984 was only 46% of the 30-year average. In 1986 it was equal to the average, while in 1988 it was 130% of the 30-year average. On one day in October 1985 the landfill received 110 mm of rainfall, nearly 12% of the annual figure, most of which must have run off. These extreme figures illustrate the uncertainties inherent in water balance calculations.

On the basis of these water balances, a time for each landfill to reach field capacity can be calculated, as can the ensuing rate of generation of leachate. The calculations showed that the Witwatersrand landfill can be expected to reach a field capacity of 60% moisture around the year 2000, after which leachate is likely to be generated at a rate of about 130 mm per year. However, as the calculations are based on conservative figures, these approach worst case estimates. The rate of 130 mm per year is equivalent to 4×10^{-7} cm/s, which is about one tenth of the coefficient of permeability of the base of the landfill. Hence any leachate that is generated will disperse into the underlying unsaturated zone.

It should also be emphasised that the result of a water balance calculation depends on the interval adopted for the calculation. For example, for Witwatersrand a weekly calculation interval gives 70% of the figures based on a daily interval.

At Cape Town, overall field capacity is predicted for 1993 whereafter the predicted rate of leachate production is 200 mm per year. As recorded earlier, this compares with the currently measured leachate production of 60 mm per year.

DIRECT SAMPLING OF LANDFILLS

A direct sampling of both the Witwatersrand and Cape Town landfills was carried out in order to examine in detail the distributions of moisture and solutes in the two landfills. An analysis of these, both at the end of the wet season and at the end of the dry season, would give an indication of the accuracy of the predictions based on water balance.

Sampling was performed from 1200-mm diameter holes augered in the landfills by means of a pile augering machine. Each hole was cased against possible collapse of the sides with a cylindrical cage made of heavy reinforcing steel. A steel collar and hinged lid was fitted to each hole to exclude surface water and retain the landfill gas. The level of each lid was kept above the surface of the landfill and surrounded by a mound, further to exclude surface water.

Two holes were drilled at the Witwatersrand site, one in a well-drained area (the South hole), the other in a slight depression where water ponds to a shallow depth (the North hole). One hole was drilled at Cape Town, adjacent to the lined leachate collection cell.

The actual sampling was carried out by descending into the holes on a bosun's chair, wearing a full-face diving mask, and extracting the samples from the sides of the holes through the apertures in the casing cage.

To analyse the soluble components of the refuse, a mass of 2 kg of refuse sample was agitated with 1500 ml of distilled water for 90 minutes. A sub-sample of the water was filtered and centrifuged to remove any suspended solids and then analysed. The dry mass of the original sample was determined. From the analysis of the water sample, the various soluble components were expressed as mg per kg of dry refuse.

The results of the analyses of the holes is shown in Fig. 6(a) (Witwatersrand North), Fig. 6(b) (Witwatersrand South), and Fig. 7 (Cape Town). Each figure shows profiles of water content, TDS (Total

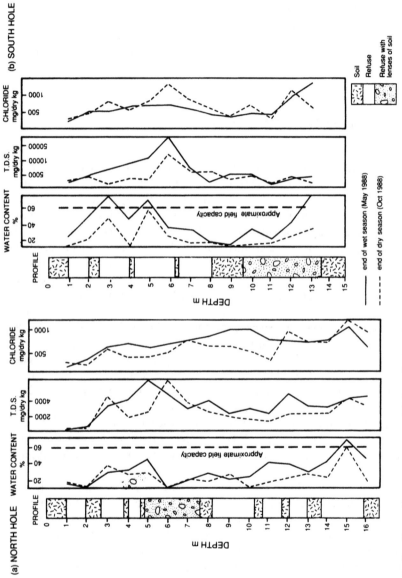

Figure 6. Profiles of water content, TDS and chlorides for Witwatersand landfill.

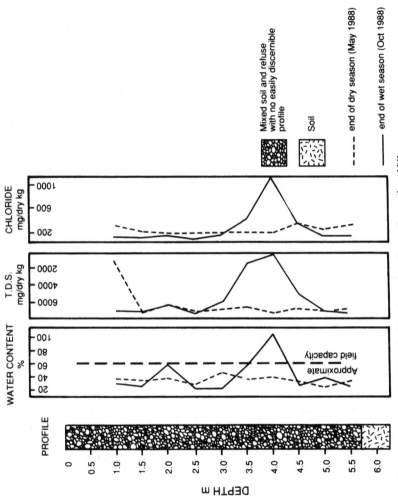

Figure 7. Profiles of water content, TDS and chlorides for Cape Town landfill.

Dissolved Solids) and chloride (both in mg per kg of dry refuse). A number of other parameters were also measured but have not been shown in Figs 6 and 7.

RESULTS OF DIRECT SAMPLING

Witwatersrand

The Witwatersrand North hole shows a definite seasonal fluctuation of water content. All the measurements except two were well below the assumed average field capacity of 60%. There appears to be an isolated wet layer at a depth of 15 m, and the water content of the soil at the base of the landfill rose to over 40% at the end of the wet season. This is well in excess of the field capacity of this soil, estimated at 15%. Hence this profile shows evidence of some leachate exiting from the base of the fill.

The South hole shows a greater fluctuation of water content down to about 5 m and also a zone in excess of the field capacity between depths of 3 and 5 m. There was also a wet season concentration of water near the base of the landfill. Both profiles are thus generally below the field capacity but show evidence of some outflow of leachate near the base.

The TDS in both profiles changed little from season to season. At about 5 m depth, the North hole shows a concentration of TDS which may be related to a corresponding peak in the water content profile. This may correspond to the 30% above average rainfall that fell in the 1988 wet season immediately preceding the sampling. This unusual precipitation may also account for the high TDS at about 6 m deep in the South hole.

Neither hole shows a tendency for TDS to accumulate at the base of the profile. Equally, neither shows a tendency for a concentration towards the top of the profile during the dry season.

The profiles of chloride at Witwatersrand show similar trends to the TDS profiles. However, there does appear to be an increase of chloride content with depth.

The overall impression is one of a fairly stable, but seasonally fluctuating water and solute profile with evidence of some outflow at the base of the landfill. Strangely enough, the profile underlying the area where ponding occurs is drier than that under the well-drained area. This may be because the cover is more pervious at the South hole, but this has not been established.

The water content profiles can be used to estimate the total moisture accumulation in the refuse since placing, if initial water contents for refuse and cover material are assumed. On average, at the end of the 1988 wet season, the profiles for the two holes had accumulated 3000 mm of water since placing. This reduced to 1750 mm at the end of the ensuing dry season. Bearing in mind that 1988 had an unusually wet season, it thus appears that the seasonal fluctuation of moisture in the profile is of the same order as the average annual precipitation of 950 mm.

Because the profile is mostly well dry of the field capacity, this moisture must mainly be moving in and out of the top cover, and the quantity of leachate escaping from the base can only be very small. Hence there is fair agreement between the results of the direct sampling and the prediction based on a water balance, with the water balance calculation likely to be pessimistic.

Cape Town

At the Cape Town site, the end of wet season water content was mostly below the field capacity, but there was a zone of material at a depth of 4 m that was above field capacity and one at 2 m that was at field capacity. There were also concentrations of TDS and chloride coinciding with this wet zone in the profile. These peak values probably represent a wetting front moving through the landfill.

The end of dry season water content profile was well below the field capacity, which must indicate that moisture moves up the profile and out through the cover during the dry season. Moisture moving downward by gravity flow could be expected to leave material at the field capacity above it.

The profile appears to have accumulated about 900 mm of water since placing at the end of the wet season and 650 mm at the end of the dry season. Hence taking run-off at 40 mm per annum, the moisture available for increasing the water content of the landfill and ultimately for producing leachate amounts to about 200 mm per year, which agrees with the prediction based on the water balance.

CONCLUSIONS

Calculations based on considerations of the water balance for a landfill can be used to predict the time likely to be taken for the landfill to reach its overall field capacity. The results of such calculations are liable to uncertainty because of the highly variable nature of the weather and the variable properties of the refuse and cover material. Uncertainties in the estimation of run-off and evapotranspiration add to the uncertainty of the prediction.

A comparison between the results of direct sampling of two landfills and predictions based on their water balance showed that the water balance method is realistic and as accurate as can be expected, given the uncertainty of the input data.

In order to predict whether or not a landfill sited in an arid or semi-arid area will produce leachate, it is essential to consider the detailed distributions of precipitation and evaporation through the year, and to carry out a full water balance calculation.

In the examples examined here, both landfills are sited in areas of nominal water deficit, with the annual deficit being between 600 and 800 mm. Neither landfill is producing very much leachate at present, but it seems that the Cape Town landfill has the potential to produce around 200 mm of leachate per year, with the landfill at Witwatersrand around 130 mm per year.

REFERENCES

Ball, J. M. & Blight, G. E. (1986). Groundwater pollution downstream of a long established sanitary landfill. In *Proceedings, International Symposium on Environmental Geotechnology*, Allentown, PA, ed. H. V. Fang, Envo Publishing, Bethlehem, PA, pp. 149–57.

Blakey, N. C. (1992). Model prediction of landfill leachate production. This book, Chapter 2.1

Blight, G. E., Vorster, K. & Ball, J. M. (1987). The design of sanitary landfills to reduce groundwater pollution. In *Proceedings, International Conference on Mining and Industrial Waste Management*. S. A. Institution of Civil Engineers, Johannesburg, South Africa, pp. 297–306.

Burns, J. & Karpinski, G. (1980). Water balance method estimates how much leachate a site will produce. *Solid Waste Management*, August, 54–84.

Campbell, D. J. V. (1983). Understanding water balance in landfill sites. *Waste Management*, November, 594–605.

Canziani, R. & Cossu, R. (1989). Landfill hydrology and leachate production. In: *Sanitary Landfilling: Process, Technology and Environmental Impact*, ed. T. H. Christensen, R. Cossu & R. Stegmann. Academic Press, London.

Fenn, D. G., Hanley, K. J. & De Geare, T. V. (1975). Use of the water balance method for predicting leachate generation for solid waste disposal sites. US Environmental Protection Agency, Report No. EPA/530/SW168, 1975.

Holmes, R. (1980). The water balance method for estimating leachate production from landfill sites. *Solid Wastes*, **LXX**(1), 20–33.

Keenan, J. D. (1986). Landfill leachate management. *Journal of Resource Management and Technology*, **14**(3), 177–88.

Lundgren, T. & Elander, P. (1987). Environmental control in disposal and utilization of combustion residues. Swedish Geotechnical Institute Report No. 28E, 1987.

Naylor, J. A., Rowland, C. D., Young, C. P. & Barber, C. (1978). The investigation of landfill sites. Water Research Centre, Technical Report TR9, 1978.

Roper, K. & Fongoqa, F. (1988). Determination of the field capacity of domestic refuse. Final year project report, Department of Civil Engineering, Witwatersrand University, 1988.

Saxton, K. E. (1983). Soil water hydrology: simulation for water balance computations. Proceedings, Workshop on New Approaches in Water Balance Computations, Hamburg, Germany, 1983.

Schulz, E. F. (1974). *Problems in Applied Hydrology*. Water Resources Publications, USA, 1974.

Thornwaite, C. W. & Mather, J. R. (1957). Instructions and tables for computing potential evapotranspiration and the water balance. *Publications in Climatology*, **X**(3), Drexel Institute of Technology, Philadelphia, PA.

2.3 Effects of Vegetation on Landfill Hydrology

MATTI ETTALA

Department of Environmental Sciences, University of Kuopio, PO Box 1627, SF-70211, Kuopio, Finland

INTRODUCTION

Of the components contributing to the hydrological balance of the landfill, the evaporation term is of major importance, since the amount of water not being evapotranspired contributes to surface run-off and leachate. Increasing the actual evapotranspiration would result in less leachate generation and as such contribute to a reduction of leachate treatment costs.

The actual evapotranspiration depends on the vegetation cover and the annual distribution of precipitation. The latter aspect may be managed by recycling leachate to the top of the landfill in the dry summer months, as is done at many landfills. The former aspect, the importance of the vegetation on the landfill, has gained only little attention as a factor to control. This supposedly is due to the many reports on the absence or scarcity of vegetation on landfill sites. However, the more dense the vegetation, the higher the evapotranspiration and the less the leachate generation is expected to be.

This chapter summarises research activities conducted in Finland studying the effect of vegetation on landfill hydrology, in particular focusing on the introduction of short-rotation plantations on landfill sites. The chapter presents a survey of natural vegetation coverage on Finnish landfills and discusses the ability of plantations to survive on landfills and their influence on the water balance of the landfill.

MATERIAL AND METHODS

Study Period and Areas

The study was carried out in 1982–1989. The vegetation coverage was examined at 40 sites and the leachate quality at 26 sites in southern Finland. Data on the survival of short-rotation tree plantations were gathered from nine sanitary and industrial landfills. Most of the measurements concerning hydrological effects and growth conditions were made at two sanitary landfills, especially at Lahti (Table 1). All the sites are described in detail by Ettala *et al.* (1988*a,b*).

Vegetation Studies

The plant species and vegetation coverage were examined using 1-m wide survey lines with a combined length of 11 km (Ettala *et al.*, 1988*a*). The quality and depth of the cover material and age of the refuse were also studied in order to find reasons for variations in the vegetation cover.

TABLE 1. Landfill Characteristics

	Characteristics of the landfill at	
	Hollola	*Lahti*
Established in	1974	1955
Location:		
latitude	60°58′	60°57′
longitude	25°30′	25°45′
Refuse:		
area (ha)	3	17
max height (m)	12	12
saturated height (m)	6–8	3–5
Waste amount (t a^{-1})	5 200	70 000
Degree of density[a]	2	5
Use of covering soil[a]	2	5
Special wastes[b]	A,D	A,B,C

[a] 1 not at all . . . 5 high
[b] A: septic sludge
  B: galvanic sludge
  C: ash from incineration plant
  D: oily wastes and solvents

A preliminary study was carried out at the Hollola landfill in 1982 to examine the possibility of establishing short-rotation tree plantations. Since the results were promising, 65 000 cuttings were then planted at six sanitary landfills to obtain a general view of the survival of the stands (Ettala, 1988*a*). In addition, 15 000 cuttings were planted at three industrial landfills. The biomass production was studied in relation to the plant species, substrate depth and quality, irrigation method and volume, leachate quality, and coppicing.

Studies on Evapotranspiration

Evapotranspiration was first estimated on the basis of the above-ground biomass production of *Salix* stands and values presented for water consumption per unit dry matter weight of *Salix viminalis* and *Salix aquatica*. The second estimate was based on the Penman–Monteith formula (Monteith, 1965), the values for heat flux and surface resistance being measured at the landfill in field conditions. The third estimate of evapotranspiration was obtained from direct measurements of precipitation, throughfall, stemflow and transpiration (Fig. 1). The last-mentioned variable was measured with an analyser based on infrared radiation.

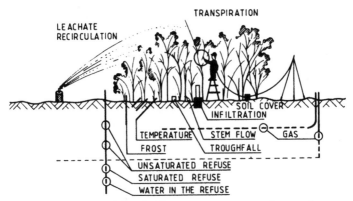

Figure 1. Arrangement of the field measurements made on short-rotation tree plantations at the Lahti landfill (Ettala, 1988*b*).

Landfill Characteristics

Measurements were made of temperatures in the refuse, heat flux from
the refuse, frost and snow depth, snow water equivalent, infiltration rate
and hydraulic conductivity. The results were examined in order to find
relations between infiltration, evapotranspiration and surface run-off,
and possibilities of influencing them. To elucidate the degradation stage
of the refuse and the conditions for plant growth at sanitary landfills in
Finland, a study was also made of the quality of the refuse, gas, leachate
and water in the refuse (Ettala *et al.*, 1988*b*).

The short-rotation tree plantations (see Fig. 1) were irrigated with

TABLE 2. Quality of the Leachate at the Lahti and Hollola Landfills

Parameter	Leachate quality at					
	Lahti landfill			Hollola landfill		
	n	$\bar{x}$	s	n	$\bar{x}$	s
Conductivity (mS m^{-1})	34	550	110	17	320	73
pH	35	7·9	0·42	17	6·6	0·10
Alkalinity (mmol litre^{-1})	29	53	12	15	27	2·8
COD$_{Cr}$ (mgO$_2$ litre^{-1})	2	330	110	2	1 700	280
COD$_{Mn}$ (mgO$_2$ litre^{-1})	33	110	41	15	150	61
BOD (mgO$_2$ litre^{-1})	34	73	100	17	2 100	3 100
Total nitrogen (mgN litre^{-1})	36	130	31	17	140	61
Ammonium (mgN litre^{-1})	35	120	30	17	130	18
Total phosphorus (mg litre^{-1})	35	0·58	1·1	17	0·74	0·16
Chloride (mg litre^{-1})	28	360	110	15	200	29
Iron (mg litre^{-1})	32	14	15	17	330	140
Zinc (mg litre^{-1})	19	0·032	0·037	8	0·27	0·14
Lead (mg litre^{-1})	17	0·017	0·027	5	0·020	0·004 6
Copper (mg litre^{-1})	14	0·015	0·012	9	0·077	0·077
Chromium (mg litre^{-1})	17	0·012	0·005 9	5	0·079	0·026
Cadmium (mg litre^{-1})	13	0·000 5	0·001 3	5	0·000 6	0·000 2
Nickel (mg litre^{-1})	14	0·037	0·009 1	5	0·046	0·008 7
Sodium (mg litre^{-1})	14	860	160	7	170	24
Potassium (mg litre^{-1})	14	230	21	7	120	27
Calcium (mg litre^{-1})	14	67	94	7	290	100
Aluminium (mg litre^{-1})	15	0·31	0·28	7	3·9	2·0
Total sulphur (mg litre^{-1})	19	6·3	3·8	14	8·8	7·4
Susp. solids (mg litre^{-1})	32	120	150	15	220	100
Manganese (mg litre^{-1})	19	0·45	0·49	13	5·8	2·8
Boron (mg litre^{-1})	12	2·7	0·15	6	1·9	0·27

leachate, the quality of which corresponded to typical Finnish leachate (Table 2).

RESULTS AND DISCUSSION

Vegetation

On landfills still in operation the vegetation coverage was low, approximately 25% as shown in Fig. 2. Altogether 306 plant species were

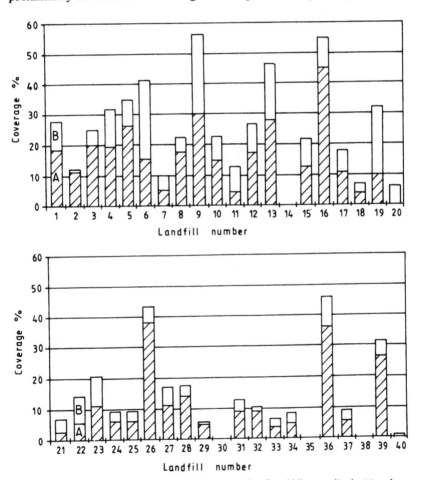

Figure 2. The coverage of vegetation at the landfills studied. Numbers 1–20 serve 3700–94 000 inhabitants, numbers 21–40 less than 3000 inhabitants. A = dominant species, B = other species (Ettala *et al.*, 1988a).

found. The 20 most abundant species, with a coverage of at least 10%, consisted of common weeds, meadow species and ruderals. Woody plants were scarce and natural afforestation took place slowly at sanitary landfills no longer in use. The main reasons for this were disturbance of the vegetation caused by vehicles and dumping of new refuse and cover soil, and unsuitable or insufficient cover material.

Good growth of grass vegetation was achieved with at least 10 cm of cover soil and a sufficiently fine-textured substrate, preferably silt or moraine-type soils. Organic material in the substrate had a growth-stimulating effect. Use of bark and sludge from a municipal sewage treatment plant as substrate supported good growth. The possibility of replacing the soil cover with waste material should be studied separately in each case.

In order to increase the present vegetation coverage of sanitary landfills, plantings are needed. Most of the short-rotation tree plantations at nine landfills developed well, even the six-year-old stands growing satisfactorily. *Salix aquatica* was the most productive species and therefore the best for increasing evapotranspiration. *Salix viminalis* also developed well, but had lower biomass production and in one case was eaten by hares. *Populus rasumowskyana* and *Betula pendula* proved to be possible alternatives for landscaping a site.

Irrigation with leachate had a beneficial effect on the growth of short-rotation plantations. The biomass production achieved is one of the highest values recorded in Finland and 10–15 times higher than that of normal Finnish forest (Table 3). Unirrigated stands lost their leaves early during the dry summer period. Growth of the stands was disturbed when leachate rich in salts was sprayed directly on the leaves or the

TABLE 3. Mean Annual Biomass Production of *Salix aquatica* at the Lahti Landfill in 1983–1987 (Ettala, 1988a)

Irrigation level mm/growing season	No. of stands	Biomass production (kg dry matter $m^{-2} year^{-1}$)		
		Leaves	Stem and branch	Total
0	2	0·32	0·73	1·05
500	3	0·55	1·71	2·26
1000	1	0·63	1·25	1·88

irrigation volume was too high. High sulphate concentrations caused growth disturbance even when leachate was applied to the ground with a hose. Growth was seldom disturbed by pests, disease or cold winters. The best growth was achieved on a substrate with a high humus content and a thickness of at least 0·2–0·3 m. Five years' irrigation with leachate raised substrate salinity. This did not cause any disturbance of growth, because salinity was kept low enough by occasionally using high-intensity irrigation to wash the substrate.

The study areas are located in southern Finland. Because of similarities in waste disposal technology, quality of refuse and hydrological conditions, short-rotation plantations can be used at sanitary landfills in Scandinavia, as is also indicated by the results of Anshelm (1986). The recommendation to use, for example, willow, poplar and birch to revegetate sites in France (Leroy, 1987) and the field studies on a reclaimed landfill site in Northern Ireland (Dawson, 1988) indicate that short-rotation tree plantations can be applied further south in Europe as well. However, the landfill characteristics should be studied in detail in each case.

Evapotranspiration

Throughfall averaged 67% of the sum of irrigation and precipitation, and stem flow 2%. The proportion of the irrigation and precipitation that was intercepted was 31% (Fig. 3).

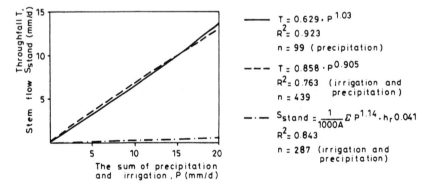

$$T = 0.629 \cdot P^{1.03}$$
$$R^2 = 0.923$$
$$n = 99 \text{ (precipitation)}$$

$$T = 0.858 \cdot P^{0.905}$$
$$R^2 = 0.763 \text{ (irrigation and}$$
$$n = 439 \qquad \text{precipitation)}$$

$$S_{stand} = \frac{1}{1000A} \varepsilon \, P^{1.14} \cdot h_r \, 0.041$$
$$R^2 = 0.843$$
$$n = 287 \text{ (irrigation and}$$
$$\text{precipitation)}$$

The sum of precipitation and irrigation, P (mm/d)

Figure 3. Throughfall and stemflow in the stands of *Salix aquatica* as a function of precipitation and irrigation at the Lahti landfill, 23 June–29 August 1986, h_r = relative height of the sprouts, A = area covered by the sprouts (m^2) (Ettala, 1988*b*).

TABLE 4. Transpiration Rate (E_t), Leaf Mass of the Stand (Y_{stand}) and per Unit Leaf Area (Y_l), Leaf Area Index (*LAI*) and Transpiration (E_{stand}) of the Stands of *Salix aquatica* at the Lahti Landfill in 1986 (Ettala, 1988*b*)

Period	Age of sprouts/ stump (years)	Irrigation level (mm/ growing season)	E_t (mmole m^{-2} s^{-1})	Y_{stand} (g dry matter m^{-2})	Y_l $\bar{x}$ (g dry matter)	s	LAI	$E_{t\,stand}$ (mm h^{-1})
14–17	2/4	0	0·44	480	47·0	11·7	10·2	0·29
June	2/4	500	2·00	740	52·9	13·9	14·0	1·81
	4/4	0	0·42	340	48·4	11·7	7·02	0·19
	4/4	500	1·09	690	47·0	10·3	14·7	1·04
12–15	2/4	0	2·31	540	70·2	8·2	7·70	1·15
July	2/4	500	2·52	830	70·7	10·9	11·7	1·92
	4/4	0	2·75	360	79·0	13·7	4·56	0·81
	4/4	500	1·98	770	76·7	11·3	10·0	1·29
1–4	2/4	0	2·10	560	59·1	12·8	9·48	1·29
Sep-	2/4	500	1·69	1010	76·2	17·8	13·3	1·45
tember	4/4	0	1·97	660	71·5	12·8	9·23	1·18
	4/4	500	1·71	410	77·0	18·7	5·32	0·59
	3/4	500	2·22	940	79·8	16·9	11·8	1·69

Irrigation increased transpiration significantly, especially during dry periods. Although the transpiration rates per unit leaf area did not depend on the irrigation level during rainy periods, transpiration from the irrigated stands was markedly higher than from unirrigated ones because of a difference in the leaf area index (Table 4).

According to the estimates of the evapotranspiration from the *Salix aquatica* stands (Fig. 4), the use of short-rotation tree plantations at sanitary landfills increases evapotranspiration and minimises leachate discharge. When the stands are irrigated with leachate, evapotranspiration can exceed the mean annual precipitation, even in Finland. The lysimeter studies of Grip (1981) with *Salix* stands agree with the evapotranspiration values achieved at the Lahti landfill.

Landfill Characteristics

As the results reported in the previous sections differ remarkably from those published elsewhere, it is necessary to examine the growth conditions at sanitary landfills in more detail.

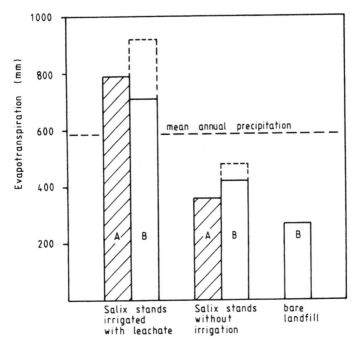

Figure 4. Calculations of evapotranspiration from stands of *Salix aquatica* based on biomass production (A) compared with values obtained using the Penman–Monteith formula (B) at the Lahti landfill in the period 1 May–30 September 1986 (Ettala, 1988*b*). The indicated variation of B is caused by different values assigned to surface resistance and aerodynamic resistance.

The quality of the refuse, gas and water in boreholes at the sites shows that the degradation stage varies in different parts of the sites. The acetic phase existed at two of the landfills studied. Besides low temperatures, other factors limiting achievement of the methanogenic phase were a low proportion of organic matter, a low moisture content and the presence of inhibiting compounds. The methane content in the landfill gas averaged 22% by volume (Ettala *et al.*, 1988*b*), being low from the point of view of gas recovery. Gas with one of the highest methane concentrations (67% vol.) was extracted from the refuse under the short-rotation tree plantations at the Lahti landfill. This, together with the good growth results, only 0·2–0·3-m deep rooting layer and observations of gas discharge from single small points, indicates that production of landfill gas does not notably interfere with revegetation measures at landfills.

The temperatures measured deeper in the refuse than 3·0 m at the seven landfills were low, mostly in the range of 10–15°C. In the upper part of the site, the seasonal variation is wide and even in fresh refuse the six-month average at a depth of 0·8 m was only 12·4°C. The results show that aerobic degradation was negligible. Moreover, temperatures below 15°C are unfavourable to the methanogenic phase, too. Increasing the depth of landfills diminishes seasonal variation in the refuse temperature and the leachate discharge in proportion to the amount of waste. The heat flux from a new, slightly compacted and covered landfill averaged 4·1 W m^{-2} and that from five-year-old refuse 1·4 W m^{-2}. The heat flux from a well-compacted and covered site was negligible (Fig. 5). The records indicated that the heat flux was favourable for the revegetation measures carried out.

The frost depth and snow water equivalent at the landfills were smaller than in natural soils. Short-rotation tree plantations seem to strengthen this effect (Ettala, 1988*c*). For this reason, the leachate discharge from a

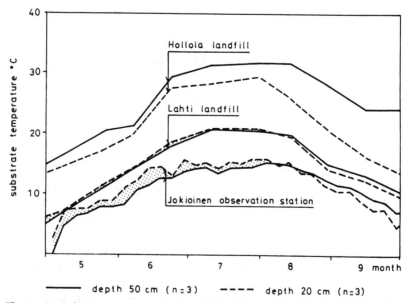

Figure 5. Substrate temperature in the stand of *Salix aquatica* irrigated with leachate at the Lahti landfill in 1986. For comparison the substrate temperature at the Hollola landfill and soil temperature at the Jokioinen observation station are presented (Ettala, 1988*b*).

sanitary landfill revegetated with willow stands will be smaller and steadier than the discharge from natural areas.

The infiltration rate in the refuse or soil cover at the landfills studied mostly exceeds 1 mm min^{-1}. In addition to the infiltration measurements, factors suggestive of low surface run-off are the low precipitation and the flatness of the sites in Finland and cracks in the soil cover at the landfills. Establishment of short-rotation tree plantations reduces the surface run-off still further.

Implications on Landfill Management

Sanitary landfills should be planned and managed in such a way that, for example, 30–50% of the site can be kept free from disposal activities for 5–10 years and revegetated with short-rotation tree plantations while the other part of the landfill is still in use.

The stands can be established when the refuse is at least two years old. The plantation should consist of several species, mainly *Salix aquatica* and *Salix viminalis*. In areas where landscaping is important, *Populus rasumowskyana* and *Betula pendula* can be used. The substrate should be rich in humus and 0·2–0·3-m thick. The final soil cover of the closed sites should be thicker, even when willow stands are used.

The stands need irrigation with leachate throughout the growing season, but this should not exceed 500 mm. Irrigation with sprinklers or water guns is preferable. Irrigation may increase the evapotranspiration by up to 400 mm per year which in some cases could outbalance the normal annual leachate generation. Short-rotation tree plantations should be coppiced a year after planting and the most suitable rotation time is then 3–4 years, the plantations consisting of stands at different stages of growth. Plantations of 50 000 cuttings per hectare can best be established gradually and the cuttings can be taken from the coppiced stands. A look-out should be kept for growth disturbance, and attention paid to leachate quality, the volume used for irrigation and its timing.

REFERENCES

Anshelm, L. (1986). Lakvattentillförsel i mark-växtsystem. Etapp rapport 2 (Application of leachate to ground-vegetation system, Report 2). Forskning och utveckling inom avfallsomr ådet. Eslöv. 72 pp. (In Swedish).

Dawson, W. M. (1988). Biomass on marginal land—an alternative energy source. *Journal of the Institute of Wastes Management*, **78**, 323–9.

Ettala, M. (1988a). Short-rotation tree plantations at sanitary landfills. *Waste Management & Research*, **6**, 291–302.

Ettala, M. (1988b). Evapotranspiration from a *Salix aquatica* plantation at a sanitary landfill. *Aqua Fennica*, **18**, 3–14.

Ettala, M. (1988c). Heat flux from a sanitary landfill. *5th International Solid Wastes Conference and Exhibition of the International Solid Wastes and Public Cleansing Association*, Copenhagen, Denmark. Academic Press, London, Vol. 1, pp. 109–14.

Ettala, M., Yrjönen, K. & Rossi, E. (1988a). Vegetation coverage at sanitary landfills in Finland. *Waste Management & Research*, **6**, 281–9.

Ettala, M., Rahkonen, P., Kitunen, V., Valo, O. & Salkinoja-Salonen, M. (1988b). Quality of refuse, gas and water at a sanitary landfill. *Aqua Fennica*, **18**, 15–28.

Grip, H. (1981). Evapotranspiration experiments in *Salix* stands. Technical report 15. Energy Forestry Project, Swedish University of Agricultural Sciences, Uppsala, Sweden, 29 pp.

Leroy, J.-B. (1987). Design criteria for landfills—the French concept. Proceedings of ISWA International Symposium on Process, Technology and Environmental Impact of Sanitary Landfills, Cagliari, Italy, 19th–23rd October 1987, Vol. II, 17 pp.

Monteith, J. L. (1965). Evaporation and environment. *Proc. Symp. Soc. Exp. Biol.*, **19**, 205–34.

2.4 Chemical and Biological Characteristics of Landfill Leachate

GIANNI ANDREOTTOLA

*CISA, Environmental Sanitary Engineering Centre,
Via Marengo 34, I-09123 Cagliari, Italy*

&

PIERO CANNAS

*CISA, Environmental Sanitary Engineering Centre,
Via Marengo 34, I-09123 Cagliari, Italy*

INTRODUCTION

Leachate composition has been the object of numerous research studies in the last twenty years because leachate treatment and disposal has been and still remains one of the main problems in sanitary landfill management. At the beginning of 1970 leachate quality studies were focused on a rough characterization of leachate components. Later on, research efforts shifted towards the identification of the main factors influencing the composition of leachate. In that period the most significant leachate quality models were developed. Recently detailed investigations have been carried out in order to identify and to evaluate hazardous components of leachate. In this chapter the main chemical, physical and biological factors which influence leachate quality are described. Monitoring data from numerous sanitary landfills all over the world are also presented.

BIOCHEMICAL PROCESSES IN SANITARY LANDFILL

The mechanisms which regulate mass transfer from wastes to leaching water, from which leachate originates, can be divided into three categories:

—hydrolysis of solid waste and biological degradation
—solubilization of soluble salts contained in the waste
—dragging of particulate matter

The first two categories of mechanism, which have greater influence on the quality of leachate produced, are included in the more general concept of waste stabilization in landfills. In this chapter the various phases of waste stabilization process are described and the mechanisms that lead to mass transfer from wastes to leachate are discussed.

Aerobic Degradation Phases

The first phase of aerobic degradation of organic substances is generally of limited duration due to the high oxygen demand of waste relative to the limited quantity of oxygen present inside a landfill. The only layer of a landfill involved in aerobic metabolism is the upper layer where oxygen is trapped in fresh waste and is supplied by rainwater. In this phase (Phase I, Fig. 1) it was observed that proteins are degraded into amino acids, thus into carbon dioxide, water, nitrates and sulphates, typical catabolites of all aerobic processes (Barber, 1979). Carbohydrates are converted to carbon dioxide and water and fats are hydrolysed to fatty acids and glycerol and are then further degraded into simple catabolites through intermediate formation of volatile acids and alkalis. Cellulose, which constitutes the majority of organic fraction of wastes, is degraded by means of extracellular enzymes into glucose which is used subsequently by bacteria and converted to carbon dioxide and water (Bevan, 1967). This stage, due to the exothermicity of reactions of biological oxidation, may reach elevated temperatures if the waste is not compacted. Usually the aerobic phase is short and no substantial leachate generation will take place.

In old landfills, when only the more refractory organic carbon remains in the landfilled wastes, a second aerobic phase will appear in the upper layer of the landfill. In this phase (Phase V, Fig. 1) the methane production rate is so low that air will start diffusing from the atmosphere, giving rise to aerobic zones and zones with redox potentials too high for methane formation (Christensen & Kjeldsen, 1989).

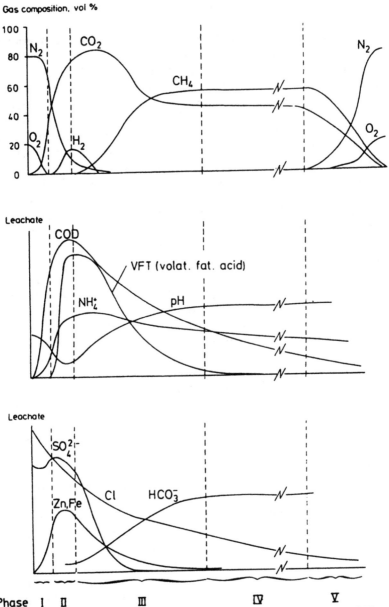

Figure 1. Illustration of developments in leachate and gas in a landfill cell (Christensen & Kjeldsen, 1989).

Anaerobic Degradation Phases

Three different phases can be identified in the anaerobic decomposition of waste. The first phase (Phase II, Fig. 1) of anaerobic degradation is acid-fermentation, which causes a decrease in leachate pH, high concentrations of volatile acids and considerable concentrations of inorganic ions (for example, Cl^-, SO_4^{2-}, Ca^{2+}, Mg^{2+}, Na^+). The initial high content of sulphates may slowly be reduced as the redox potential drops. The generated sulphides may precipitate iron, manganese and heavy metals that were dissolved by the acid-fermentation (Christensen & Kjeldsen, 1989). The decrease in pH is caused by the high production of volatile fatty acids and the high partial pressure of CO_2. The increased concentration of anions and cations is due to lixiviation of easily soluble material including that originally available in the waste mass and that made available by degradation of organic substances. Initial anaerobic processes are elicited by a population of mixed anaerobic microbes, composed of strictly anaerobic bacteria and facultative anaerobic bacteria. The facultative anaerobic bacteria aid in the breakdown of materials and reduce the redox potential so that methanogenic bacteria can grow. In fact, the latter are extremely sensitive to the presence of oxygen and require a redox potential below $-330\,mV$ in order to carry out their function (Christensen & Kjeldsen, 1989). Leachate from this phase is characterized by high BOD_5 values (commonly $>10\,000\,mg/litre$), high BOD_5/COD ratios (commonly >0.7) and acidic pH values (typically 5–6) and ammonia (often 500–$1000\,mg/litre$) (Robinson, 1989), the latter due to hydrolysis and fermentation of proteinous compounds in particular.

The second phase (Phase III, Fig. 1) is intermediate anaerobiosis and starts with slow growth of methanogenic bacteria. This growth may be inhibited by an excess of organic volatile acids which are toxic to methanogenic bacteria at concentrations of 6000–$16\,000\,mg/litre$ (Stegmann & Spendlin, 1989). The methane concentration in the gas increases, whilst hydrogen, carbon dioxide and volatile fatty acids decrease. Moreover, the concentration of sulphate decreases owing to biological reduction. Conversion of fatty acids causes an increase in pH values and alkalinity with a consequent decrease in solubility of calcium, iron, manganese and heavy metals. The latter are probably precipitated as sulphides. Ammonia is released and is not converted by an anaerobic environment.

The third phase (Phase IV, Fig. 1) of anaerobic degradation is characterized by methanogenic fermentation elicited by methanogenic bacteria. The pH range tolerated by methanogenic bacteria is extremely limited and ranges from 6 to 8. At this stage, the composition of leachate is characterized by almost neutral pH values, low concentrations of volatile acids and total dissolved solids whilst biogas presents a methane content which is generally higher than 50%. This confirms that solubilization of the majority of organic components has decreased at this stage of landfill operation, although the process of waste stabilization will continue for several years.

Leachates produced during this phase are characterized by relatively low BOD values and low ratios of BOD/COD. Ammonia continues to be released by the first stage acetogenic process.

To partially summarize this paragraph, Table 1 reports leachate parameters which, according to the value reached, indicate the degree of stabilization reached by the landfill.

TABLE 1. Leachate Indicator Parameters (Pohland, 1989)

Parameter identity	*Utility for phase description*
Physical	
pH^a	Acid–base/stabilization phase indicator
ORP^a	Oxidation–reduction/stabilization phase indicator
Conductivity	Ionic strength/activity indicator
Temperature	Reaction indicator
Chemical	
COD^a, TOC, TVA^a	Substrate indicators
TKN^a, NH$_3$-N^a, PO$_4$-P^a	Nutrients indicators
SO$_4$/S^a, NO$_3$/NH$_3$^{a}	Stabilization phase indicators
TS, chloride	Dilution/mobility indicator
Total alkalinity^a	Buffer capacity indicator
Alkali/alkaline earth metals	Toxicity/environmental effects indicators
Heavy metals	Toxicity/environmental effects indicators
Biological	
BOD$_5$	Substrate/biodegradability
Total/faecal coliforms	Health effect indicators
Faecal streptococci	Health effect indicators
Viruses	Health effect indicators
Pure/enrichment cultures	Stabilization phase indicators

^{a} Parameters frequently used for evaluation.

CHEMICAL COMPOSITION OF LEACHATE

Factors Affecting Leachate Composition

The chemical composition of leachate depends on several parameters, including those concerning waste mass and site localization and those deriving from design and management of the landfill. These latter factors are discussed in Chapter 2.7. Of the former the main factors influencing leachate quality are:

Waste composition. The nature of the waste organic fraction influences considerably the degradation of waste in the landfill and thus also the quality of the leachate produced. In particular, the presence of substances which are toxic to bacterial flora may slow down or inhibit biological degradation processes with consequences for the leachate. The inorganic content of the leachate depends on the contact between waste and leaching water as well as on pH and the chemical balance at the solid–liquid interface. In particular, the majority of metals are released from the waste mass under acid conditions.

pH. pH influences chemical processes which are the basis of mass transfer in the waste leachate system, such as precipitation, dissolution, redox and sorption reactions. It will also affect the speciation of most of the constituents in the system. Generally, acid conditions, which are characteristic of the initial phase of anaerobic degradation of waste, increase solubilization of chemical constituents (oxides, hydroxides and carbonated species), and decrease the sorptive capacity of waste.

Redox potential. Reducing conditions, corresponding to the second and third phases of anaerobic degradation, will influence solubility of nutrients and metals in leachate.

Landfill age. Variations in leachate composition and in quantity of pollutants removed from waste are often attributed to landfill age, defined as time measured from the deposition of waste or time measured from the first appearance of leachate. Landfill age obviously plays an important role in the determination of leachate characteristics governed by the type of waste stabilization processes. It should be underlined that variations in composition of leachate do not depend exclusively on landfill age but on the degree of waste stabilization and volume of water which infiltrates into the landfill. The pollutant load in leachate generally reaches maximum values during the first years of operation of a landfill (2–3 years) and then gradually decreases over following years. This trend is generally applicable to organic components, main indicators of organic pollution (COD, BOD, TOC), microbiological population and to main inorganic ions (heavy metals, Cl, SO_4, etc.).

TABLE 2. Chemical Composition of Landfill Leachate (Ehrig, 1989; Andreottola *et al.*, 1990)

Parameter	Range
COD (mg/litre)	150–100 000
BOD_5 (mg/litre)	100–90 000
pH	5·3–8·5
Alkalinity ($mgCaCO_3$/litre)	300–11 500
Hardness ($mgCaCO_3$/litre)	500–8 900
NH_4 (mg/litre)	1–1 500
N_{org} (mg/litre)	1–2 000
N_{tot} (mg/litre)	50–5 000
NO_3 (mg/litre)	0·1–50
NO_2 (mg/litre)	0–25
P_{tot} (mg/litre)	0·1–30
PO_4 (mg/litre)	0·3–25
Ca (mg/litre)	10–2 500
Mg (mg/litre)	50–1 150
Na (mg/litre)	50–4 000
K (mg/litre)	10–2 500
SO_4 (mg/litre)	10–1 200
Cl (mg/litre)	30–4 000
Fe (mg/litre)	0·4–2 200
Zn (mg/litre)	0·05–170
Mn (mg/litre)	0·4–50
CN (mg/litre)	0·04–90
AOX[a] (μgCl/litre)	320–3 500
Phenol (mg/litre)	0·04–44
As (μg/litre)	5–1 600
Cd (μg/litre)	0·5–140
Co (μg/litre)	4–950
Ni (μg/litre)	20–2 050
Pb (μg/litre)	8–1 020
Cr (μg/litre)	30–1 600
Cu (μg/litre)	4–1 400
Hg (μg/litre)	0·2–50

[a] Adsorbable organic halogen.

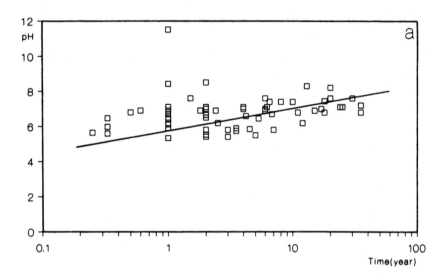

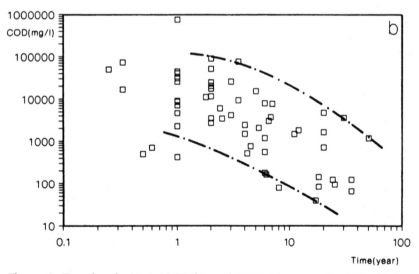

Figure 2. Trends of pH(a), COD(b) and BOD₅(c) versus landfill age for several landfill leachates (Andreottola *et al.*, 1990).

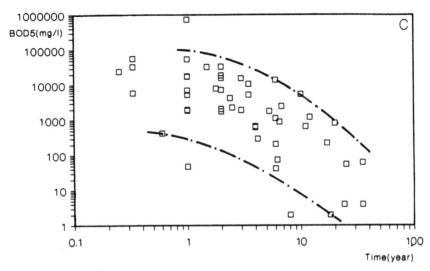

Figure 2—*contd.*

Leachate Chemical Characteristics

This summary of leachate quality is based on data from the technical literature concerning more than 70 municipal solid waste sanitary landfills in Europe and the USA (Andreottola *et al.*, 1990).

Besides the previously discussed factors, the procedures used for sampling, conservation and pretreatment, and analytical methods used to characterize leachate, can also have a considerable influence (see Chapter 2.9).

Table 2 reports concentration ranges for the main parameters of landfill leachate. Figures 2–5 show trends of more significant parameters according to landfill age. Time trends of organic compound concentrations (BOD_5, COD) clearly show that passing from acid phase to methanogenic phase leads to a notable decrease in concentrations. Similarly, pH shows a tendency to increase according to evolution of the biochemical processes previously described.

The trend of organic compounds present in leachate versus landfill age, expressed as a percentage of TOC, is shown in Fig. 6. This figure shows that the production of volatile acids, corresponding to the first stage of anaerobic degradation, is dominant in the first years of landfill operation. The hydroxyl aromatic compounds present in humic and

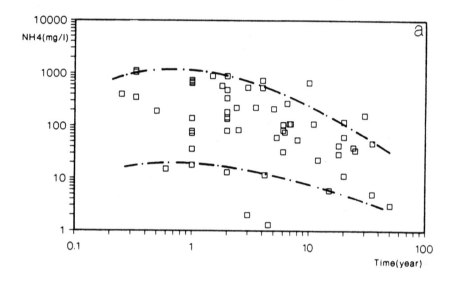

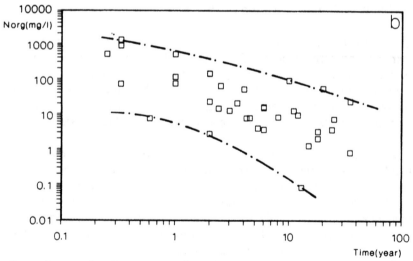

Figure 3. Trends of NH$_4$ (a) and N$_{org}$ (b) versus landfill age for several landfill leachates (Andreottola *et al.*, 1990).

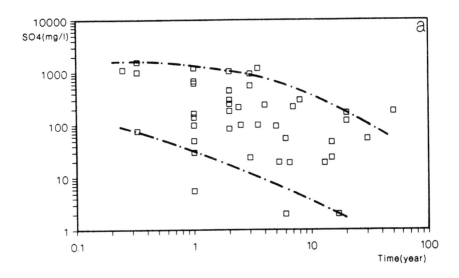

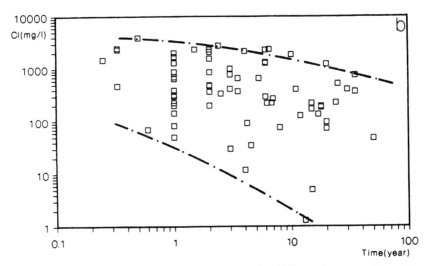

Figure 4. Trends of SO₄ (a) and Cl (b) versus landfill age for several landfill leachates (Andreottola *et al.*, 1990).

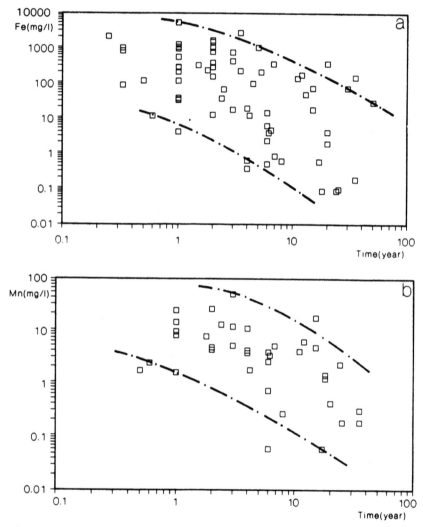

Figure 5. Trends of Fe(a), Mn(b) and Zn(c) versus landfill age for several landfill leachates (Andreottola *et al.*, 1990).

fulvic-like fractions of leachate organics show a slight decrease according to landfill age but represent, however, the greater portion of TOC observed in older landfills.

The concentrations of ammonia are high in leachate following hydrolysis and fermentation of the proteic fraction of the biodegradable

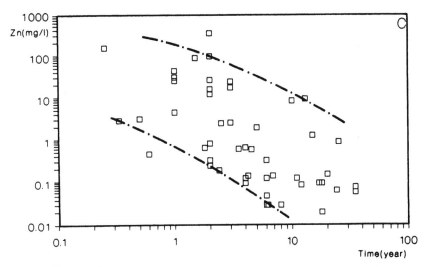

Figure 5—*contd.*

substrate. At the onset of the second intermediate anaerobic phase, the concentration of ammonia tends to decrease slowly in agreement with the theoretical trend reported in Fig. 1. Moreover, Fig. 3 shows the trend for organic nitrogen and the tendency to decrease can be noted.

Figure 4 shows concentrations of sulphates and chlorides in leachate as a function of time. With regard to sulphates, concentrations observed in a young landfill are high. Reduction to sulphides induced in an anaerobic environment determines a gradual decrease of sulphate content in the landfill. The production of sulphides may induce precipitation of various heavy metals contained in the leachate. The high initial concentration of chlorides decreases with landfill age owing to the washing phenomena.

Figure 5 reports the concentrations of iron, zinc and manganese, chosen as the more representative of metals present in leachate. During the first phase of landfill operation there is a high degree of metal solubilization due to low pH values caused by the high production of organic acids whilst, as landfill age increases, the consequent increase in pH causes a minor degree of solubilization as shown in the trend reported in Fig. 5.

Stabilization of organic compounds influences lixiviation of trace metals. Many organic compounds containing nitrogen, oxygen and sulphur can form soluble complexes with the metals and thus increase the metal concentrations (see, for example, Lun & Christensen (1989)).

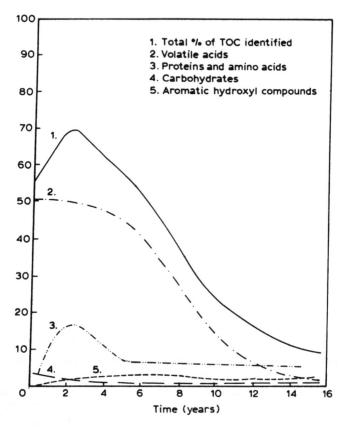

Figure 6. Trend in the identified fraction of leachate TOC versus landfill age (Chian & DeWalle, 1977).

Humic and fulvic acids are considered strong complexing ligands and may play an important role in the long-term release of heavy metals from landfills. Also inorganic ligands, in particular, chloride, may complex heavy metals.

The phenomena of adsorption and complexation are probably those mainly responsible for attenuation and mobilization of trace metals in the waste mass. Under oxidizing conditions, adsorption may regulate the concentration of a constituent to below the concentrations controlled by chemical precipitation. Solid lignin-type compounds may sorb trace metals from the leachate.

TABLE 3. Specific Organic Compounds Identified in Landfill Leachates (Shridharan & Didier, 1988, modified)

Parameter	Range	Median
Acenaphthene	13·9–21·3	17·60 µg/litre
Benzene	1–1 630	11·10 mg/litre
Bis-2-ethylhexyl phthalate	91–7 900	1 050 µg/litre
Butyl benzyl phthalate	10–64·1	37·05 µg/litre
Carbon tetrachloride	3–995	28·00 µg/litre
Chlorobenzene	3–188	25·20 µg/litre
Chloroethane	2–730	17 µg/litre
Chloroform	4·4–16	7·14 µg/litre
Di-N-butyl phthalene	13–540	28·70 µg/litre
Di-N-octyl phthalene	16·1–542	110 µg/litre
Dibromochloromethane	22–160	91 µg/litre
Dichlorodifluoromethane	100–242·1	171·05 µg/litre
Dichloromethane	27·6–58 200	483 µg/litre
Diethyl phthalate	12–230	44 µg/litre
Ethylbenzene	1–1 680	43·50 µg/litre
Fluoranthene	9·56–723	39·10 µg/litre
Fluorene	21–32·6	26·80 µg/litre
Fluorotrichloromethane	1–183	34 µg/litre
Formaldehyde	1–1·4	1·20 mg/litre
Halogen, total organic	0·0039–33 400	623·50 µg/litre
Isophorone	3·18–520	76 µg/litre
Methyl ethyl ketone	2 100–37 000	19 550 µg/litre
Naphthalene	4·6–186	33·75 µg/litre
p-dichlorobenzene	2–250	14 µg/litre
Phenanthrene	8·1–1 220	50·70 µg/litre
Phenol	1·1–2 170	174 µg/litre
Phenolics, total	0·052–19 000	619 µg/litre
Tannin and lignin, combined	0·12–264	1·94 mg/litre
Tetrachloroethylene	1–232	16·30 µg/litre
Tetrahydrofuran	410–1 400	730 µg/litre
Toluene	1–11 800	360 µg/litre
Trichloroethylene	1–372·2	19 µg/litre
Vinyl chloride	10–3 000	230 µg/litre
Xylene	9·4–240	72·50 µg/litre

Aside from the leachate constituents presented in Table 2, many other organic compounds have been identified in landfill leachates. Tables 3 and 4 present, respectively, the organic compounds and classes reported by several investigations. Particular care must be given to the presence of hazardous classes like AOX and PAH. In particular, AOX have been found in all tested German MSW landfills (Lemmer, 1986).

TABLE 4. Organic Classes and Substances Identified in Landfill Leachates (Lu *et al.*, 1982 modified)

Author	Compound
Burrows & Rowe (1975)	Acetone
	Short chain alcohols
	Short chain acids
Khare & Dondero (1977)	Alkanes
	Ketones (acetone, 2-butanone)
	PAH
	Short chain alcohols
	Short chain amines
	Short chain acids
Robertson *et al.* (1974)	Phthalate esters
	PAH
	Alcohols
	Methylpyridine
	Ethers
	Short chain acids
Engers (1978)	Short chain acids
	(C1–C7)
Shridharan & Didier (1988)	PAH
	AOX

MICROBIOLOGICAL COMPOSITION OF LEACHATE

Contrary to the chemical characteristics of leachate which have been widely reported, data on the microbiological composition are scarce.

Municipal solid waste (MSW) contains a large microbial population and may be contaminated by pathogenic micro-organisms. Wastes often contain animal excrement, animal carcasses, diapers, sewage sludges and sometimes hospital waste (in those countries where disposal of this type of waste is still permitted) which act as a vehicle for pathogenic micro-organisms and thus represent a health risk.

The first microbiological studies carried out on sanitary landfills investigated and emphasized hygienic and sanitary considerations focusing on micro-organisms, especially bacteria and viruses of faecal origin, present in waste and in leachate (e.g. Engelbrecht *et al.*, 1974). More recent studies have started to closely examine counts and physiological activity of micro-organisms involved in stabilization of MSW, including the extracellular enzymatic activity and rapid chemical methods to estimate the number of methanogenic micro-organisms.

Bacteria

Various studies observed a significant leachate-connected bacterial population with varying entity and composition according to landfill age (Donnelly *et al.*, 1981; Lu *et al.*, 1982; Barlaz *et al.*, 1989; Sleat *et al.*, 1989). The presence of faecal streptococci indicates faecal contamination. The presence of faecal coliforms indicates contamination caused by warm-blooded animals. Total coliforms have also been used to indicate the possible presence of pathogenic constituents.

It has been observed that survival of bacteria in a sanitary landfill is inversely proportional to the temperature in the landfill. Bacterial growth and survival are inhibited at temperatures of 60°C and above. The bacterial activity decreases on a par with pH and the synergic effect of temperature and pH further accelerates this phenomenon (Engelbrecht & Amirhor, 1975).

The concentration of faecal indicator bacteria generally decreases as landfill age increases and ultimately reaches levels which are no longer detectable. In lysimeter tests containing sewage sludge and MSW, the presence of streptococci was no longer observed after two years (Donnelly *et al.*, 1981).

Viruses

Contrary to bacteria, viruses are parasitic organisms which are unable to multiply outside the host. The presence of enteroviruses in leachate can be attributed to the presence of faecal material of varying origin in MSW. Studies carried out on leachate from MSW landfills only reported the presence of enteroviruses in extremely rare cases (Engelbrecht, 1973; Cooper *et al.*, 1974; Sobsey, 1975, 1978).

Fungi

Little information is available in the literature on the presence of fungi in leachate. Species observed by various researchers are mainly saprophytes such as *Aspergillus, Penicillum* and *Fusarium* which are not pathogenic.

The only pathogenic fungus observed is *Allescheriaboydii* which can cause madura foot abscesses (Cook *et al.*, 1967; Donnelly *et al.*, 1981).

Parasites

There are no data in the literature concerning the presence of parasites in leachate. Parasites such as protozoa, helminths and nematodes could be observed in leachate in the presence of animal and human faeces in landfills and sewage sludges. Parasitic cysts and eggs are extremely resistant faecal micro-organisms. Conditions which deactivate bacteria and viruses are generally ineffective for parasites, in particular for nematodes and helminths (Hays, 1977).

LEACHATE QUALITY MODELS

The development of mathematical models for the prediction of leachate quality can lead to the following advantages (Lu *et al.*, 1982):

—to assess the potential impacts of leachate on receiving waters;
—to aid in the design of a leachate treatment system;
—to help estimate the concentration of contaminants entering an underlying groundwater system so as to determine leachate concentration in that system;
—to assess the possible effects of co-disposal of various liquid and/or semi-solid wastes with municipal solid waste in landfills.

Modelling efforts which describe and predict leachate quality always include a sub-model, more or less complex, for the prediction of leachate quantity. There is, in fact, a strong correlation between leachate quality and the total volume of leachate produced. There are generally two approaches used when attempting to model leachate quality as a function of time. The first approach is to quantitatively describe the physical, chemical and biological processes which occur (or are presumed to occur) within the landfill. The second approach is based on regression of historical data.

Mechanistic Models

Mechanistic models are generally very complex, because they have to describe a dynamic system, and need therefore some simplifying assumptions. Several authors have developed models using both theoretically and empirically derived mass-transfer equations (Qasim & Burchinal, 1970; Phelps from Lu *et al.*, 1982; Straub & Lynch, 1982*a,b*). Some models consider the landfill as a packed column where a solution percolating through the column picks up contaminants from the solid phase at a rate proportional to the difference in concentration between the compound in the liquid phase and the compound adsorbed on the solid particles (Qasim & Burchinal, 1970; Phelps from Lu *et al.*, 1982).

Qasim & Burchinal tested their model on chloride using experimental columns. The model is responsive to physical parameters such as the depth of the waste, waste compaction, and cumulative leachate volume. The concentrations of various leachate components (acidity, alkalinity, hardness, BOD, N, P, Ca, Fe, Mg, K, Na, S, tannin and lignin) were estimated on the basis of chloride behaviour. Experimental and theoretical concentrations of leachate constituents in leachate samples showed fair agreement, with maximum deviations of 30%. Theoretical calculations tended to underestimate chloride concentrations as leaching continued.

The model developed by Phelps, built upon both theoretical and empirical parameters, was tested on the following leachate components: BOD, COD, total residue, Ca, Fe and Cl. Their concentrations in leachate were described as a function of time, leachate flow rate and depth of waste. Correlation coefficients between the calculated and observed curves for concentration versus time exceeded 0·9 for young leachate (not older than twice the time of its first appearance). For older leachate the observed concentration values were consistently above those calculated.

Finally, a series of models have been developed by Straub & Lynch (1982*a,b*) which describe separately the leaching processes of organic and inorganic constituents of leachate. Unlike the previous models, however, the model complexity is greatly increased, because of the formulation of equations describing hydraulic, physical/chemical and microbial processes. The processes described by the model include hydraulic behaviour, inorganic leachate composition and organic leachate composition. The concentrations and relative strength of inorganic contaminants, represented by total solids, are estimated by two sub-models. One sub-model presents the landfill as a single well-mixed reactor, providing mass balances for both moisture and contaminant yields. The leaching pattern of inorganic constituents is described as a function of the net infiltration rate of moisture, the volume and field capacity of the waste, and the rate of generation of dissolved substances. The other is a vertically distributed model of contaminant transport in an unsaturated porous medium. Model parameters include leachable contaminant mass, waste moisture content, volume, depth, density and field capacity of waste and moisture application rates.

Three models were developed for the description of the organic leachate composition. They are based on an unsaturated flow of moisture, contaminant generation and transport, oxygen exchange and microbial activity. The roles of aerobic and anaerobic bacteria are simulated using conventional kinetic formulations. The first model, similar to that used for inorganic leachate components, assumes the landfill as a single well-mixed reactor. The second is a vertically-cascaded well-mixed reactor model. The third incorporates unsaturated flow and contaminant transport modelling to simulate organic leachate strength. Each of the reactor models is responsive to landfill volume, void volume, moisture content at field capacity, moisture flow rates, concentrations of substances in moisture entering the landfill, ultimate mass of leachable constituents per unit bulk volume of waste and an assumed maximum constituent concentration. The three models have been tested simulating the leachate COD concentration. Each of them closely approximates experimental COD data, but the unsaturated flow and contaminant transport model is the most precise.

Regression Models

The second approach is based on empirical equations fitting, in the best way, historical data sets on leachate quality. These fitting models are simpler than the mechanistic models and are generally effective only in modelling the landfill from which the data are derived, because leachate quality is so strongly affected by waste composition and landfill operational schemes. For this reason, this type of model is not strictly predictive, because it needs site-specific leachate data to be developed.

The first regression model was developed by Revah & Avnimelech (1979) modelling the concentration of leachate constituents (TOC, VA, TKN, NH_4, NO_3, Fe and Mn) in waste-filled columns. The empirical equation describing the quality of leachate is a simple exponential one

$$C = a \cdot b^t$$

where

C = concentration of the leachate constituent

t = time

a = concentration at $t = 0$

b = exponential base

The verification of this model showed a general similarity with field data.

A second fitting model was proposed by Wigh (1979), who proposed a more complex empirical equation to describe the relationship between leachate concentration of various constituents and the total volume of leachate produced

$$C = k_1 . k_2 . M/(k_2 - k_1) . (e^{-k_1} . V - e^{-k_2} . V)$$

where

C = pollutant concentration

M = total leachable mass per unit surface area

V = volume of leachate collected per unit surface area

k_1, k_2 = empirically fitted rates constants

The model fitted well to the historical leachate quality data set.

With regard to long-term evaluations, a model for evolution of landfills has been proposed (Belevi & Baccini, 1989) in which the general trend of leachate versus time is evident. The parameter 'specific flow', used as a key element, represents the yearly flow of a compound per mass unit of waste. On the basis of data from four landfills, the oldest of which has been operating for 10 years, the model proposed is able to supply specific flows of compound versus time. The concept of stabilization should be taken as the situation in which leachate concentrations do not exceed freshwater concentrations by more than a factor of ten. Referring to Swiss law, Belevi & Baccini (1989) claim that organic substances constitute the more important species to keep under observation; they also observe that, for the landfills examined, all metals analysed reach 'stabilization' within 10 years.

CONCLUSIONS

Although the basic biochemical processes which occur inside a sanitary landfill have been identified, it is not yet possible to precisely correlate leachate quality with the type of waste disposed, methods of management and age of landfill.

Leachate quality is strictly linked to physical, chemical and biological processes which occur in sanitary landfills. In particular the leachate from the stable methanogenic phase in landfill exhibits decreased pollutant concentrations.

Currently, the validity of models on leachate quality is strictly limited to the case studied, because of the existing lack of knowledge of the description of the basic processes occurring in the landfill environment and of the high variability of many important factors (waste composition, operation schemes, etc.), not only from landfill to landfill but also within the same facility. For these reasons leachate quality models are generally used for research purposes only.

REFERENCES

Andreottola, G., Cannas, P. & Cossu, R. (1990). Overview on landfill leachate quality. CISA, Environmental Sanitary Engineering Centre, Technical Note No. 3.

Barber, C. (1979). Behaviour of wastes in landfills. Review of processes of decomposition of solid wastes with particular reference to microbiological changes and gas production. Water Research Centre, Stevenage Laboratory Report LR 1059, Stevenage, UK.

Barlaz, M. A., Schaefer, D. M. & Ham, R. K. (1989). Bacteria population development and chemical characteristics of refuse decomposition in a simulated sanitary landfill. *J. Applied and Environmental Microbiology*, **55**, 55–65.

Belevi, H. & Baccini, P. (1989). Water and elements fluxes from sanitary landfills. In: *Sanitary Landfilling: Process Technology and Environmental Impact*, ed T. H. Christensen, R. Cossu & R. Stegmann. Academic Press, London.

Bevan, R. E. (1967). Notes on the science and practice of controlled tipping. Institute of Public Cleansing, London.

Burrows, W. D. & Rowe, R. S. (1975). Ether soluble constituents of landfill leachate. *Journal of Water Pollution Control Federation*, **47** (5), 921.

Chian, E. S. K. & DeWalle, F. B. (1977). Evaluation of leachate treatment, Volume 1; characterization of leachate. EPA 600/2-77-186a. US EPA, Cincinnati, OH.

Christensen, T. H. & Kjeldsen, P. (1989). Basic biochemical processes in landfills. In: *Sanitary Landfilling: Process, Technology and Environmental Impact*, ed. T. H. Christensen, R. Cossu & R. Stegmann. Academic Press, London.

Cook, H. H., Cromwell, D. C. & Wilson, H. A. (1967). Microorganisms in household refuse and seepage water from sanitary landfill. *Proc. W. Va. Acad. Sci.*, **39**, 107–14. (From Lu *et al.*, 1982.)

Chemical and Biological Characteristics of Landfill Leachate   87

Cooper, R. C., Potter, J. L. & Leong, C. (1974). Virus survival in solid waste treatment systems. In: *Virus Survival in Water and Wastewater Systems*, ed. J. F. Malina & B. P. Sagik. Center for Research in Water Resources, University of Texas, Austin, TX, pp. 218–32. (From Lu *et al.*, 1982.)

Donnelly, J. A., Scarpino, P. V. & Brunner, D. (1981). Recovery of fecal-indicator and pathogenic microbes from landfill leachate. Land Disposal: Municipal Solid Waste. US EPA-600/9-81-0029, US EPA, Cincinnati, OH.

Ehrig, H. J. (1989). Leachate quality. In: *Sanitary Landfilling: Process, Technology and Environmental Impact*, ed. T. H. Christensen, R. Cossu & R. Stegmann. Academic Press, London.

Engelbrecht, R. S. (1973). Survival of viruses and bacteria in a simulated sanitary landfill. NTIS/PB-234 589, Springfield, VA.

Engelbrecht, R. S. & Amirhor, P. (1975). Inactivation of enteric bacteria and viruses in sanitary landfills leachate. NTIS/PB-252 973/AS, Springfield, VA.

Engelbrecht, R. S., Weber, M. J., Amirhor, P., Foster, D. H. & LaRossa, D. (1974). Biological properties of sanitary landfill leachates. In: *Virus Survival in Water and Wastewater Systems*, ed. J. F. Mallina & B. P. Sagik. Center for Research in Water Resources, University of Texas, Austin, TX, pp. 210–17. (From Lu *et al.*, 1982.)

Engers, L. W. (1978). Mineralization of organic matter in the subsoil of a waste disposal site: a laboratory experiment. *Soil Science*, **126**, 22–8.

Hays, B. D. (1977). Potential for parasitic disease transmission with land application of sewage plant effluents and sludges. *Water Research*, **11**(7), 583–95.

Khare, M. & Dondero, N. C. (1977). Fractionation and concentration of volatiles and organics on high vacuum system: examination of sanitary landfill leachate. *Environmental Science and Technology*, **11**(8), 814–19.

Lemmer, F. (1986). Möglichkeiten der Behandlung von Sickerwasser durch verfahren nach dem Stand der Technik. ATV Dokumentation 4, Deponiesickerwasser—ein Problem der Abwassertechnik?, pp. 35–41.

Lu, J. C. S., Eichenberger, B. & Stearns, R. J. (1982). Production and management of leachate from municipal landfills: summary and assessment. Office of Research and Development, US EPA, Cincinnati, OH.

Lun, X. Z. & Christensen, T. H. (1989). Cadmium complexation by solid waste leachates. *Water Research*, **23**, 81–7.

Phelps, D. H. (1982). Solid waste leaching model. Department of Civil Engineering, University of British Columbia, Canada. (From Lu *et al.*, 1982.)

Pohland, F. G. (1989). Leachate recirculation for accelerated landfill stabilisation. Sardinia '89 Symposium, Porto Conte, Italy, 9–13 October.

Qasim, S. R. & Burchinal, J. C. (1970). Leaching of pollutants from refuse beds. *J. Sanitary. Eng. Div. ASCE*, **96**(SA1), 49–58.

Revah, A. & Avnimelech, Y. (1979). Leaching of pollutant from sanitary landfill models. *J. Water Pollution Control Federation*, **51**(11), 2705–16.

Robertson, J. M., Toussaint, C. R. & Jorque, M. A. (1974). Organic compounds entering ground water from landfill. US EPA 660/2-74-077, Washington, DC.

Robinson, H. D. (1989). Development of methanogenic conditions within landfill. Sardinia '89 Symposium, Porto Conte, Italy, 9–13 October.

Shridharan, L. & Didier, P. (1988). Leachate quality from containment landfills in Wisconsin. *ISWA 88 Proceedings*, ed. L. Andersen & J. Moller, Vol. 2. Academic Press, London.

Sleat, R., Harries, C., Viney, L. & Rees, J. F. (1989). Activities and distribution of the key microbial groups in landfill. In *Sanitary Landfilling: Process, Technology and Environmental Impact*, ed. T. H. Christensen, R. Cossu & R. Stegmann. Academic Press, London.

Sobsey, M. D. (1975). Studies on the survival and fate of enteroviruses in experimental model of a municipal solid waste landfill and leachate. *J. Applied Microbiology*, **30**(12), 565–74. (From Lu *et al.*, 1982.)

Sobsey, M. D. (1978). Field survey of enteric viruses in solid waste landfill leachate. *Am. J. Public Health*, **68**(9), 858–63. (From Lu *et al.*, 1982.)

Stegmann, R. & Spendlin, H. H. (1989). Enhancement of degradation: German experiences. In: *Sanitary Landfilling: Process, Technology and Environmental Impact*, ed. T. H. Christensen, R. Cossu & R. Stegmann. Academic Press, London.

Straub, W. A. (1980). Development and application of models of sanitary landfill leaching and landfill stabilization. RP.#259. Resource Policy Center, Thayer School of Engineering, Dartmouth College, Hanover, NH. (From Lu *et al.*, 1982.)

Straub, W. A. & Lynch, B. R. (1982a). Models of landfills and leaching: moisture flow and inorganic strength. *J. Environ. Eng. Div.*, ASCE, **108**(EE2), 231–49.

Straub, W. A. & Lynch, B. R. (1982b). Models of landfills and leaching: organic strength. *J. Environ. Eng. Div. ASCE*, **108**(EE2), 251–68.

Wigh, R. J. (1979). Boone county fields site interim report, test cells 2A, 2B, 2C, 2D. EPA-600/2-79-058, U.S. EPA, Cincinnati, OH.

2.5 Ecotoxicological Characteristics of Landfill Leachate

PREBEN KRISTENSEN

Water Quality Institute, 11 Agern Allé, DK-2970 Hørsholm, Denmark

INTRODUCTION

Leachate most often contains a complex variety of organic and inorganic compounds. Its constituents have the potential to leach to both ground-water and surface waters with associated contamination of drinking water resources and deleterious effects on aquatic ecosystems. Comprehensive chemical analysis programmes may provide detailed information on specific chemicals within the mixture and the fate and effects of these chemicals may be evaluated by obtaining information on their environmental properties from the literature.

Many analytical methods are however relatively limited in their applicability. Even 'high-power' equipment like GC/MS, an instrument widely applied in leachate analysis, is incapable of detecting about 80% of all synthetic organic compounds (US-EPA, 1988). The complexity of the leachate may also reduce the possibility of identifying a number of chemicals, as peak overlapping will be a matter of concern. Comparing results of chemical summary parameters like NVOC (non-volatile organic carbon) and TOX (total organic halides) with the sum of specific substances identified, less than 50% of the content of total organic carbon can often be accounted for. Failure to identify a component does not mean that the chemical is not toxic or otherwise harmful to the environment.

Another major problem associated with the specific analytical approach is that for relatively few chemicals sufficient fate and toxicity data will be available, and for a number of these, data may not include information relevant to the specific environment of the particular site.

Within the last decade an increasing number of studies have been reported which include in their approach screenings for the 'inherent' toxicity of leachates by means of biological testing methods. Comparing the toxicity measured by such screening methods with toxicity evaluations of specific chemicals, the risk of obtaining false-positive as well as false-negative results concerning the hazardous properties of the leachate is reduced.

In this chapter, biological screening methods (biotests), their limitations and advantages, are described and recently published studies are referred to in order to illustrate their uses and the ecotoxicological characteristics of landfill leachate.

PRINCIPLES OF ECOTOXICOLOGICAL METHODS (BIOTESTS)

The general approach for measuring toxicity is incubation of a number of organisms in a dilution series of the sample with a spacing not exceeding a factor of 3·2. After a defined incubation period the number of organisms suffering a defined effect (e.g. death, growth) is recorded. The percentage of organisms affected (response) is plotted against the concentrations which ideally give a sigmoidal curve as theoretically shown in Fig. 1.

This curve may be fitted by linear regression using probit transformed response-data and concentrations on a logarithmic scale and limits of confidence calculated as shown in Fig. 1.

The most frequently used toxicity figure is the concentration which gives 50% response (LC_{50}: concentration lethal to 50% of the organisms; EC_{50}: effect-concentration for 50% of the organisms). Also the lowest concentration showing a significant effect is often reported (LOEC, 10 or 20% response). The no observed effect concentration (NOEC) is defined as the highest concentration tested having no significant effect. Limits of confidence for sigmoid shaped curves will always show the least variation at the 50 per cent fractile. Therefore, the LC_{50} or EC_{50} value will be the best statistical estimate of toxicity.

Two major groups of test types may be defined: acute and chronic toxicity tests. Acute toxicity tests are designed to evaluate the toxicity

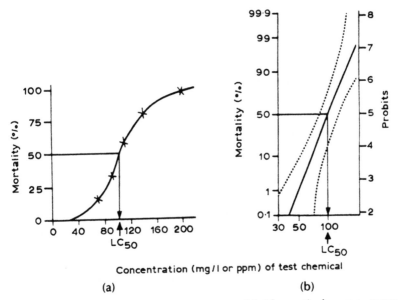

Figure 1. Concentration–response curves. (a) Theoretical curve representing mortality versus concentration; (b) same as (a) but mortality on a probit scale. Shown by dotted lines are 95% confidence limits of the linearized curve (from Rand & Petrocelli, 1984).

after a short-term exposure related to the life span of the test organism (fish, crustaceans: 24–96 h; algae: 4–6 h; bacteria: 5–30 min). Common effect parameters (endpoints) for acute toxicity are mortality (fish), immobility (crustaceans) and reduced photosynthesis (algae).

In chronic toxicity tests, adverse effects are studied under conditions of long-term exposure at concentrations proven to be non-lethal in acute toxicity tests. For organisms having a relatively short life-cycle (e.g. bacteria, algae, small crustaceans) the test normally includes one to several generations of the organism. For organisms with longer life-cycles (e.g. bivalves, fish) the studies conducted cover especially sensitive life-stages and toxicity endpoints (e.g. reproduction, embryo and larval growth and survival).

Two major objectives of the use of biotests in leachate investigations may be identified:

—to detect the 'inherent' toxicity potential of the complete mixture as a complementary parameter to chemical–physical measurements;

—to evaluate the potential of the mixture to intoxicate a target (aquatic) environment.

It is important to separate these two objectives, as the type of biotests selected will most often depend on the objectives of the investigation.

SCREENING METHODS FOR 'INHERENT' TOXICITY

Methods applicable for investigating the 'inherent' toxicity of leachate are most often short-term acute toxicity tests. A list of commonly applied test methods is shown in Table 1.

Included in Table 1 are also test methods designed to pick up substances with mutagenic potentials (Ames tests).

TABLE 1. Examples of Screening Methods Often Applied in Leachate Investigations

Trophic level	Organism	Effect parameter	Test duration
Bacteria	Photobacterium phosphoreum (Microtox)	Inhibition of light emission	5–30 min
	Salmonella typhimurium (Ames test)	Revertants	48 h
Algae	Selenastrum capricornutum	Inhibition of photosynthesis	6 h
	Nitzschia palea Skeletonema costatum	Inhibition of cell growth	72 h
Plants	Lemma minor (Duckweed)	Inhibition of growth (chlorophyll *a* and weight)	96 h
	Radish Sorghum	Reduced germination or seedling growth	24–96 h
Crustacea	Daphnia magna Ceriodaphnia dubia Mysidopsis	Reduced survival of larvae	24–96 hr
Fish	Zebra fish Guppy Rainbow trout Fathead minnow	Reduced survival of larvae or fingerlings	96 h

It is a prerequisite that biotests applied to measure the inherent toxicity of substances/mixtures as far as possible meet the following requirements:

—standardized methods having wide international acceptance;
—high degree of reproducibility;
—test organisms which may be cultured in the laboratory and are thus available at all times of year;
—sensitivity to a wide range of chemicals;
—low cost and easy to conduct with clearly identifiable effect parameters.

Thus, the demands of such screening tests are very similar to those of chemical analysis.

As the results of screening tests are not primarily meant to be predictable for impact to specific environments, 'standardized' test organisms are applied under 'standardized' conditions. Thus, zebra fish might be applied to screen the potential toxicity of leachate samples provided this species is at least as sensitive as other commonly utilized laboratory-reared fish species. For an impact assessment on a marine environment, however, marine species of relevance to the impacted area would be a better choice for biotesting.

A brief description of principles of frequently used screening methods is given below.

Effect Studies on Fish

Immature fish (zebra fish, guppies, fathead minnow) or fingerlings (rainbow trout) are exposed for a period of 96 h. Daily readings of mortality are the basis of estimating primarily the LC_{50} value after 96 h (OECD, 1981). An increased sensitivity may often be obtained when larvae of fish are used for the test (Nordberg-King, 1989).

Effect Studies on Crustaceans

Larvae of *Daphnia magna* or *Ceriodaphnia dubia* (small shelled crustaceans) less than 24 hours old are most often utilized (OECD 1981; ASTM, 1988). The test duration is 24–48 h. As the mortality of these small organisms is difficult to register exactly, immobility is applied as an effect parameter. Also the saltwater species *Mysidopsis bahia* and *Nitocra spinipes* have occasionally been used.

Effect Studies on Bacteria

The single most frequently used method of estimating toxicity to bacteria is the Microtox™ test, developed by Beckmann Instruments (1982). According to this method, light diminution of bioluminescent bacterial cells (*Photobacterium phosphoreum*) is measured by a 'Microtox toxicity analyser' (Beckmann Instruments, Inc.). The measurements are performed after 5, 15 and 30 min of incubation.

Bacteria cells are also used to screen for the content of mutagenic substances (*Salmonella typhimurium*, Ames test) (Ames *et al.*, 1974).

Effect Studies on Micro-algae and Plants

Screening methods used on unicellular algae have been adopted for the measurement of either the effect on photosynthesis (^{14}C assimilation) as a short-term (6 h) acute test (Kusk & Nyholm, in press) or the effect on algal growth (ISO, 1987; Nyholm & Källkvist, 1989). The growth inhibition is measured after three days of incubation. As this time duration allows for several cell-multiplications, this test is considered to measure chronic effects on the algae.

The effects on plants are often measured as the inhibition of the germination and early growth of seeds or seedlings (US-EPA, 1975; OECD, 1981). Often applied species are sorghum or radish. Also the effects on growth of duckweed (*Lemna minor*) have been applied as a test method (US-EPA, 1975).

METHODS OF PREDICTING ENVIRONMENTAL IMPACTS

In Table 2 examples of methods applicable to the prediction of impact on freshwater and marine aquatic environments are given. Many of the aspects discussed for the screening methods also apply to the methods in Table 2; further discussion is beyond the scope of this chapter.

INFLUENCE OF SAMPLE PREPARATION ON THE RESULTS OF BIOTESTS

A number of parameters connected with the sample preparation may influence the results of biotests. These problems are primarily connected

TABLE 2. Applied Biotests for Impact Assessment on Resident Species

Trophic level	Organism	Effect parameter	Test duration
Bacteria	Mixed resident population	Inhibition of community respiration	<3 h
Algae	Single species or mixed resident population	Photosynthesis, cell growth and shift in diversity	6 h–3 days
Crustacea	Resident single species *Daphnia magna Nitocra spinipes Acartia tonsa Gammarus* sp. a.o.	Survival of larvae or premature organisms, reproduction	48 h–21 days
Fish	Resident single species: Rainbow trout Stickleback Eel Plaice a.o.	Survival of larvae, juveniles or other premature stages, reproduction, growth	96 h–60 days

with:

—sampling and storage
—particulate materials
—pH

A comprehensive discussion of the influence of these parameters has been reported previously (Epler *et al.*, 1980).

For samples taken for chemical analysis, sample degradation is often prevented by changing the physical/chemical environment (e.g. acidification, chelation). For samples to be subjected to biotests, the storage method having the least effect is keeping the sample below 5–8°C for short-term storage and below −18°C for long-term storage (Epler *et al.*, 1980). As deep-freezing may also change the toxicity, biotesting should be performed as quickly as practically possible.

Only chemicals dissolved in water will be bioavailable for the organisms and thus responsible for toxic effects. Thus the bioavailable concentrations of substances having a high tendency to sorb to particulate materials (heavy metals, lipophilic organic substances) may be changed by altering the concentration of particulates in relation to the sample volume. To investigate the content of absorbable compounds, a

number of simple manipulations of parallel subsamples involving pH-changes prior to filtration may be performed (US-EPA, 1988).
Most test methods prescribe a relatively narrow pH-interval for the test medium (e.g. pH 7·5 ± 0·5). As leachate will often deviate from this range, pH-corrections are necessary prior to testing. For a number of substances, bioavailability will be affected by changes in pH, e.g. the sorption of heavy metals and lipophilic substances to particulates, precipitation of ions, and thus the toxicity potential will be affected. Changes of pH may also affect the toxicity potential of acid/bases, as the noncharged substance is most often the most toxic compared to the ionic compound (e.g. ammonia compared to ammonium).

RELATIVE SENSITIVITY OF BIOTESTS

Often differences of several orders of magnitude exist between the least sensitive and most sensitive species when they are exposed to a particular toxicant (Sloof *et al.*, 1983*a*; LeBlanc, 1984). For complex mixtures of substances like leachate and sewage effluents, this difference is generally less than that seen for single compounds, but still differences of one order of magnitude are often seen (Sloof *et al.*, 1983*b*). Since the measured toxicity of leachate will often be caused by unknown constituents, the relative sensitivity will also be unknown. Therefore, proper leachate toxicity analysis requires tests with a range of different species. Cost is most often balanced against the optimal investigation programme and thus the level of accepted uncertainty. In this respect it is often recommended to use three species representing three different trophic levels, e.g. algae/plant, crustacean, and fish (Kimerle *et al.*, 1984; OECD, 1986; US-EPA, 1987).

An example of the deviation in sensitivity between species is seen in Fig. 2, representing an investigation of leachate from eight sanitary landfills in Sweden (Naturvårdsverket, 1989).

The general range of sensitivity for this study was algae > crustacean > fish > plant > Microtox, with algae as the generally most sensitive. However, no organism was the most sensitive in all samples.

Investigation of leachate from a landfill containing household (40%) and industrial (60%) waste using four different test species (fathead minnow, *Daphnia*, algae and Microtox) also showed high deviations in sensitivity (Plotkin & Ram, 1984). The sensitivity range was algae > Microtox > *Daphnia* > fish (Table 3).

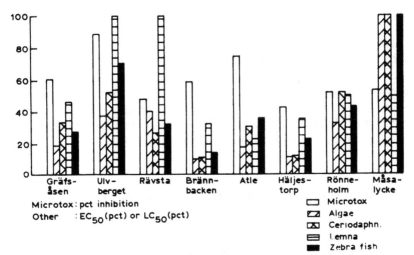

Figure 2. Ecotoxicological analysis of leachate from eight different sanitary landfills. The species tested were Microtox™, *Selenastrum capricornutum* (algae), *Ceriodaphnia dubia* (crustacean), *Lemna minor* (duckweed) and zebra fish. Effect concentrations are given as percentage of leachate responsible for 50% effect (EC or LC). For Microtox, percentage inhibition in 100% leachate is applied (from Naturvårdsverket, 1989).

TABLE 3. The Results of Testing Sanitary Landfill Leachate by Four Biotests (after Plotkin & Ram, 1984)

Test	Treatment	Toxicity values (% leachate)	Range of dissolved oxygen over test duration (mg/litre)
Microtox assay:	Unfiltered	5 min. $EC_{50} = 14$	4·0
Photobacterium phosphoreum	Filtered	5 min. $EC_{50} = 17$	4·0
Algal assay: *Selenastrum capricornutum*[a]	Filtered	$1 < EC_{50} < 10$	—
Macroinvertebrate assay: *Daphnia magna*	Filtered	$48LC_{50} = 62-66$ $ILC_{50} = 37$	4·8–7·4
Fish bioassay:	Unfiltered	$96LC_{55} = 100$	5·0–8·3
Pimephales promelas[b]	Filtered	$96LC_{15} = 100$	4·0–7·9

[a] As determined by chlorophyll *a* analysis.
[b] 96LC values were based solely upon survival in 100 percent leachate after 96 h of exposure. High survival rate precluded an exact LC_{50} value determination.

Differences in species-sensitivity dependent on the composition of leachate were observed for *Daphnia magna* and Microtox on testing electroplating sludge leachate (heavy metals) and pesticide waste leachate, respectively. For both types of leachate *Daphnia magna* was more sensitive than Microtox, the difference being 2–26 times for the leachates predominantly containing heavy metals and 42–3000 times for the pesticide-containing leachates (Calleja *et al.*, 1986).

CHEMICAL ANALYSIS IN RELATION TO BIOTESTS

The overall advantage of including biotests in leachate characterization schemes is that these methods will complement chemical measurements

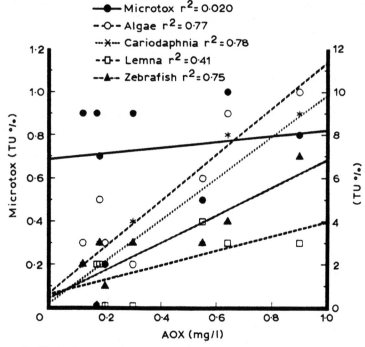

Figure 3. Biotesting results versus AOX of leachate from eight sanitary landfills. TU = 100 · (LC(EC)$_{50}$, % leachate)$^{-1}$. Linear correlations and correlation coefficients (r^2) are indicated (recalculated from Naturvårdsverket, 1989).

and reduce the risk of false-negative results for the characterization. In other words, if chemical analytical programmes were comprehensive enough to pick up all substances hazardous to life, and if the toxicity of these substances were known, there would be less need to apply biotests as an additional screening tool.

In the above-mentioned study on leachate from eight Swedish sanitary landfills, specific analysis for 24 metals and metalloids and AOX-measurements were performed in addition to biotests on five different trophic levels (Naturvårdsverket, 1989). On the basis of the results it was concluded that except for the Microtox, relatively good correlations between the results of each biotest and chemical analysis were obtained. In Figs 3 and 4 the results from AOX-measurements and heavy metals

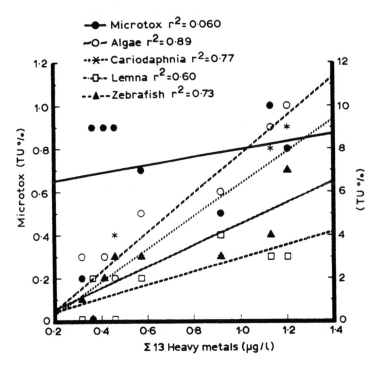

Figure 4. Biotest results versus the concentrational sum of 13 heavy metals for leachate from eight sanitary landfills. TU: $100 \cdot (LC(EC)_{50}, \%$ leachate$)^{-1}$. Linear correlations and correlation coefficients (r^2) are indicated (recalculated from Naturvårdsverket, 1989).

analysis, respectively, have been compared with the results of the biotests performed.

To allow a comparison for all biotests, the results have been recalculated to 'toxicity units' (TU)

$$TU = \frac{100}{\% \text{ leachate corresponding to } LC_{50} \text{ or } EC_{50}}$$

This way of expressing toxicity is preferred as TU is numerically directly proportional to concentration, i.e. TU is increasing with increasing concentration, whereas the LC_{50}-value has an inverse relationship to concentration. Thus, if the mixture concentration that induces an LC_{50} is 25% (250 ml/litre), this mixture has a TU-value of four.

The results indicate that toxicity to *Ceriodaphnia*, algae and zebra fish is correlated to the degree of pollution, measured either as AOX or as the sum of heavy metals. The toxicity to duckweed (*Lemna minor*) is only slightly correlated and the results with Microtox are not correlated at all to the measured chemical parameters. Substances other than those included in the analysis may have influenced the correlation (e.g. ammonia). The most sensitive biotests are shown to be *Ceriodaphnia* and algae, while duckweed (and Microtox) were relatively non-sensitive. The same trend in biotest data is seen for the comparison with both AOX and heavy metals, although as expected duckweed and algae correlated more significantly to heavy metals than to AOX.

The non-correlation of Microtox data to the chemical analysis may be due to parameters not included in the analytical programme possessing specific toxicity to the Microtox bacteria. This unknown parameter may not necessarily be a substance of major hazardous concern.

The low sensitivity of Microtox compared to the other applied biotests would be expected for leachate predominantly composed of metals, as most metals have been shown to be relatively low-toxic to *Photobacterium* (Qurestri *et al.*, 1982).

Analysis of landfill leachate by acute toxicity screening (*Daphnia magna*) and a number of chemical parameters has been performed at 40 different landfills in Finland (Assmuth & Melanen, 1988). Acute toxicity of leachate was observed in 30% of the landfills. Mean LC_{50} (48 h) for *Daphnia magna* was 450 ml/litre (45%) for these landfills.

The data obtained on toxicity and AOX (absorbable organohalides) are shown in Fig. 5.

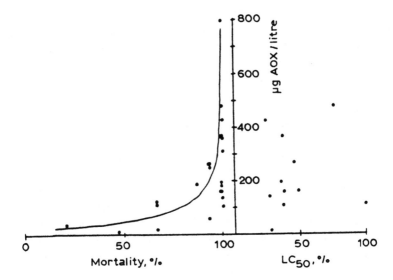

Figure 5. Relations between AOX (absorbable organohalides) and acute toxicity to *Daphnia magna* (24 h mortality in 100% leachate and 48 h, LC_{50} data) (from Assmuth & Melanen, 1988).

It is concluded from this study that an incubation in pure leachate followed by a registration of the percentage of dead organisms after 24 hours may be a favourable choice for relatively rapid screening of leachate samples. It is seen (Fig. 5) that samples with low AOX may possess a high degree of toxicity and vice versa. Both parameters are thus necessary to pick up samples possessing major hazard potential.

IDENTIFICATION OF COMPOUNDS RESPONSIBLE FOR TOXICITY

The results of biotests applied on leachate may be interpreted by comparing the concentrations of specific compounds identified in the leachate with literature toxicity data. The benefit of this exercise will, however, be dependent on the extent and success of the analytical programme and on relevant data being available in the literature. Other problems concerned with such comparisons are that literature data are most likely obtained under different experimental conditions (e.g. hardness, pH, alkalinity, particulate matter), which may affect the availability of the substance and hence the concentration producing the toxic effect. Also interaction phenomena in the leachate mixture may influence the comparison.

Comparisons of the leachate toxicity to the toxicity of single constituents should therefore only be performed qualitatively or quantitatively on orders of magnitudes.

Plotkin & Ram (1984) qualitatively compared the results of biotests (four trophic levels) on leachate with literature-based effect-concentrations on single chemicals. The specific analysis programme contained 13 heavy metals, 27 pesticides, 8 organic chemicals, ammonia, cyanide and iron. They concluded that all observed concentrations of pesticides as well as other organic substances were several orders of magnitude below literature LC_{50} values. Ammonia and to a lesser extent silver, mercury, lead, cadmium and manganese were assumed to be the primary contributors to the observed toxicity.

In a study on leachate from three sanitary landfills, toxicity to *Daphnia magna* (24-h LC_{50}) was compared to the content of metals (Fe, Pb, Cd, Zn, Cu, Ni, Mn and Cr) (Mezzanotte *et al.*, 1988). The 24-h LC_{50} measured was 2·3, 7 and 35%, respectively, for the three landfill leachates. Although the concentration of iron exceeded the threshold (LOEC) for toxicity to *Daphnia magna*, no single element could account for the measured overall toxicity. Assuming that the toxicity of the individual elements is additive (although below the threshold concentration for toxicity each individual element contributes additively to the toxicity of the mixture) a LC_{50} value of 10% (100 ml/litre) may be calculated based on the data from Mezzanotte *et al.* (1988) compared to the measured toxicity of leachate, showing a 48-h LC_{50} of 2·3% (23 ml/litre). Thus elements other than those included in the analytical programme may have contributed to the toxicity.

The identification of (at least) the groups of chemicals responsible for an observed toxicity may be performed following a testing strategy developed for industrial and municipal wastewater by US-EPA (1988). According to this method, toxicity testing is conducted on fractions of the mixture that have been separated using physical/chemical laboratory fractionation techniques (e.g. via solid, reverse phase extractor columns eluted with different concentrations of methanol/water) and techniques involving relatively simple sample manipulations (e.g. filtration/evaporation at different pH-levels, complexing with EDTA, oxidation).

CONCLUSION

Characterization of landfill leachate may be conducted cost-effectively by combining chemical analytical methods with screening for the inherently toxic properties of the leachate mixture to living organisms. Such combined procedures will most likely reduce the risk of false-negative monitoring results, i.e. that a reduced chemical programme has led to a characterization of samples of leachate as 'non-polluting' even though they may still contain hazardous substances. It should be remembered, though, that biotests will only pick up substances having toxic properties. Other substances of environmental concern, persistent or bioaccumulative substances will only be identified by the chemical analytical approach.

REFERENCES

Ames, B. N., McCann, J. & Yamasaki, E. (1974). Methods for detecting carcinogens and mutagens with the *Salmonella* Microsome Mutagenicity Test. *MUT. Res.*, **31**, 347–64.

Assmuth, T. & Melanen, M. (1988). Screening toxicants in waste deposit runoff. In *ISWA 88, 5th International Solid Waste Conference, Proceedings*, vol. 2. Academic Press, London, pp. 35–40.

ASTM (1988). Standard guide for conducting acute toxicity tests on aqueous effluents with fishes, macroinvertebrates and amphibians. E-1192-88, American Society for Testing and Materials, Philadelphia, PA.

Beckmann Instruments (1982). Toxicity testing of hazardous waste. Microtox Application Notes M105. The Microtox System Product Development Bulletin, Beckmann Instruments, 6964 Carlsbad, CA.

Calleja, A., Baldasano, J. M. & Mulet, A. (1986). Toxicity analysis of leachates from hazardous wastes via Microtox and *Daphnia magna*. *Toxicity Assessment: An International Quarterly*, **1**(1), 73–83.

Epler, J. L., Larimer, F. W., Rao, T., Burnett, E. P. & Griest, W. H. (1980). Toxicity of Leachates. EPA-600/2-80-059, PB 80-179328, US Environmental Protection Agency, Washington, DC.

ISO (1987). Water Quality—Algal Growth Inhibition Test. International Standardization Organisation, Geneva, Switzerland.

Kimerle, R. A., Wesser, A. F. & Adams, W. J. (1984). Aquatic hazard evaluation principles applied to the development of water quality criteria. In *Aquatic Toxicology and Hazard Assessment*, Seventh Symposium. ASTM, STP 854, Philadelphia, PA.

Kusk, O. & Nyholm, N. (in press). Evaluation of a phytoplankton toxicity test method based upon ^{14}C-assimilation as a biotest for water pollution assessment and control. *Arch. Environ. Contam. Toxicol.*

Le Blanc, G. A. (1984). Interspecies relationships in acute toxicity of chemicals to aquatic organisms. *Env. Tox. and Chem.*, 3(1), 47–60.

Mezzanotte, V., Sora, S., Viganò, L. & Vismara, R. (1988). Using bioassays to evaluate the toxic and mutagenic properties of landfill leachate. In *ISWA 88, 5th International Solid Waste Conference, Proceedings*, vol. 1. Academic Press, London, pp. 131–6.

Naturvårdsverket, Sweden (1989). Biological and chemical characterization of leachate. DEPÅ 90, Avfallsdeponering på 90-talet. Rapport no. 3702, 89:9 (in Swedish).

Norberg-King, T. J. (1989). An evaluation of the fathead minnow seven-day subchronic test for estimating chronic toxicity. *Env. Tox. Chem.*, 8, 1075–89.

Nyholm, N. & Källquist, T. (1989). A critical review on methodologies for growth inhibition toxicity tests with freshwater algae. *Env. Tox. Chem.*, 8, 689–703.

OECD (1981). OECD guidelines for testing of chemicals. Organization for Economic Cooperation and Development, Paris, France.

OECD (1986). Environmental Committee, Water Management Policy Group. The use of biological tests for water pollution assessment and control. Organization for Economic Cooperation and Development, Paris, France.

Plotkin, S. & Ram, N. M. (1984). Multiple bioassays to assess the toxicity of a sanitary landfill leachate. *Arch. Environ. Contam. Toxicol.*, 13, 197–206.

Qurestri, A., Flood, K. W., Thompson, S. R., Janhurst, S. M., Inniss, C. S. & Rolosh, D. A. (1982). Comparison of a luminescent bacterial test with other bioassays for determining toxicity of pure compounds and effluents. *Aquatic Toxicology and Hazard Assessment*: Fifth Cont., ed. J. G. Pearsen, R. B. Foster & W. E. Bishop. ASTM, STP 766, Philadelphia, PA, pp. 179–95.

Rand, G. M. & Petrocelli, S. R. (1984). *Fundamentals of Aquatic Toxicology. Methods and Applications.* Hemisphere Publ. Corp., Washington, DC.

Sloof, W., Canton, J. H. & Hermens, J. L. M. (1983a). Comparison of the susceptibility of 22 freshwater species to 15 chemical compounds. I: (sub)acute toxicity tests. *Aquatic Toxicology*, 4, 113–28.

Sloof, W., de Zwart, D. & van de Kerkheft, J. F. J. (1983b). Monitoring the rivers Rhine and Meuse in the Netherlands for toxicity. *Aquatic Toxicology*, 4, 189–98.

US-EPA (1975). Test methods for assessing the effects of chemicals on plants. EPA 560-17-75-008, US Environmental Protection Agency, Washington, DC.

US-EPA (1987). Permit Writers Guide to Water Quality-based Permitting for Toxic Pollutants. EPA 440-4-87-005, US Environmental Protection Agency, Washington, DC.

US-EPA (1988). Methods for Aquatic Toxicity Identification Evaluations: Toxicity Characterization Procedures. EPA-600/3-88/034, US Environmental Protection Agency, Washington, DC.

2.6 Treatability Characteristics of Landfill Leachate

NORBERT MILLOT* & PATRICK COURANT

*France Dechets, F. D. Conseil, 71 rue Henri Brettonet,
F-78970 Mezieres-sur-Seine, France*

INTRODUCTION

Landfill leachate composition varies widely among landfills, as shown by
Johansen & Carlson (1976) and Robinson & Marris (1979), supposedly
due to factors such as waste types landfilled, climate, hydrogeological
structure of the landfill, operational conditions and age of landfill. This
variation of leachate composition makes a thorough characterization of
the leachate mandatory for each landfill before appropriate treatment
schemes can be defined. Basic physico-chemical parameters for the
leachate may provide a basis for determining the organic load and its
general degradability, but due to the complex composition of the organic
fraction of the leachate, this may not be sufficient information for
defining a treatment scheme.

The application of conventional analytical techniques for characterization of the organic fraction, e.g. gas chromatography, may, in some
cases, be difficult to apply due to the complexity of the leachate. For this
reason, the gel permeation technique (GPC) has been introduced. Chian
(1977) and Harmsen (1983) used this technique to demonstrate the
importance of compounds with high molecular weights (humic-like
compounds) in older landfills and that of volatile fatty acids in more
recent landfills.

* Present address: Labo Services, Route de la Centrale, F-69700 Givors, France.

A treatability diagnosis method has been developed combining conventional leachate analysis with GPC analysis and for different leachates compared to results obtained in pilot scale treatment experiments by Granet *et al.* (1985) and Millot (1986).

In this chapter, the diagnosis methodology is presented and its application to two actual cases demonstrated: Villeparisis Landfill and Val Saint Germain Landfill in France.

DIAGNOSIS METHODOLOGY

The methodology is based on:

—a physical-chemical analysis of the leachate, which allows determination of the organic loading, the biodegradability, the salinity and possible toxic elements present;

—gel permeation chromatography (GPC) in order to determine the distribution of dissolved organic compounds according to their molecular weight.

The synthesis of this information allows definition of a suitable method of treatment for this leachate and forecasting of the specific efficiency of each step of the treatment line.

Procedure

The characterization of the leachate is based mainly on an analysis of the conventional physical-chemical parameters used in waste water analysis. The analyses examine global parameters (COD, BOD_5, TOC, TKN, pH) and also specific parameters (ammonia, nitrate, carboxylic acids, metals). The carboxylic acids are measured by gas chromatography in a Carbopack C, 0·3% Carbowax 20 M column filled with 0·1% H_3PO_4; helium vector gas: 40 ml/min, temperature programmed from 100 to 180°C at 4°C/min, flame ionization detector, column length: 2 m.

Before proceeding to the GPC column, the samples are filtered through a 0·45 μm membrane. To obtain comparable profiles and in particular to avoid saturating the gel column, the concentration of all leachates is corrected to between 100 and 200 mg/litre TOC, either by dilution with demineralized water or by concentration in a rotary evaporator. A 10 ml part of the sample is placed on the surface of the gel and the successive fractions are collected at the outlet from the column.

To obtain satisfactory separation of the compounds (Granet *et al.*, 1985) and an acceptable analysis time, a gel (Sephadex G 25 Fine, Pharmacia, Sweden) is used for the GPC. Compounds with an apparent molecular weight of more than 5000 are excluded and therefore eluted first. The column is 2·7 cm in diameter and 90 cm in height. Demineralized water (Millipore) at a flow rate of 140 ml/h is used as eluent. A fraction collector (LKB 2070 Ultrarac II) is used to collect the 10 ml fractions at the outlet of the column. The TOC in these samples is measured (Carbon Analyzed Dohrmann DC 80), as is the UV absorbance at 260 nm and 220 nm (Perkin–Elmer Lambda 3 UV/Vis spectrophotometer).

Diagnosis Scheme

Based on the analysis of 20 different French landfill leachates and their tested treatability, a diagnosis scheme has been developed (Millot, 1986; Millot *et al.*, 1987) operating with three classes as shown in Table 1.

Leachate belonging to Group I has a high organic loading primarily consisting of carboxylic acids (>80% of organic carbon) and a high metal content (in particular iron) due to the low pH. Group III contains

TABLE 1. Diagnosis Scheme for Landfill Leachate Treatability Together with Proposed Treatability Scheme

I	II	III
pH < 6·5	6·5 < pH < 7·5	pH > 7·5
BOD/COD > 0·3	0·3 > BOD/COD > 0·1	BOD/COD < 0·1
GPC:	GPC:	GPC:
—little exclusion peak	—exclusion peak	—very high exclusion
—very high fatty acids	equivalent to fatty	peak
peak	acids peak	—no fatty acids peak
Suggested treatment	*Suggested treatment*	*Suggested treatment*
line	*line*	*line*
(pre-treatment to	(pretreatment)	coagulation–flocculation
remove metals)	⋮	
↓	↓	↓
extended aeration	extended aeration	activated carbon
↓	↓	
coagulation–flocculation	coagulation–flocculation	
↓	↓	
(activated carbon)	(activated carbon)	

leachates with a moderate organic load with hardly any carboxylic acids present and extremely reduced metal concentrations due to the increased pH. Group II represents an intermediate stage. At present, the diagnosis method has been used for determining full-scale treatment schemes at 10 French landfills.

APPLICATION OF THE DIAGNOSIS METHOD: TWO CASES

In order to illustrate the applicability of the methods, the diagnosis method is demonstrated on the leachates of Villeparisis and Val Saint Germain landfills.

Presentation of the Two Landfills
The landfills of Villeparisis and Val Saint Germain are located near Paris; the first one is 20 km north-east and the other 40 km south.

The Villeparisis landfill has been in use since 1978. Its surface area is 40 ha and its capacity is 4 million cubic metres. This landfill receives approximately 1200 tonnes of waste per day. The major part (80%) of this waste comprises industrial waste which is disposed of separately in a specific area. The site is geologically leakproof and equipped with a drainage system for collection of leachates which then flow, by gravity, to a retention basin and thereafter to the leachate treatment plant. This landfill is also equipped with a collection system for biogas (120 wells connected in a network, 3500 m³/h, 45% methane).

The Val Saint Germain landfill was in operation between 1979 and 1982. Its surface area is 5 ha and its capacity is 500 000 cubic metres. The majority of waste comprises domestic waste which was compacted in small cells. The site has since been recovered and revegetated. The leachates are collected in a retention basin which is fed by a drainage system. The Val Saint Germain landfill is also equipped with a biogas collection system which produces 600 m³/h of biogas with a content of 55% methane).

Composition of Leachates
The two leachates were sampled in the retention basins which receive the leachates collected by the drainage systems. The results of chemical analysis of these leachates are presented in Table 2.

TABLE 2. Analysis of the Leachates at the Val Saint Germain Landfill and the Villeparisis Landfill

Parameters	Val Saint Germain	Villeparisis
pH	6·8	7·2
Conductivity	12 000	25 000
COD	20 000	5 600
BOD	8 500	2 100
TOC	6 600	1 900
TKN	1 300	950
N.NH₄	1 180	910
N.NOₓ	0·1	0·1
P.tot	1·2	1·0
Chloride	1 800	4 900
Acetic acid	3 950	1 280
Propionic acid	2 600	780
Butyric acid	2 950	860
Valeric acid	1 850	360
Iron	530	72
Zinc	22	8
Manganese	25	6

All parameters in mg/litre except pH (pH unit) and conductivity (μS/cm).

These results indicate, in both cases, a high organic load with BOD/COD ratio of about 0·4. This high biodegradable organic content may be explained by the high concentration of fatty acids which represents a theoretical TOC of about 85% of the measured TOC. The Total Kjeldahl Nitrogen content is high for the two leachates and is mainly represented by ammonia nitrogen. The heavy metal content consists primarily of iron.

GPC Profiles of Leachates

The two GPC profiles are quite similar (Figs. 1 and 2). The leachate of Val Saint Germain is diluted by 1/50 and that of Villeparisis by 1/10.

These GPC profiles are characterized by a low UV absorbance and also by a very important TOC peak; analysis showed this peak to correspond to fatty acids. The high molecular weight peak is masked by the high dilution rate, mainly for Val Saint Germain leachate.

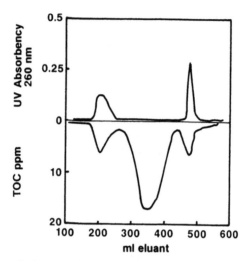

Figure 1. GPC profile of Villeparisis leachate—dilution 10.

Identification on Treatment Scheme

The diagnosis methodology indicates that the most suitable treatment schemes for these two leachates are the following since the leachates are both characterized as Group I although pH is slightly above pH 6·5:

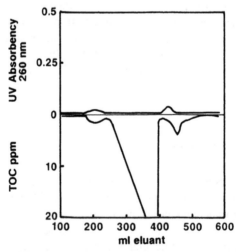

Figure 2. GPC profile of Val Saint Germain leachate—dilution 50.

—First step: biological treatment by aerated lagoon which reduces the organic load by more than 90%.

—Second step: physical–chemical treatment by coagulation flocculation.

—Third step: if necessary, final treatment by adsorption on activated carbon.

Leachate Treatment Plants

The leachate treatment plants at the two sites include (as illustrated by the scheme of the Villeparisis treatment plant shown in Fig. 3) a homogenization basin which allows the mixing of leachates from different sections

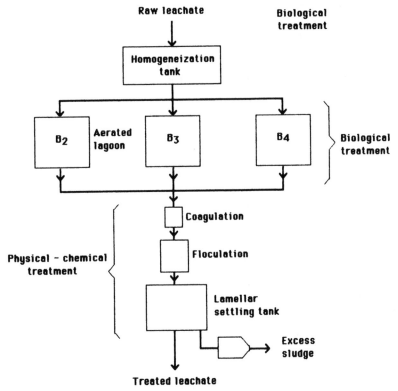

Figure 3. Scheme of Villeparisis leachate treatment plant.

and biological treatment by aerated lagoon. The volume of this basin is designed with an average retention time of 50 days. The average sludge load is 0·12 kg BOD_5/kg MLVSS (Mixed Liquor Volatile Suspended Solids) per day. Oxygen is supplied by surface aerators in Villeparisis and by small bubble air diffusors in Val Saint Germain. For nutrient supply phosphorus (KH_2PO_4) is added periodically. The final step is a physical-chemical treatment by coagulation–flocculation with ferric chloride.

The efficiencies of these treatment plants are shown in Table 3 and Fig. 4.

These results indicate that:

—biological treatment is highly efficient on organic loading, on nitrogen content and also on iron content;
—physical-chemical treatment is efficient mainly for the removal of the organic loading;
—biological treatment allows a treated effluent with a constant quality to be obtained.

The GPC profiles of the two biologically treated leachates (Figs 5 and 6), obtained without dilution, confirm the removal of carboxylic acids.

TABLE 3. Results of the Leachate Treatment Plants

Parameters	Villeparisis			Val Saint Germain		
	Raw leach-ate	After bio-logical treat-ment	Treated leachate	Raw leach-ate	After bio-logical treat-ment	Treated leach-ate
pH	7·2	7·8	6·6	6·8	7·8	6·5
COD (mg/litre)	5 600	520	230	20 000	710	320
TOC (mg/litre)	1 900	190	70	6 600	260	105
BOD_5 (mg/litre)	2 100	<10	—	8 500	<10	—
TKN (mg/litre)	950	15	14	1 300	28	25
$N.NH_4$ (mg/litre)	910	10	10	1 180	25	24
Fe (mg/litre)	72	0·8	1·0	530	1·2	1·1

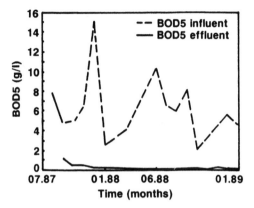

Figure 4. Influent and effluent concentrations of BOD_5 at the Val Saint Germain leachate treatment plant.

The biological treatment is ineffective or inoperative on compounds with a molecular weight >5000. The GPC profile (Fig. 7) after coagulation–flocculation shows that this treatment removes high molecular weight compounds. Experiments on a laboratory scale showed that activated carbon would eliminate the compounds remaining after biological treatment and coagulation–flocculation (Millot, 1986).

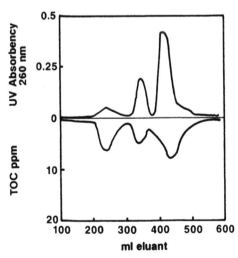

Figure 5. GPC profile of biologically treated Villeparisis leachate.

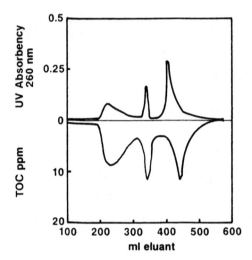

Figure 6. GPC profile of biologically treated Val Saint Germain leachate.

CONCLUSION

A rapid treatability diagnosis methodology has been developed accounting for the complex composition of the organic fraction of landfill leachate. The methodology is illustrated by application to the leachates

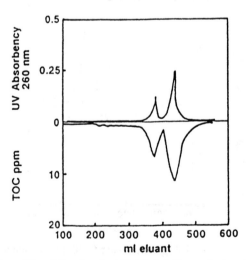

Figure 7. GPC profile of the treated Val Saint Germain leachate.

from two actual landfills. At present, the methodology has been used in determining appropriate leachate treatment schemes at 10 actual landfills in France.

REFERENCES

Chian, E. S. K. (1977). Stability of organic matter in landfill leachates. *Water Research*, **11**, 225–32.

Granet, C., Millot, N., Wicker, A. & Navarro, A. (1985). Application de la chromatographie de permeation sur gel aux lixiviats de décharge contrôlée (Application of gel permeation chromatography to landfill leachate). *Tech. Scien. Munic. l'Eau*, **80**(5), 223–9.

Harmsen, J. (1983). Identification of organic compounds in leachate from a waste tip. *Water Research*, **17**, 699–705.

Johansen, O. J. & Carlson, D. A. (1976). Characterization of sanitary landfill leachates. *Water Research*, **10**, 1129–34.

Millot, N. (1986). Les lixiviats de décharge contrôlée. Caractérisation analytique. Etude des filières de traitement. (Sanitary landfilling leachates: analytical characterization studies on treatment schemes.) PhD thesis, Institut National des Sciences Appliquées, Lyon, n. 86 ISAL 0011.

Millot, N., Granet, C., Wicker, A. & Navarro, A. (1987). Application of GPC processing system to landfill system. *Water Research*, **21**, 709–15.

Robinson, H. D. & Marris, P. J. (1979). Leachate from domestic waste. Generation, composition and treatment: a review. Technical report TR 108, Water Research Council, Medmenham, UK.

2.7 Effects of Landfill Management Procedures on Landfill Stabilization and Leachate and Gas Quality

THOMAS H. CHRISTENSEN, PETER KJELDSEN

Department of Environmental Engineering, Technical University of Denmark, Building 115, DK-2800 Lyngby, Denmark

&

RAINER STEGMANN

Institute of Waste Management, Technical University of Hamburg-Harburg, Harburger Schlossstrasse 37, D-2100 Hamburg 90, Germany

INTRODUCTION

The stabilization of landfills receiving substantial amounts of organic wastes is primarily governed by the microbial degradation processes developing in the landfill cells. A rapid degradation of the waste resulting in conversion of organic carbon into gases and neutral pH conditions in the waste is considered beneficial for the leachate composition and gas utilization potential, e.g. Ehrig (1989).

Many factors affect the microbial degradation processes as discussed in the review by Christensen & Kjeldsen (1989). They evaluated the influence on the degradation processes of oxygen, hydrogen, pH and alkalinity, sulphate, nutrients, inhibitors, temperature and moisture/water content of the waste. Although many of the factors influence the degradation process, the pH of the waste environment seems to be the most critical parameter in obtaining an effective methanogenic degradation of the landfilled waste (Christensen & Kjeldsen, 1989). However, many of the above mentioned abiotic factors cannot be controlled individually. The actual management procedures are operationally defined and will simultaneously influence several of the basic factors and the resulting effect may be difficult to predict from knowledge of the individual abiotic factors.

119

TABLE 1. Summary of Reported Investigations on the Effects of Various Landfill Management Procedures on Waste Degradation as Measured by Leachate and Gas

Reference	Scale of experiment[a]	Management procedure investigated[b,c]								Effects measured on[d]			
		WC	SS	B	S	C	SC	LR	PC	GP	GC	LP	LC
Barlaz et al. (1987)	LAB			P				Z, P		×	×	×	
Buivid (1980)	LAB		P	P	N	P		P		×	×		
Chian et al. (1977)	LAB				N					×	×		
Fungaroli & Steiner (1979)	LAB				P						×		×
Klink & Ham (1982)	LAB									×	×	×	×
Leuschner & Melden (1983)	LAB		P	P				P, N		×	×	×	×
Scharf (1982)	LAB		Z							×	×		
Stegmann (1983), Stegmann & Spendlin (1986)	LAB	P, N		Z	Z			P		×	×	×	×
Augenstein et al. (1976)	LYS							P		×	×		×
Doedens & Cord-Landwehr (1984)	LYS							P		×	×	×	×
Eifert (1976), Eifert & Schwartzbaugh (1977)	LYS				N						×		×
EMCON Associates (1975)	LYS				N			P			×	×	×
Gandolla (1982)	LYS		P							×	×		
Ham & Bookter (1982)	LYS			P	N		N			×	×	×	×
Kinman et al. (1987)	LYS		P	P				P/N	×		×	×	×

Reference	Type										
Leckie *et al.* (1979)	LYS	N			P	× ×	× ×	× ×	× ×	× ×
Pohland (1980)	LYS				P			×	×	×
Pohland & Kang (1974)	LYS	P			P	× ×	× ×	× ×	× ×	× ×
Stamm *et al.* (1984)	LYS	P Z				× ×	× ×		× ×	× ×
Stegmann (1983), Stegmann & Spendlin (1986)	LYS			P						
Tittlebau (1982)	LYS	P Z			P				×	
Barber & Maris (1984), Robinson & Maris (1985)	LAN				P				×	×
Beker (1987)	LAN	P	N		N/P	× ×	× ×	× ×	× ×	× ×
Christiansen *et al.* (1985)	LAN				P	× ×		×	×	×
Doedens & Cord-Landwehr (1984)	LAN			P	P				×	×
Ehrig (1982)	LAN			P	P	× ×	× ×	× ×	× ×	× ×
Stegmann (1983), Stegmann & Spendlin (1986)	LAN			P	P	× ×	× ×	× ×	× ×	× ×

[a] LAB: Laboratory scale (<0·2 m³), LYS: Lysimeters, LAN: Actual landfill cells.
[b] WC: Waste composition, SS: Sewage sludge addition, B: Buffer addition, S: Shredding, C: Compaction, SC: Daily soil cover; LR: Leachate recirculation; PC: Aerobic pre-composting.
[c] Symbols in the table indicate: M: Measured or controlled, P: Positive effect; N: Negative effect, Z: Zero effect.
[d] GP: Gas production, GC: Gas composition, LP: Leachate production, LC: Leachate composition.

This chapter reviews the literature with respect to the effects on landfill stabilization of various landfill management procedures: composition of waste landfilled, addition of sewage sludge, addition of buffer, shredding, compaction of waste, use of daily soil covers, leachate recirculation and use of pre-composting procedures at the landfill.

SUMMARY OF REPORTED INVESTIGATIONS

Several investigations have been reported in the literature on the effects of various landfill management procedures on leachate composition and gas generation. Table 1 summarizes the general results of the individual investigations. It contains in addition to the bibliographic references three columns: (1) the scale of the reported experiment, (2) the investigated landfill management procedures and (3) the measured resulting effects in terms of gas and leachate.

(1) The scale of the reported experiments varies greatly and for this review three categories have been assigned: laboratory experiments (LAB) employing reactors less than $0.2\,m^3$; lysimeter experiments (LYS) employing reactors of substantial volume; and full-scale experiments (LAN) involving controlled landfill cells (slightly reduced in scale compared to normal landfill cells).

(2) The investigated landfill management 'procedures' are categorized as waste composition (WC), sewage sludge addition (SS), buffer addition (B), shredding (S), compaction (C), soil cover (SC), leachate recirculation (LR), and aerobic pre-composting (PC).

(3) The resulting effects have been measured in terms of gas production rate (GP), gas composition (GC), leachate production rate (LP), and leachate composition (LC).

In Table 1 symbols are used to give information about the results obtained in the investigations: 'M' indicates that a parameter or factor has been measured or controlled, but not purposely manipulated, 'P' indicates a positive effect of an increase in the parameter, 'N' indicates a negative effect of an increase in the parameter, while 'Z' indicates zero effect. For example a 'P' in column 'SS' shows that the reported investigation found an increased waste degradation rate caused by the addition of sewage sludge. An increase in a governing factor may in some cases mean that the actual management procedure was accomplished as compared to the control, where it was not accomplished. For example an

'N' in the column concerning shredding ('S') means that shredding had a negative effect on the waste decomposition as compared to no shredding of the waste.

Table 1 shows that most of the reported investigations deal with only a few of all the potential factors and that most investigations have involved experiments in reduced scale ('LAB' or 'LYS'). The relatively small scale employed has obvious advantages as to short response times and easily controllable experimental conditions, but some of the landfill management procedures, e.g. shredding, compaction and leachate recirculation, may be difficult to study in small-scale experiments. Comprehensive full-scale experiments involving many factors are very few.

Table 1 indicates that pre-composting and in most cases also buffer addition and recirculation of leachate have a significant positive effect on landfill stabilization. The results obtained for the other managing procedures appear to be somewhat contradictory or at least inconclusive ('N' and 'P' in the same column). Further scrutiny is needed to elucidate these apparent contradictions and to understand the functioning of the managing procedures.

DISCUSSION OF INDIVIDUAL MANAGEMENT PROCEDURES

The effects of individual management procedures on waste degradation are discussed in the following paragraphs, in a sequence identical to the sequence applied for Table 1.

Waste Composition

The composition of the waste received at a landfill is often governed more by the disposal needs of the community it serves than of the needs of the landfill management and in most reported investigations the composition of the waste has been taken for granted. Very few investigations have focused on the composition of landfilled waste although this must be an important factor controlling the degradation processes.

German investigations by Stegmann & Spendlin (1986) and Wolffson (1985) evaluated the effects of the waste paper fractions, wet kitchen garbage, and garden debris. An increased content of newspaper did not affect the gas generation significantly, while an increased content of magazine printing increased the gas production. The wet organic matter from kitchen and garden delayed the methane production, supposedly due to a more intensive acid generation. However, as shown in Fig. 1, in the long run the total gas production was increased in the presence of the putrescibles.

Synthetic organic matter in the form of plastic material makes up a significant fraction of most municipal and industrial wastes. Although parts of the plastic, in the long run, may be biologically degradable (e.g. Haxo, 1977), the rate is so slow that the plastic fraction should not be included in the organic fraction when estimating methane production potentials.

Since large concentrations of sulphate increase the redox potential and sulphate reduction may compete for organic carbon (Christensen & Kjeldsen, 1989), the presence of large amounts of sulphate-containing wastes, e.g. demolition waste and incinerator slags, may (where relatively little organic matter is present and from a theoretical point of view) decrease methane formation and increase carbon dioxide production.

It has to be respected that waste composition in different countries is different (Carra & Cossu, 1990), so that investigations in various countries cannot always be compared. This is especially true in the case of lysimeter studies from the USA and European countries, since the paper content in the USA waste is higher and the amount of putrescible organics is lower than found in Europe (see also Leuschner, 1989).

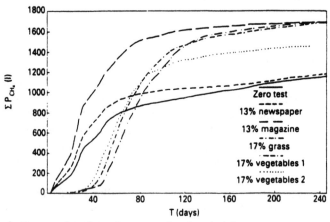

Figure 1. Accumulated methane production in laboratory experiments as a function of time for various waste compositions (Wolffson, 1985).

Sewage Sludge

The effects of sewage sludge additions on waste degradation and its consequences for leachate quality and gas quality have been studied in several experiments, but without a simple conclusion (Table 1): reports are found on positive effects, on negative effects, and on no effects. The positive effects are, however, all found for anaerobically digested sewage sludge, while the negative effects are found for septic tank sludges. Potential positive effects of sewage sludge addition could be attributed to increasing water content, supply of readily available nutrients and supply of an active anaerobic biomass, factors that all would speed up the methanogenic phase of the degradation. The beneficial effects of this potential, however, seem to be conditioned by two factors. Firstly, if the methanogenic consortia are already established or the waste environment is advantageous to methanogenic consortia, additions of sewage sludge may have limited effect. Sewage sludge addition can initially give rise to increased concentrations in the leachate and to an increased total amount of gas product, if the amount of organic matter added as sewage sludge is substantial. Secondly, the influence of sewage sludge on the pH of the waste seems crucial to the methane formation. Sludge low in pH, e.g. septic sludges, may have a negative effect on the methane formation (Leckie *et al.*, 1979; Leuschner & Melden, 1983) while neutral, well-buffered sewage sludge may have positive effects (Buivid, 1980; Leuschner & Melden, 1983; Kinman *et al.*, 1987; Leuschner, 1989).

It appears that addition of sewage sludge to the landfill will not always have beneficial effects on the degradation process. However, if methane formation is slow because of deficient water content and nutrients, addition of a sludge with a good buffer capacity may improve the methane formation substantially. However, this will also increase the amount of organics in the leachate. If methane formation is slow because of too low pH-values, addition of a well-buffered sludge may significantly increase the methane formation and decrease the organic content of the leachate.

Buffer

The detrimental effects on methane formation of low pH-values caused by an active acid phase have frequently led to the concept of adding buffer to the landfill, by either incorporating a solid buffer material such as calcium carbonate or adjusting the leachate before recirculation by addition of base (e.g. NaOH) or buffer solution (e.g. NaHCO$_3$). In most of the reported cases, the addition of buffer has had a positive effect on the degradation (Table 1); only Stegmann & Spendlin (1989) reported negative effects of pH-adjustment by NaOH addition. In this experiment, the reference reactor having no buffer addition exhibited an active methanogenic environment and the NaOH addition may have inhibited methane formation by the increased sodium content. Other non-published experiments by Stegmann & Spendlin show also the positive effect of lime addition.

The methane formation is by itself producing buffer capacity and increasing pH, and only where this process cannot outweigh the acidity produced by the acid phase has buffer addition had an effect. This also indicates that buffer addition may be very advantageous when the acid phase is very vigorous. It should be noted that buffer addition primarily has been investigated in small-scale experiments. Such experiments tend to have high reaction rates due to elevated temperatures, high content of water, organic matter or small particle sizes leading to fast hydrolysis and acid formation and are thus susceptible to buffer addition. In actual landfills, however, the reaction rates may be slower and some of the disposed waste (e.g. demolition waste) and the daily soil cover may provide sufficient buffer capacity to outweigh the acid phase acidity.

Buffer addition seems to have a beneficial effect on waste degradation judging from the reported experiments, but it does not seem justified to conclude that active buffer addition needs to be accomplished at all landfills. However, if a landfill has failed to generate methane due to low pH-values, buffer addition is an obvious measure to help establish methanogenic conditions.

Shredding

Shredding of solid waste prior to disposal in a landfill has been investigated in several experiments. The arguments for shredding are that it increases the homogeneity of the waste by size reduction and mixing, increases the specific surface area of the waste, removes water barriers caused by plastic bags and foil, and improves the water content and distribution in landfilled waste. However, the main reason for shredding has usually been the possibility of increasing the landfill capacity.

Most of the reported investigations on the effect of shredding on waste degradation show a negative effect (EMCON Associates, 1975; Eifert, 1976; Chian *et al.*, 1977; Buivid, 1980). Apparently the shredding intensifies the acid phase resulting in increasing production of carbon dioxide, low pH, and high content of organic carbon in the leachate resulting in no or reduced methane formation. If the negative effects of shredding in the initial phase of the degradation could be circumvented, shredding could prove beneficial in the later stages of the degradation process by improving the hydrolysis and acid formation. Both Barlaz *et al.* (1989) and El-Fadel *et al.* (1989) state that hydrolysis reactions tend to limit the degradation processes in the later stages. This is supported by observations by Ham & Bookter (1982) finding in lysimeter experiments that shredding, although initially resulting in a vigorous acid phase and high leachate concentrations, in the long run reduced the amount of carbon leached from the waste.

Based on current knowledge, the benefits of shredding the waste prior to landfilling seem limited. If shredding is performed, experience shows that measures must be taken to control the vigorous acid phase that will develop and eventually inhibit methane formation. Thus shredding alone cannot be recommended. Pre-composting of shredded MSW may however result in an enhanced degradation process (refer to the later section on 'Pre-composting').

Compaction

Compaction as performed by a compactor at the landfill front is a very common operation argued by the need for optimum use of the landfill capacity and for obtaining geotechnical stability of the landfilled waste. Thus it is expected that all waste sooner or later will be compacted at a modern landfill. The question is merely how an initial or delayed compaction may affect the degradation processes and the leachate and gas composition. A thorough compaction of the waste also leads to some homogenization and mixing of the waste landfilled, and in practice these factors cannot be evaluated separately.

If the upper refuse layer is supposed to undergo some aerobic degradation (pre-composting), a moderate, uncompacted or a thin, compacted layer must be established. Ehrig (1982) argues that an uncovered compacted layer of 0·5 m of refuse may allow for aerobic processes and similar effects can be obtained in 2 m of uncompacted, uncovered waste layers.

Compaction may also affect the anaerobic decomposition processes. At high water content, Rees & Grainger (1982) showed that increasing the dry weight density from 0·2 to 0·47 tonnes/m³ decreased the gas production, supposedly due to an overstimulation of the acid phase. The same investigation showed that, if the refuse is relatively dry (21%), increasing the dry weight density from 0·32 to 0·47 tonnes/m³ significantly increased the gas production. This latter effect is explained by the relatively higher moisture content at the high density, enhancing the availability of nutrients and contact between substrate and biomass. These observations are in accordance with laboratory experiments reported by Buivid (1980) and Buivid et al. (1981).

Experimental data on the effects of compaction on waste degradation are scarce and distinct conclusions cannot be justified. However, the few results available may indicate that the time of compaction of the upper refuse layer may be seen as a possible control of the acid phase and hence of the initiation of the methanogenic degradation.

Soil Cover

Application of daily soil cover is often prescribed to improve the hygienic and aesthetic standard of the landfill. However, adequate soils are often not available for this purpose, and soil covers at many landfills are considered to be an unnecessary depletion of the limited landfill capacity. These considerations, naturally, do not apply to the final soil cover established on top of completed landfill sections.

Negative effects of soil covers may be expected if the upper refuse layer is supposed to undergo aerobic degradation. A soil cover will here decrease the diffusion of oxygen into the waste layer and thus decrease the composting rate. Ham & Bookter (1982) found that the leachate concentrations were higher from shallow lysimeters having soil covers than from lysimeters without soil cover. The heavier the soil, the more adverse the effect expected. Use of heavy, clayish soils as covers may at a later stage of the landfill cause heterogeneous water distribution in the landfill, eventually perched water tables in the waste or very dry zones below soils of low permeability (Lee et al., 1986).

Positive effects of soil cover may be expected if the soil provides important buffer capacity to the landfill environment avoiding low pH-values inhibitory of methane formation. As previously discussed, buffering may not always be needed to obtain methanogenic conditions, but the soil cover may act as a prophylactic buffer in the landfill.

Recirculation of Leachate

Recirculation is the most investigated landfill management procedure. Reports are found on positive effects of leachate recirculation, on negative effects, and on no effects. The reported arguments for introducing leachate recirculation are: reduction of the organic content of the leachate and consequently the cost for leachate treatment; reduction of the amount of leachate to be treated by increasing evapotranspiration; and enhancement of the degradation of the waste by increasing the water content of the waste, supply and distribution of nutrients and biomass, and by dilution of locally high concentrations of inhibitors.

Recirculation is often practised with leachate containing high concentrations of organic matter by applying the leachate to the cell from which it originates. This indicates that the waste has not reached stable methanogenic conditions, since the leachate still contains high concentrations of organic matter. If the development of the methanogenic environment is inhibited by low pH values, which often will be the case, feeding back an acid phase leachate may not improve the situation; in fact it may further inhibit methane formation. This has led in many experiments to pH-adjustment and buffering of the leachate prior to recirculation resulting in enhancement of gas production and decreases in the organic content of the leachate (e.g. Buivid, 1980; Tittlebau, 1982; Leuschner, 1983; Leuschner & Melden, 1983; Kinman *et al.*, 1987; Leuschner, 1989). If buffer addition is avoided the effect of recirculation is less predictable: e.g. Doedens & Cord-Landwehr (1984) found positive effects of recirculation while Buivid (1980) found negative effects.

Recirculation increases the content and distribution of water in the waste cells which theoretically should benefit the degradation processes (see, for example, Christensen & Kjeldsen, 1989). However, Barlaz *et al.* (1987) found no positive effects for water contents above 20%. Besides increasing the water content, recirculation also induces a water flux in the waste cell that potentially could be beneficial. Klink & Ham (1982) showed in laboratory batch experiments that the water flux reactors gave 25–50% higher methane production compared to reactors having the same water content but no flux.

Figure 2 shows the concentrations of COD and BOD_5 in the leachate at an actual German landfill showing the positive effects of leachate recirculation. However, Doedens & Cord-Landwehr (1989) reporting on the experiences at 13 large-scale landfills in Germany found no clear

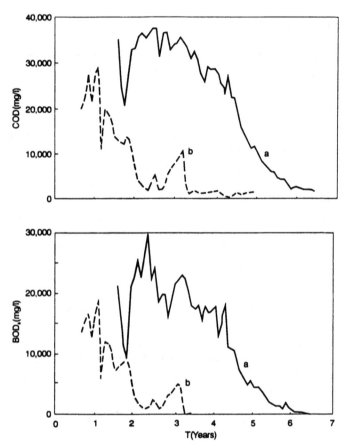

Figure 2. Organic contents in leachate from a full-scale landfill with (b) and without (a) leachate recirculation (adapted from Stegmann & Spendlin, 1986).

evidence that recirculation enhanced waste stabilization and gas production, although a faster decrease in the organic concentrations in the leachate was observed for landfills that had practised recirculation from the beginning. Although experience with recirculation of landfill leachate is not quite straightforward, a properly balanced view of its possibilities as a management procedure supposedly must pay attention to leachate composition, annual surplus precipitation and waste dryness.

In a relatively wet climate, e.g. in central Europe, recirculation is supposedly only beneficial in the first years of the landfill life by improving the water content and distribution in the waste cells, returning incompletely-stabilized leachate to the biologically active waste cells and reducing the leachate production by increasing the evapotranspiration from the uncompleted landfill sections in the dry season. However, if the leachate exhibits low pH-values and the methanogenic conditions are not well established in the cells, recirculation may be detrimental to the waste degradation unless the leachate is pH-adjusted and buffered. When the waste cells have proper water content and the leachate is low in BOD_5, recirculation supposedly has very little effect on leachate quality.

In dry climates, recirculation may be more beneficial in terms of enhancing the waste degradation and reducing the leachate amounts, and recirculation supposedly should be continued for a longer period. But also here the recirculation should be handled carefully so as not to inhibit the methanogenic consortia.

Only in very dry climates might evapotranspiration of recycled leachate outweigh leachate production, and in most landfills, diversion of part of the leachate must be done to avoid build-up of excessive water and perched water tables in the landfill.

Pre-composting

In recent years, the concept of in-situ pre-composting of the bottom layer of the landfill in order to prevent a too vigorous acid phase degradation has developed in Germany (Stegmann, 1983). The basis for this concept is to allow part stabilization of the waste through aerobic processes and thereby modify the acid phase and more quickly obtain a balance between the acid phase and methane phase microbial consortia, when the waste environment is made anaerobic by compaction and additional waste layers (see also Stegmann & Spendlin, 1989).

The pre-composting of the bottom layer may be established by:

- A thin (0·5 m) layer of compacted refuse left uncovered for several months (6–12 months). This is repeated 2–3 times.
- A 1-m loose layer of waste (homogenized by the compactor) established at the bottom of the fill and left for several months (6–12 months). This is repeated once.
- A 2–3-m layer of loose waste placed at the bottom of the fill and supplied with perforated pipes for temperature-induced venting (Spillmann & Collins, 1981).

When the composting process has progressed sufficiently (the BOD_5 content of the leachate or the temperature curve of the composting refuse may be used for indicating process progression), the waste layers are compacted and additional layers of waste established on top. This leads to a rapid development of methanogenic conditions in the bottom layer which in the future will act as an effective anaerobic filter for the acids developed in the layers above. Doedens & Cord-Landwehr (1984) showed by calculation that the capacity of the anaerobic filter in the bottom layer of the landfill hardly could be exceeded. This also indicates

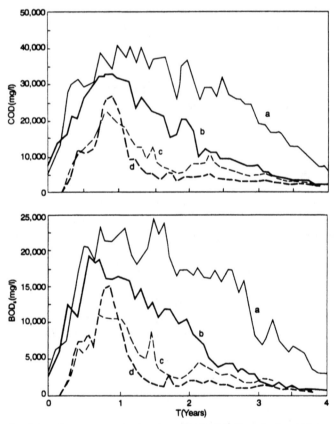

Figure 3. Organic contents in leachate from a full-scale landfill employing compaction of the waste (a) and recirculation (b) and as alternatives pre-composting (c) and pre-composting combined with leachate recirculation (d) (adapted from Stegmann & Spendlin, 1986).

that the methods for establishing the additional layers may not be of major importance, since the bottom layer will handle any leachate load from the layers above.

A well-developed pre-composting bottom layer may exhibit elevated temperatures (Rees & Grainger, 1982) that, by an effective compaction and insulation by waste layers on top, may be preserved and maintained by the temperature-enhanced methane formation. The anaerobic process produces much less heat than the composting process but, at the elevated temperatures, the heat flux generated suffices, outweighing the heat loss through a thick waste cap (Rees, 1980).

The positive effects on waste degradation of establishing pre-composting of the bottom waste layer are illustrated in Fig. 3, which shows the organic content of the leachate at a full-scale German landfill.

Pre-composting outside the landfill as a pre-treatment step is discussed in Germany. Advantages may be better landfill operation (no dust, improved aesthetics), higher compaction, less settling and enhanced degradation. On the other hand, the addition of a complete treatment step is costly and also produces emissions.

CONCLUSION

Several landfill management procedures are potentially available for establishing a well-engineered landfill with good degradation of the waste, an effective gas production and a moderate amount and composition of the leachate. However, most of the management procedures influence several of the basic factors controlling the landfill environment and the consequences of one combination of procedures compared to other combinations cannot always be predicted. Although much has been learned in recent years about the consequences on leachate and gas of landfill management procedures, it must be concluded that the current knowledge primarily is at the 'explanatory' level. We are often able to explain or pose a likely explanation for observations at actual landfills in terms of leachate and gas composition, but we still have not developed and verified predictive models founded on basic mechanisms taking place in the landfill. Less advanced predictive tools may eventually also suffice for practical purposes. Apparently the control of the acid phase, to prevent it being too vigorous, seems to be a key element in further developing our understanding of the effect of landfill management procedures on waste degradation, gas production and leachate quantity and quality. Finally it shall be emphasized that the knowledge we have from different sources about enhancement of landfills should be tested in intensively monitored full-scale experiments.

REFERENCES

Augenstein, D. C., Wise, D. L. & Wentworth, R. L. (1976). Fuelgas recovery from controlled landfilling of municipal wastes. *Resource Recovery and Conservation*, **2**, 103–17.
Barber, C. & Maris, P. J. (1984). Recirculation of leachate as a landfill management option: benefits and operational problems. *Quarterly Journal of Engineering Geology*, **17**(1), 19–29.
Barlaz, M. A., Milke, M. W. & Ham, R. K. (1987). Gas production parameters in sanitary landfill simulators. *Waste Management & Research*, **5**, 27–40.
Barlaz, M. A., Schaeffer, D. M. & Ham, R. K. (1989). Bacterial population development and chemical characteristics of refuse decomposition in a simulated sanitary landfill. *Applied and Environmental Microbiology*, **55**(1), 55–65.
Beker, D. (1987). Control of acid phase degradation. International Symposium Process, Technology and Environmental Impact of Sanitary Landfill, Cagliari, Italy, 19–23 Oct. 1987.
Buivid, M. G. (1980). Laboratory simulation of fuel gas production enhancement from municipal solid waste landfills. Dynatec R & D Co., Cambridge, MA.
Buivid, M. G., Wise, D. L., Blanchet, M. J., Remedios, E. C., Jenkins, B. M., Boyd, W. F. & Pacey, J. G. (1981). Fuel gas enhancement by controlled landfilling of municipal solid waste. *Resources and Conservation*, **6**, 3–20.
Carra, J. S. & Cossu, R. (1990). *International Perspectives on Municipal Solid Wastes and Sanitary Landfilling*. Academic Press, London.
Chian, E. S. K., DeWalle, F. B. & Hammerberg, E. (1977). Effect of moisture regime and other factors on municipal solid waste stabilization. In: *Management of Gas and Leachate in Landfills. Proceedings of the Third Annual Municipal Solid Waste Research Symposium*. EPA-600/9-77-026, US Environmental Protection Agency, Cincinnati, OH, pp. 73–86.
Christensen, T. H. & Kjeldsen, P. (1989). Basic biochemical processes in landfills. In: *Sanitary Landfilling: Process, Technology and Environmental Impact*, ed. T. H. Christensen, R. Cossu & R. Stegmann. Academic Press, London.
Christiansen, K., Prisum, M. & Skov, C. (1985). Undersøgelse af lossepladsers selvrensende effekt ved recirkulering af perkolat. Enviroplan A/S, Lynge, Denmark.
Doedens, H. & Cord-Landwehr, K. (1984). Sickerwasserkreislaufführung auf Deponien—neue Erkenntnisse und betriebliche Varianten. *Müll und Abfall*, **16**, 68–77.

Doedens, H. & Cord-Landwehr, K. (1989). Leachate recirculation. In *Sanitary Landfilling: Process, Technology and Environmental Impact*, ed. T. H. Christensen, R. Cossu & R. Stegmann. Academic Press, London.

Ehrig, H.-J. (1982). Auswirkungen der Deponietechnik auf die Umsetzungsprozesse im Deponiekörper—Einführung in die Thematik sowie Untersuchungen der Deponie Venneberg/Lingen. In: *Gas- und Wasserhaushalt von Mülldeponein*. Internationale Fachtagung 29.9–1.10.1982 Braunscheweig, pp. 124–44. TU Braunschweig. (Veröffentlichungen des Instituts für Stadtbauwesen, Heft 33).

Ehrig, H.-J. (1989). Leachate quality. In: *Sanitary Landfilling: Process, Technology and Environmental Impact*, ed. T. H. Christensen, R. Cossu & R. Stegmann. Academic Press, London.

Eifert, M. C. (1976). Variations in gas and leachate production from baled and non-baled refuse. In: *Gas and Leachate from Landfills: Formation, Collection and Treatment*. Proceedings of a research symposium held at Rutgers University, New Brunswick, NJ, March 1975. EPA-600/9-76-004, US Environmental Protection Agency, Cincinnati, OH, pp. 71–82.

Eifert, M. C. & Schwartzbaugh, J. T. (1977). Influence of municipal solid waste processing on gas and leachate generation. In: *Management of Gas and Leachate in Landfills*. Proceedings of the Third Annual Municipal Solid Waste Research symposium. US Environmental Protection Agency, Cincinnati, OH, pp. 55–72.

El-Fadel, M., Findikakis, N. & Leckie, J. O. (1989). A numerical model for methane production in managed sanitary landfills. *Waste Management & Research*, **7**, 31–42.

EMCON Associates (1975). Sonoma County Solid Waste Stabilization Study. EPA SW-65d.1, PB 239 778. US Environmental Protection Agency, Cincinnati, OH.

Fungaroli, A. A. & Steiner, R. L. (1979). Investigation of Sanitary Landfill Behaviour. EPA-600/2-79-053 US Environmental Protection Agency, Cincinnati, OH.

Gandolla, M. (1982). Ergebnisse von Lysimetern auf der Deponie Croglio, Schweiz. In: *Gas- und Wasserhaushalt von Mülldeponien*. Internationale Fachtagung 29.9.–1.10.1982, Braunschweig, pp. 163–82. TU Braunschweig. (Veröffentlichungen des Instituts für Stadtbauwesen, Heft 33).

Ham, R. K. & Bookter, T. J. (1982). Decomposition of solid waste in test lysimeters. *Journal of Environmental Engineering Division ASCE*, **108**, 1147–70.

Haxo, H. E. (1977). Compatibility of liners with leachate. In: *Management of Gas and Leachate in Landfills*. Proceedings of the Third Annual Municipal Solid Waste Research Symposium, St. Louis, Missouri, 14–16 March. EPA-600/9-77-026, US Environmental Protection Agency, Cincinnati, OH.

Kinman, R. N., Nutini, D. L., Walsh, J. J., Vogt, E. G., Stamm, J. & Richabaugh, J. (1987). Gas enhancement techniques in landfill simulators. *Waste Management & Research*, **5**, 13–26.

Klink, R. E. & Ham, R. K. (1982). Effects of moisture movement on methane production in solid waste landfill samples. *Resources and Conservation*, **8**, 29–41.

Leckie, J. O., Pacey, J. G. & Halvadakis, C. (1979). Landfill management with moisture control. *ASCE, Journal of Environmental Engineering Division*, **105**, 337–55.

Lee, G. F., Jones, R. A. & Ray, C. (1986). Sanitary landfill leachate recycle. *Biocycle*, **27**, 36–8.

Leuschner, A. P. (1983). Feasibility study for recovering methane gas from the Greenwood Street sanitary landfill, Worchester, MA, Vol. I. Task 1— Laboratory feasibility. Dynatech R/D Company, Cambridge, MA.

Leuschner, A. P. (1989). Enhancement of degradation: laboratory scale experiments In: *Sanitary Landfilling: Process, Technology and Environmental Impact*, ed. T. H. Christensen, R. Cossu & R. Stegmann. Academic Press, London.

Leuschner, A. P. & Melden, Jr., H. A. (1983). Landfill enhancement for improving methane production and leachate quality. Presented at the 56th Annual Conference of the Water Pollution Control Federation, 2–7 October, Atlanta, GA.

Pohland, F. G. (1980). Leachate recycle as landfill management option. *ASCE, Journal of Environmental Engineering Division*, **106**, 1057–69.

Pohland, F. G. & Kang, S. J. (1974). Sanitary landfill stabilization with leachate recycle and residual treatment. *AIChE Symposium Series*, **71**(145), 308–18.

Rees, J. F. (1980). Optimisation of methane production and refuse decomposition in landfills by temperature control. *Journal of Chemical Technology & Biotechnology*, **30**, 458–65.

Rees, J. F. & Grainger, J. M. (1982). Rubbish dump or fermenter? Prospects for the control of refuse fermentation to methane in landfills. *Process Biochemistry*, **17**(6), 41–4.

Robinson, H. D. & Maris, P. J. (1985). The treatment of leachates from domestic waste in landfill sites. *Water Pollution Control Federation Journal*, **57**, 30–8.

Scharf, W. (1982) Untersuchungen zur gemeinsamen Ablagerung von Müll und Klarschlamm im Labormasstab. In: *Gas- und Wasserhaushalt von Mülldeponien*. Internationale Fachtagung 29.9–1.10.1982, Braunschweig, pp. 83–98. TU Braunschweig. (Veröffentlichungen des Instituts für Stadtbauwesen, Heft 33).

Spillmann, P. & Collins, H. J. (1981). Das Kaminzug-Verfahren. *Forum Städtehygiene*, **32**, 15–24.

Stamm, J., Vogt, G. & Walsh, J. (1984). Evaluation of the impacts of sludge landfilling. US Environmental Protection Agency, Cincinnati, OH.

Stegmann, R. (1983). New aspects on enhancing biological processes in sanitary landfills. *Waste Management & Research*, **1**, 201–11.

Stegmann, R. & Spendlin, H.-H. (1986). Research activities on enhancement of biochemical processes in sanitary landfills. *Water Pollution Research Journal Canada*, **21**(4).

Stegmann, R. & Spendlin, H.-H. (1989). Enhancement of degradation: German experiences. In: *Sanitary Landfilling: Process, Technology and Environmental Impact*, ed. T. H. Christensen, R. Cossu & R. Stegmann. Academic Press, London, UK.

Tittlebau, M. E. (1982) Organic carbon content stabilization through landfill leachate recirculation. *Water Pollution Control Federation Journal*, **54**, 428–33.

Wolffson, C. (1985). Untersuchungen über den Einfluss der Hausmüllzusammensetzung auf die Sickerwasser- und Gasemissionen—Untersuchungen im Labormasstab. In: *Sickerwasser aus Mülldeponien—Einflüsse und Behandlung*. Fachtagung März 1985, Braunschweig, pp. 119–46. Technische Universität Braunschweig, Deutschland. (Veröffentlichungen des Instituts für Stadtbauwesen, Heft 39).

2.8 Managing Co-disposal Effects on Leachate and Gas Quality

FREDERICK G. POHLAND

Department of Civil Engineering, 949 Benedum Hall, University of Pittsburgh, Pittsburgh, Pennsylvania 15261, USA

INTRODUCTION

Most municipal landfills exist as dynamic, microbially-mediated and operationally-influenced treatment systems with inherent capacities to attenuate and assimilate a variety of waste constituents. This capacity is driven by the availability and sufficiency of moisture and nutrients, and is reflected by the nature of the leachate and gas produced as the landfill matures.

Attenuation may be defined (Pohland *et al.*, 1988) as a change in waste components and a concomitant lessening of their impacts as they are released to the leachate and gas transport phases, while assimilation more specifically involves the processes of conversion and removal within the landfill system. Hence, attenuation may be caused by a simple washout or dilution, perhaps preceded by assimilation, whereas the latter involves the collective effects of biochemical and physico-chemical alterations leading to a particular condition. Consequently, when considering the totality of the inorganic and organic fractions constituting the waste mass, potential transformations may be induced by the combination of biotic and abiotic processes operative throughout the phases of landfill stabilization. If operational control over the waste input and resultant leachate and gas generation is exercised, attenuation can be optimized and made more predictable, while adverse environmental effects are curtailed.

This chapter describes and discusses the results of landfill simulator experiments conducted to study the effect of organic and inorganic hazardous waste co-disposal on landfill leachate and gas quality and its implications for landfill management.

MUNICIPAL LANDFILL CO-DISPOSAL PRACTICE

Refuse, the primary solid waste input to municipal landfills, contains a variety of common materials (Table 1). The normal input of 'hazardous wastes' is much less and often undetectable, whether derived from the household (Table 2) or from small quantity industrial generators (Table 3). In 1986, about 83% of the municipal solid waste generated in the USA was managed in landfills, without a distinction being made between the hazardous and nonhazardous fractions. Separate studies, however, have indicated between 0·0015 and 0·4% hazardous waste by weight depending on location (US-EPA, 1988).

LANDFILL SIMULATOR EXPERIMENTS

Because of the many uncertainties associated with waste inputs to municipal landfills and the associated effects on in-situ processes of

TABLE 1. Temporal Trends in Municipal Refuse Composition in the United States (US-EPA, 1988)

Refuse category	*Composition, % by weight*		
	1970	*1986*	*2000*
Paper and paperboard	32·4	35·6	39·1
Yard waste	20·6	20·1	19·0
Metal	12·0	8·9	8·5
Food waste	11·4	8·9	7·3
Glass	11·1	8·4	7·1
Wood	3·6	4·1	3·6
Plastic	2·7	7·3	9·2
Rubber and leather	2·7	2·8	2·3
Textile	1·8	2·0	2·0
Miscellaneous	1·7	1·8	1·9

TABLE 2. Major Household Hazardous Wastes by Activity (US-EPA, 1988)

Household activity	Major consituents
Automotive maintenance	Refrigerants; waxes, polishes and cleaners; lubricating, starter and radiation fluids, solvents; oil, fuel, radiator and transmission additives
Home maintenance	Adhesives; paints, paint thinners, strippers and removers; stains; varnishes; sealants; polishes; cleaners; disinfectants
Yard and garden care	Fungicides, herbicides and pesticides; wood preservatives
Miscellaneous	Batteries and electronic items; cosmetics; drugs; swimming pool chemicals

stabilization, pilot-scale simulations have been devised to study attenuating mechanisms under controlled conditions. Early studies firmly established the sequential nature of landfill stabilization, and emphasized the importance of the acid formation and methane fermentation phases (Pohland & Harper, 1985). These two stabilization phases were described in terms of leachate and gas parameters, as influenced by design and operational circumstance. In this regard, the controlled influx of moisture, coupled with leachate containment and recycle, served to accelerate the microbially-mediated stabilization processes and made the onset and duration of each phase more predictable. As stabilization progressed, waste components released from the waste mass and transferred to the leachate and gas transport phases were subjected to in-situ attenuation and assimilation.

TABLE 3. Representative Sources of Small Quantity Generator Hazardous Wastes[a] (Adapted from US-EPA, 1988)

Analytical and clinical laboratories	Motor freight terminals
Chemical manufacturing and formulation	Paper and paper products industry
Construction and demolition	Pesticide application and end use
Educational and vocational institutions	Photography and printing
Furniture manufacturing and refinishing	Textile manufacturing
Laundries and dry cleaning establishments	Vehicle and equipment maintenance
Metals and metal products manufacturing	Wholesale and retail establishments

[a] Less than 100 kg of hazardous waste/month generated by >50% of the industrial or commercial category.

Experimental Procedures

The pilot-scale landfills used in these simulations generally consisted of containment structures with leachate and gas management capabilities configured similarly to those indicated in Fig. 1. The experiments were devised to examine single pass leaching, often in comparison with leachate recycle, in duplicate units containing the base refuse and possible admixtures of test waste components. In studies on co-disposal of refuse with hazardous wastes, the organic and/or inorganic species were added to the test units in batches to provide contact opportunity as well as to represent probable conventional loading practices. Consequently, the control units simulated the attenuation/assimilation potential of municipal refuse, including possible inputs of household hazardous wastes, whereas the test units simulated this same potential together with the effects of hazardous wastes additions similar to those from small

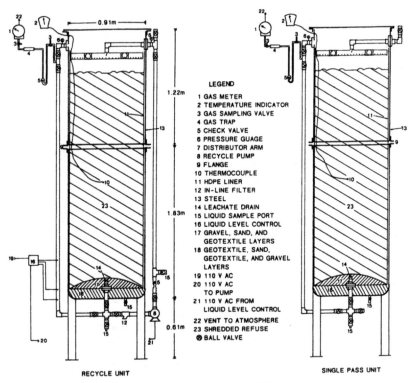

LEGEND

1 GAS METER
2 TEMPERATURE INDICATOR
3 GAS SAMPLING VALVE
4 GAS TRAP
5 CHECK VALVE
6 PRESSURE GUAGE
7 DISTRIBUTOR ARM
8 RECYCLE PUMP
9 FLANGE
10 THERMOCOUPLE
11 HDPE LINER
12 IN-LINE FILTER
13 STEEL
14 LEACHATE DRAIN
15 LIQUID SAMPLE PORT
16 LIQUID LEVEL CONTROL
17 GRAVEL, SAND, AND GEOTEXTILE LAYERS
18 GEOTEXTILE, SAND, GEOTEXTILE, AND GRAVEL LAYERS
19 110 V AC
20 110 V AC TO PUMP
21 110 V AC FROM LIQUID LEVEL CONTROL
22 VENT TO ATMOSPHERE
23 SHREDDED REFUSE
⊗ BALL VALVE

RECYCLE UNIT   SINGLE PASS UNIT

Figure 1. Schematic diagram of simulated landfill columns.

quantity generators. With this approach, the nature and significance of mechanisms responsible for attenuation and assimilation at landfills receiving varying inputs of hazardous wastes co-disposed with municipal refuse could be assessed by 10 pilot-scale landfill columns, loaded and operated as indicated in Table 4. By adding water to the columns to satisfy indicated field capacity and to promote stabilization and leaching,

TABLE 4. Column Identity and Loading Characteristics for Co-disposal Investigations with Simulated Landfills

Column identity	Mode of operation	Shredded refuse/ waste loading		Admixed test constituents	
		Initial loading height (cm) (above underdrain)	Compacted density (kg/m³)[a]	Inorganic[b]	Organic[c]
1 CR	Recycle	29	313	None	None
2 C	Single pass	30	301	None	None
3 O	Single pass	29	309	None	Yes
4 OL	Single pass	28	327	Low	Yes
5 OM	Single pass	30	305	Medium	Yes
6 OR	Recycle	28	317	None	Yes
7 OLR	Recycle	29	309	Low	Yes
8 OH	Single pass	30	305	High	Yes
9 OMR	Recycle	29	313	Medium	Yes
10 OHR	Recycle	31	293	High	Yes

[a] For shredded refuse as placed; 42 individual 9-kg batches to each column or 378 kg total.
[b] Added as augmented metal plating sludge, divided into three equal portions and placed on the compacted refuse at locations 30 cm above the underdrain, at mid-depth, and 30 cm below the refuse surface.

Low: (0·11%)			Medium: (0·22%)			High: (0·44%)		
	Cd	35 g		Cd	70 g		Cd	140 g
	Cr	45 g		Cr	90 g		Cr	180 g
	Hg	20 g		Hg	40 g		Hg	80 g
	Ni	75 g		Ni	150 g		Ni	300 g
	Pb	105 g		Pb	210 g		Pb	420 g
	Zn	135 g		Zn	270 g		Zn	540 g

[c] Portions of 120 g each of bis-2-ethylhexyl phthalate, 1,4-dichlorobenzene, 1,2,4-trichlorobenzene, dibromomethane, γ-1,2,3,4,5,6-hexachlorocyclohexane, hexachlorobenzene, 2,4-dichlorophenol, 2-nitrophenol, naphthalene, nitrobenzene and trichloroethene, and 30 g dieldrin all placed as a mixture at the surface of the first 30 cm of compacted refuse in each column; 1·35 kg total or 0·36%.

samples could be obtained for initial analysis as well as throughout the experimental period. Routine moisture additions equivalent to a rainfall intensity of 123 cm/year (single pass columns), or from leachate recycle with accumulated quantities sufficient to sustain recycle and replace sampling volumes (recycle columns), were provided. Gas sampling and analysis were also conducted to permit examination of corresponding data from the associated gas transport phase.

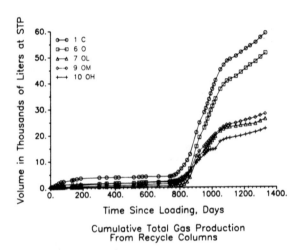

Cumulative Total Gas Production
From Recycle Columns

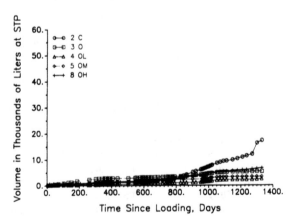

Cumulative Total Gas Production
From Single Pass Columns

Figure 2. Changes in gas production and leachate COD, TVA and pH during co-disposal investigations with simulated landfills.

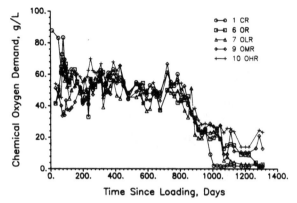

Chemical Oxygen Demand Concentration
in Leachate From Recycle Columns

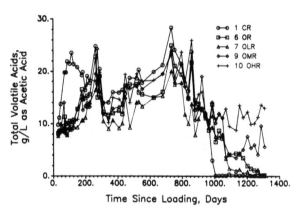

Total Volatile Acid Concentration in
Leachate From Recycle Columns

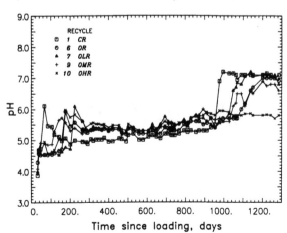

Figure 2—*contd.*

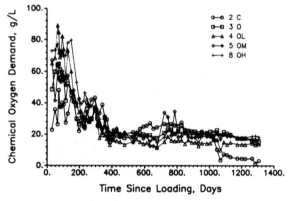

Chemical Oxygen Demand Concentration
in Leachate From Single Pass Columns

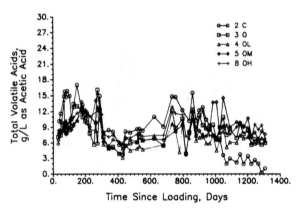

Total Volatile Acid Concentration in
Leachate From Single Pass Columns

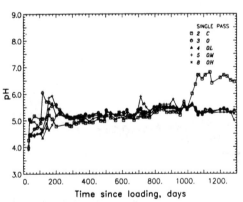

Figure 2—*contd.*

Experimental Results

By selecting gas production to complement leachate chemical oxygen demand (COD), total volatile acids (TVA) and pH as major indicator parameters for both the single pass and recycle columns, the phasic nature of landfill stabilization, as well as the impacts of the hazardous waste loadings, could be ascertained. As indicated in Fig. 2, high leachate COD and TVA concentrations at low pH and minor gas production indicated the early onset of acid formation and a strong and chemically aggressive aqueous phase. As a consequence, enhanced release of both inorganic and organic test species occurred (Figs 3–6).

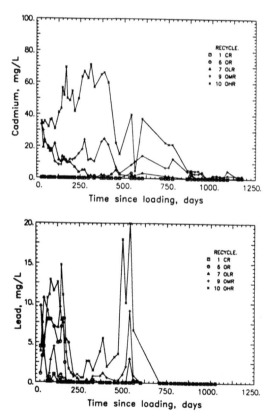

Figure 3. Changes in leachate cadmium, lead, nickel and zinc during co-disposal investigations with simulated landfills.

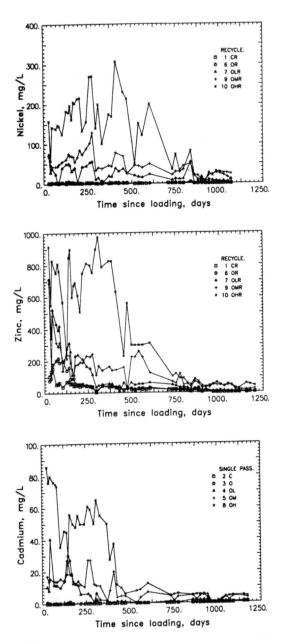

Figure 3—*contd.*

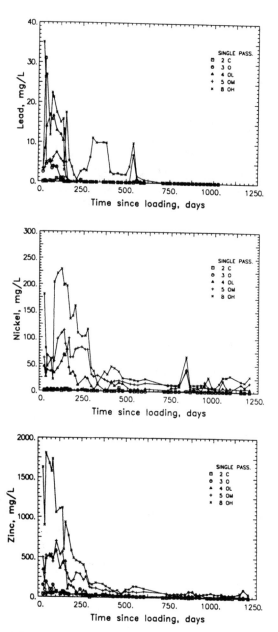

Figure 3—*contd.*

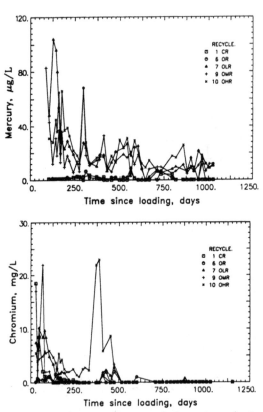

Figure 4. Changes in leachate mercury and chromium during co-disposal investigations with simulated landfills.

This was apparent especially for the heavy metals and, to a lesser extent, for the more mobile of the test organics, respectively. In the latter case it was likely that attenuation was already occurring, with a suggestion of partitioning depending upon the physical-chemical properties of the waste matrix and specific compounds in question. Indeed, four of the test compounds essentially remained within the waste matrix without release to the leachate, even with leachate recycle (Table 5).

As gas production increased with the onset of methane fermentation (Fig. 2), leachate COD and TVA concentrations decreased with increasing pH, particularly for the control columns and the test columns with the lower inorganic (heavy metals) loadings. This behavior is consonant

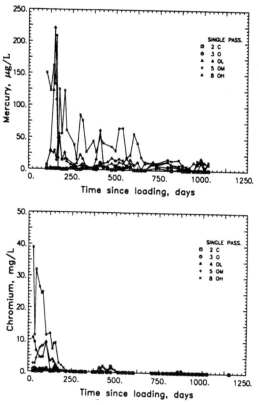

Figure 4—*contd.*

with the microbially-mediated conversion of organic intermediates to methane (CH_4) and carbon dioxide (CO_2), and the acknowledged influence of heavy metals on methane fermentation. Accordingly, as stabilization progressed, associated attenuation and assimilation mechanisms dampened or eliminated the inhibitory influences of the heavy metals. In the final analysis, these observations could be quantified and used to determine possible loadings of co-disposed heavy metals which would not impose a lasting detrimental effect on the progress of stabilization of the readily decomposable organic refuse constituents. Whether possible bioconversion of the test organic compounds would be similarly affected was not as evident at the relatively low leachate concentrations detected.

Comparison of results between the single pass and recycle columns also emphasized the importance of leachate management by recycle in terms of optimizing attenuation and assimilation capacities. Whereas both organic and inorganic waste constituents released to the leachate and gas transport phases were removed *in situ*, much less microbially-mediated removal occurred in the single pass columns, with reductions in leachate concentrations being mainly accountable to physical washout. Therefore, leachate recycle prohibited potential migration by washout and enhanced the opportunity for in-situ removal. In contrast, single pass leaching encouraged washout, diminished opportunities for in-situ

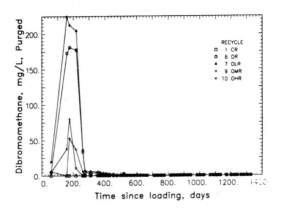

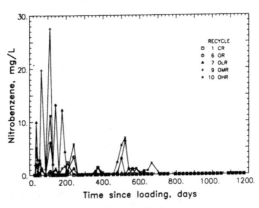

Figure 5. Changes in leachate dibromomethane, nitrobenzene, trichloro-benzene and 2-nitrophenol during co-disposal investigations with simulated landfills.

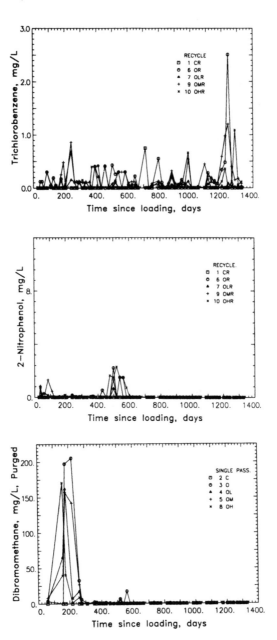

Figure 5—*contd.*

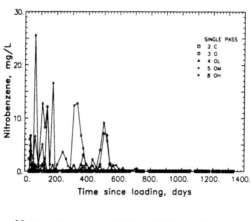

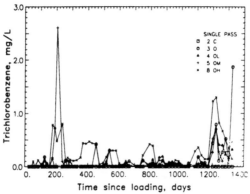

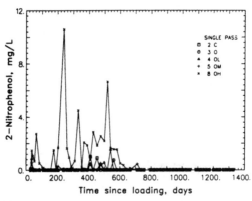

Figure 5—*contd.*

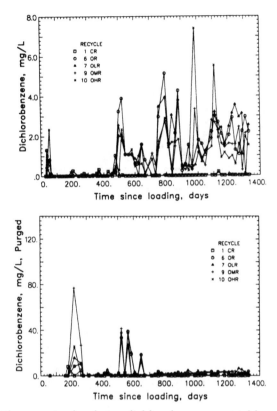

Figure 6. Changes in leachate dichlorobenzene, trichloroethene and dichlorophenol during co-disposal investigations with simulated landfills.

removal, contributed to the loss of leachate constituents, and incurred the probable need for external leachate management and treatment.

MECHANISMS OF ATTENUATION AND ASSIMILATION

The data from the pilot-scale co-disposal investigations also provided information sufficient to establish the significance of the various mechanisms responsible for in-situ attenuation and assimilation of inorganic and organic waste constituents. In this regard, there is no doubt that the readily available and reactive waste constituents are more effectively (and

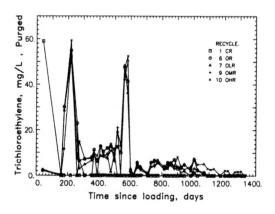

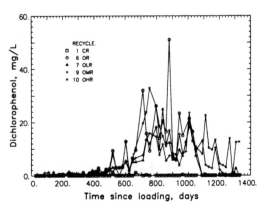

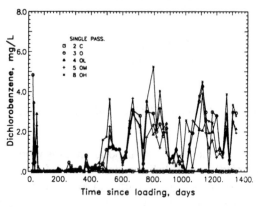

Figure 6—*contd.*

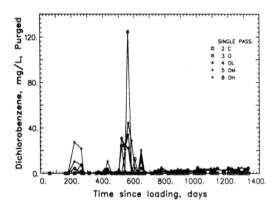

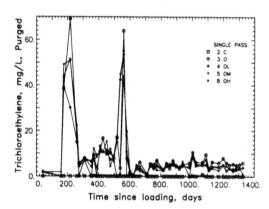

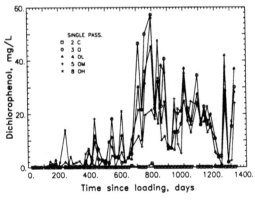

Figure 6—*contd.*

TABLE 5. Mass of Admixed Test Organic Constituents Contained in Recycled Leachate from Simulated Landfill Columns[a]

Test constituents	Column identity[b]				
	1 CR	*6 OR*	*7 OLR*	*9 OMR*	*10 OHR*
Dibromomethane	ND	54 000	84 000	17 000	26 000
		(45·0)	(70·0)	(14·1)	(21·7)
Trichloroethene	ND	17 000	15 500	16 600	17 700
		(14·2)	(12·9)	(13·8)	(14·8)
2,4-Dichlorophenol	ND	17 600	9 500	11 400	15 300
		(14·7)	(7·9)	(9·5)	(12·8)
Naphthalene	1 100	3 600	2 600	2 900	2 000
	(0·0)	(2·1)	(1·2)	(1·5)	(0·8)
Nitrobenzene	ND	520	4 500	1 000	3 100
		(0·4)	(3·8)	(0·8)	(2·6)
1,4-Dichlorobenzene	ND	1 400	1 600	1 400	1 700
		(1·2)	(1·3)	(1·2)	(1·4)
2-Nitrophenol	ND	760	180	350	670
		(0·6)	(0·2)	(0·3)	(0·6)
1,2,4-Trichlorobenzene	ND	360	70	400	130
		(0·3)	(0·06)	(0·3)	(0·1)
Lindane[c]	ND	ND	ND	1·6	0·4
				(0·001)	(<0·001)
Hexachlorobenzene	ND	ND	ND	ND	ND
Bis-2-ethylhexyl phthalate	ND	ND	ND	ND	ND
Dieldrin	ND	ND	ND	ND	ND

[a] Reported in mg and (%) of co-disposed mass transferred to leachate; ND = not detected.
[b] From Table 4; naphthalene (%) corrected for mass in control column (1 CR) leachate.
[c] γ-1,2,3,4,5,6-Hexachlorocyclohexane.

predictably) converted under the influence of leachate recycle than with single pass leaching. The microbially-mediated production and conversion of leachate COD and TVA to gas during the acid and methane phases of landfill stabilization are prime examples, but the similarly mediated reduction of inorganic sulfate to sulfide is no less significant. Indeed, processes such as these tend to condition the chemical environment and actually direct the mechanisms responsible for overall attenuation. Therefore, since the results from the recycle columns were more conclusive and unaffected by significant washout effects, these were selected for further evaluation of mechanisms.

Attenuation of Heavy Metals

It is apparent that the principal attenuating mechanisms for the heavy metals contained within the simulated landfill columns, either as part of the refuse input or added intentionally, were influenced by leachate management and the chemical character of the landfill environment during a particular phase of stabilization. Moreover, since the appearance or disappearance of the heavy metals in the leachate is largely determined by solubility equilibria, special scrutiny of potential precipitating, complexing and immobilizing mechanisms was considered appropriate.

Precipitation and oxidation–reduction mechanisms. In terms of the heavy metals added to the landfill columns (Table 4) and appearing in the leachate (Figs 3 and 4), chromium (Cr^{3+}), with its low solubility ($pK_{so} = 30·5$), would precipitate as the hydroxide, even at pH levels as low as 5·0. As methane fermentation commenced and the pH became more neutral (Fig. 2), leachate chromium concentrations for the recycle columns decreased below detectable levels.

In contrast to chromium, several of the other test metals were affected by the reducing potential prevailing under the anaerobic conditions of the simulated landfills. Persistently negative oxidation–reduction potentials (ORP) ranging from -50 to -500 mV (E_c) were indicative of requirements for microbially-mediated reduction of sulfates to sulfides. Since sulfides form very sparingly soluble precipitates with many heavy metals, and ORP levels became even more negative during methane fermentation, removal of heavy metals by sulfide precipitation decreased not only the sulfate pool (Fig. 7), but the leachate heavy metal concentrations as well (Fig. 3). Accordingly, formation of cadmium, lead, nickel and zinc sulfides ($pK_{so} = 27·2$, 23·8, 26·6 and 24·8, respectively) reduced the leachate concentrations of these metals to very low levels.

At higher leachate metal concentrations during acid formation, when elevated sulfate levels prevailed without significant reduction to sulfides, control of metal solubility may have involved other generally abundant anions such as sulfate, chloride, carbonate and possibly phosphate. However, within a landfill environment, these potential precipitants may have only transient significance, particularly if leachate recycle with accelerated stabilization is employed. The more rapid onset and intensity of reducing conditions, the formation of excess sulfides, and the attenuation of inhibiting levels of metals by the resultant mechanisms of

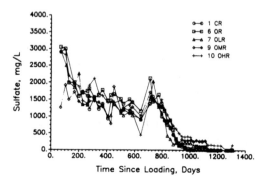

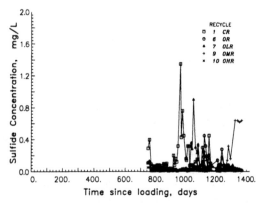

Figure 7. Changes in recycled leachate sulfate, sulfide, chloride and bromide concentrations during co-disposal investigations with simulated landfills.

precipitation, filtration, and localized containment are important features of such a landfill management option.

Another direct impact of reducing conditions was apparent for mercury, with a rapid initial decrease to leachate concentrations below 100 μg/litre and an eventual leveling to about 10 μg/litre during both the acid formation and methane fermentation phases of landfill stabilization (Fig. 4). Even with sulfate reduction and the exceptionally low solubility of mercuric sulfide ($pK_{so} = 52 \cdot 0$), no additional decrease in concentration was observed. Therefore, the suggested mechanism for persistence of low leachate mercury levels is reduction of the mercuric ion to metallic mercury, which has been shown to have an aqueous solubility of 5–30 μg/litre (Hughes, 1957).

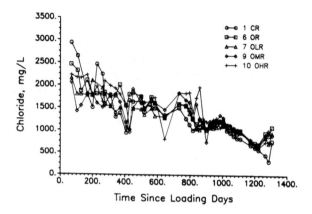

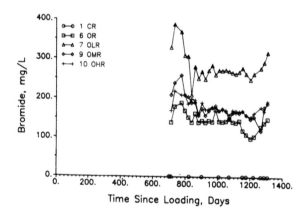

Figure 7—*contd.*

Heterogeneous physical-chemical mechanisms. The manner by which heavy metals are co-disposed with municipal refuse is important when investigating possible attenuating mechanisms. If mixed as household or as small quantity generator hazardous wastes, the relative homogeneity or heterogeneity of the mixture would likely dictate the potential for reaction and release. In either case, the creation of localized physico-chemical microenvironments within the waste matrix would provide opportunities for the development of a complex array of heterogeneous (liquid/solid) interactions.

Co-disposal of municipal refuse with the heavy metal sludges added to the simulated landfill columns probably established such microenvironments at the location of loading. Accordingly, the highly basic waste sludge provided a source of localized acid-neutralizing capacity (alkalinity) which would increase the pH and reduce metal mobility at the sludge/leachate interface. Moreover, the abundance of hydroxide in the sludge would enhance chemical reaction with leachate anions (sulfates, sulfides and phosphates) and the formation of surficial encapsulating layers capable of impeding further dissolution of sludge metals into the leachate. As a consequence, these alkaline microenvironments effectively contributed to in-situ attenuation during the acid formation phase when it was most needed, whereas this contribution to the overall attenuation of heavy metals likely diminished with encapsulation and when other attenuating mechanisms were promoted and gained dominance.

The refuse matrix also provides abundant surfaces for sorptive interactions with leachate (and gas) constituents. Potential mechanisms include physical adsorption, ion exchange (particularly with soil), chemisorption upon complexation with insoluble ligands, and mechanical containment in transiently stagnant void volume liquid. In addition, complexation of metals by soluble ligands, such as the moderate to high molecular weight humic-like substances, although initially enhancing mobility, may eventually promote sorption by formation of relatively hydrophobic molecules.

Attenuation of Organic Contaminants

The landfill environment provides an equally significant setting for the attenuation of organic hazardous contaminants, although the behavior and fate of these constituents are not as pronounced, even with leachate recycle. Moreover, whether a specific mechanism can be defined is concentration dependent; the larger and more concentrated the inputs, the more likely the possibility of evaluation. Scrutiny of the leachate data from the co-disposal investigations suggests that detection and/or mobility of the 12 test compounds (Table 4) in the leachate or gas phases varied depending upon the physical-chemical characteristics of the compound and the opportunity for reaction within the waste mass.

Microbially-mediated mechanisms. The extended opportunity for contact and acclimatization to the organic contaminants benefited the potential for microbially-mediated attenuation. Although the general lack of discrete tracers and corresponding identification of reaction products precluded specific conclusions, presumptive evidence for such attenuation was available. As an example, the presence in the leachate samples of bromide (Fig. 7) coincident with the decrease in dibromomethane (Fig. 5) strongly suggested such a mechanism. Likewise, headspace gas analysis for vinyl chloride indicated low but measurable concentrations ($\sim$300 μg/litre) at the same time leachate trichloroethene was being attenuated (Fig. 6), again suggesting partial microbially-mediated conversion to this recognized intermediate. Unfortunately, the high leachate chloride concentrations (Fig. 7) tended to mask the use of chloride as an effective tracer of possible dehalogenation for this and other chlorinated compounds.

Physical-chemical mechanisms. In addition to microbially-mediated conversions, other potential attenuating mechanisms can be perceived as operative within the landfill environment. In terms of concentration or mass (Table 5), the potential for release of a particular compound was influenced by equilibrium conditions between the gas, liquid and solid phases.

Headspace gas analysis indicated that three compounds (trichloroethene, naphthalene and dichlorobenzene) were present in the gas phase, and that the opportunity for release to the leachate was dependent upon solubility and the mobilizing influences of other leachate constituents on the more hydrophobic compounds. Otherwise, physical sorption could be selected as a primary attenuating mechanism. Indeed, a strong correlation has been shown to exist with the octanol/water partition coefficient (Reinhart, 1989), suggesting a selective fractionation depending upon the degree of hydrophobicity. Therefore, sorption of the test compounds could be attributed to hydrophobic interactions and removal from the aqueous phase onto nonspecific surfaces of the materials constituting the waste matrix.

Collectively, these results indicate that only the soluble, less hydrophobic and more refractory test organic compounds would be expected to elute from the columns, and that these would emerge in the approximate order of increasing affinity for the refuse constituents. However, the type and extent of landfill stablilization and the operational procedure employed influences this pattern. Hence, within an operational perspective, the completion of accelerated stabilization through methane fermentation and into final maturation provides an opportunity to remove residual leachate and to impart a lasting interruption of continued fractionation of the less mobile test contaminants from the waste matrix. In controlled landfills incorporating leachate containment and recycle capabilities, such an operational strategy can be easily accommodated.

SUMMARY AND CONCLUSIONS

Municipal landfills possess finite capacities to attenuate and assimilate both nonhazardous and hazardous wastes at loadings potentially encountered in conventional practice. Inorganic heavy metals co-disposed with refuse tend to be attenuated by in-situ microbially-mediated physico-chemical processes, including reduction, precipitation, sorption and waste matrix containment or interaction. Similarly, co-disposed organic contaminants tend to fractionate according to their physical and chemical properties, and are removed primarily by sorption and/or bioconversion, with the possible generation of identifiable reaction products.

Microbial acclimatization and degradation of more recalcitrant compounds are enhanced by the extended contact time and reaction opportunity provided within the landfill environment. Leachate management by recycle and final removal augments these processes by retaining leached waste constituents, accelerating stabilization and promoting increased treatment efficiency. Such controlled landfill management provides added assurances that releases of hazardous constituents will be minimized and adverse health and environmental impacts will be concomitantly curtailed.

REFERENCES

Hughes, W. L. (1957). A physicochemical rationale for the biological activity of mercury and its compounds. *Ann. New York Acad. Sci.*, **65**.
Pohland, F. G. & Harper, S. R. (1985). Critical review and summary of leachate and gas production from landfills. PB. 86-240 181/AS, NTIS, Springfield, VA.

Pohland, F. G., Cross, W. H., Gould, J. P. & Reinhart, D. R. (1988). Assimilative capacity of landfills for solid and hazardous wastes. *ISWA 88 Proceedings,* Vol. 1, ed. L. Andersen & J. Moller. Academic Press, London, pp. 101–108.

Reinhart, D. R. (1989). Fate of selected organic pollutants during landfill codisposal with municipal refuse. PhD thesis, Georgia Institute of Technology, Atlanta, GA.

US-EPA. (1988). Report to Congress—Solid waste disposal in the United States, vol. II. EPA/530-SW-88-011B, US Environmental Protection Agency, Washington, DC, Chapter 4.

2.9 Analytical Methods for Leachate Characterization

ULLA LUND, LIS RASMUSSEN, HENRIK SEGATO &
PREBEN ØSTFELDT

Water Quality Institute, Agern Allé 11, DK-2970 Hørsholm,
Denmark

INTRODUCTION

The analysis of leachate is a new discipline in environmental analytical chemistry, compared to, for example, wastewater analysis. Leachate analysis is to a great extent performed using the methods of wastewater analysis. In most cases this is straightforward, but for some parameters the different composition and often greater complexity of leachate compared to wastewater gives rise to analytical problems. Due to the comparative newness of the field of leachate analysis these problems are not always well understood, and the literature on leachate analysis is scarce.

This chapter describes methods for the analysis of leachate emphasizing instances where wastewater methods are known to create problems when used for leachate characterization. Furthermore some cautionary notes are given, where the composition of leachate would be expected to influence the accuracy or precision of the analysis.

NUTRIENTS

The main problem of sampling and analysing nutrients is that they convert both chemically and biologically into other substances of their group. This conversion may take place very rapidly. It is, therefore, very important that the samples are stored in a cold and dark place immediately after sampling. Furthermore, the analysis should be performed as soon as possible. Sample containers for nutrient analysis should be made of polyethylene or similar material. Immediate analysis is preferable, but short-term storage is possible in most cases upon acid preservation.

Analysis of most of the nutrients (i.e. nitrite, nitrate, organic nitrogen, total nitrogen, relative and total phosphorus) can be made using methods developed for wastewater analysis, e.g. Clesceri *et al.* (1989). Ammonia may, however, give rise to analytical problems.

The two most commonly used methods for the analysis of ammonia are the phenate method (Clesceri *et al.*, 1989, p. 4-120A) and the titrimetric method (Clesceri *et al.*, 1989, p. 4-121E). Both methods have advantages and disadvantages. The titrimetric method has poor precision at low concentrations and is only to be used at concentration levels higher than 1 mg/litre. With a certain degree of modification the phenate method may detect 5 μg/litre, but the reaction is performed at alkaline pH and problems arising from precipitation of metal hydroxides are often encountered, especially for samples with high iron concentrations.

Ion chromatographic methods have not been taken into consideration in this chapter because experience with these methods is still limited with respect to leachate. There are, however, indications that ion chromatography will be one of the commonly used methods in the future.

Both Danish and international experience suggest that laboratories with a reasonable general quality are able to perform nutrient analyses. However, the variation between laboratories is so great that long-term investigations should be made by a single laboratory. In this way the considerable variation between laboratories does not cloud trends in the data.

METALS

In this section metals cover sodium (Na), potassium (K), calcium (Ca), magnesium (Mg), iron (Fe), manganese (Mn), zinc (Zn), arsenic (As), cadmium (Cd), cobalt (Co), chromium (Cr), copper (Cu), mercury (Hg), nickel (Ni), lead (Pb) and strontium (Sr).

Sample bottles should be made of linear polyethylene or polypropylene, except for determination of Hg where sample bottles made of borosilicate glass or quartz are preferable. A special acid cleaning technique should be used for bottles intended for collection of trace element samples. Bottles should thus be delivered from the laboratory in question.

The sample must be acidified for preservation, preferably with nitric acid to a pH value <2. However, for determination of Hg the sample should be preserved with nitric acid to a pH value <1. In both cases use high-purity acid. Furthermore, the specific laboratory should be consulted because the selection of the method for preservation depends on the methods of analysis to be performed. Storage of the acid-preserved samples can take place at room temperature.

Sample pre-treatment may start in connection with sampling. Before collecting a sample, it must therefore be decided what fraction is to be analysed (dissolved, suspended, total, or acid-extractable). This decision will determine in part whether the sample is acidified with or without filtration and the type of digestion required.

If the sample is to be filtered it must be done directly after sample collection; optimum is on-line filtration minimizing oxidation of the sample. A 0·45 μm membrane filter is to be preferred. It should be noticed that a small amount of suspended material in the sample can significantly influence the result, as the concentration of metals is generally orders of magnitude larger in the suspended sediment than dissolved in the leachate. If iron precipitates after sample collection, other metals will co-precipitate. After acid preservation the precipitated metals will be dissolved again.

In some cases, depending on the fraction to be determined, it is necessary to digest the sample before analysis. For determination of mercury it is always necessary to digest the sample before analysis.

The majority of metals can be determined using atomic absorption spectroscopy (AAS), atomic emission spectrometry (AES), or emission spectrometry using an inductively coupled plasma source (ICP), which, if no other pre-treatment than digestion of the sample is used, all give the total content of the specific metal. It should however, be noticed that the determination of metals by ICP yields higher detection limits for many metals than determination by the use of electrothermal atomic absorption (HGAAS). Anodic stripping voltammetry may also be used, but the method is more dependent on the preceding digestion than HGAAS techniques.

With respect to both the fate and toxicity of the metal, it is of interest in many cases to evaluate not only the total content, but also the different species of the element. Much work is being performed to develop methods for speciation analysis suited to use also in routine analysis. Depending on the element, different techniques are described for speciation study in the literature, e.g. anodic stripping voltammetry. It is beyond the scope of this chapter to go into details on this subject.

Colorimetric measurements are applicable to determine metal content in leachate when interferences are known to be within the capacity of the method, but great care should be taken because leachate often contains substances that may interfere in colorimetric measurements.

The analysis of trace metals with AAS techniques necessitates the use of either background correction or a double-beam instrument with absorption at two different wavelengths. To avoid the influence of salts, which are often found in high concentrations in leachate, on the determination of the trace metals, it is suggested that the method of standard addition for determination of every sample be used.

The different AAS/AES techniques are described in Clesceri *et al.* (1989).

It is necessary to contact the specific laboratory to learn about the precision and accuracy to be expected, because accuracy depends on the method of analysis and the instrument in use. A round-robin analysis performed in the USA and Canada (DeWalle *et al.*, 1979) showed that variation between laboratories is much larger than within-laboratory variation. In the same intercomparison study, percentage recoveries were between -138 and 328. This emphasizes the necessity of taking matrix effects into account, for example by the use of standard addition techniques.

OTHER INORGANICS: PHYSICO-CHEMICAL PARAMETERS

This group of parameters covers a broad range of parameters and related analytical problems that is difficult to subdivide and describe in sub-groups. Therefore, parameters, preferred containers, preservation, maximum suggested storage time and methods of analysis are listed in Table 1.

The general experience is that leachate is stable for periods much longer than the suggested storage time.

Normally, the samples should always be kept in a cold and dark place. Certain sample types tend to form precipitates, which is why special precautions must be taken (storage in a dark place at sampling temperature and analysis must be performed as soon as possible).

TABLE 1. Sample Treatment and Analysis for Inorganic and Physico-Chemical Parameters.

Parameter	Preferred container	Preservation	Max. suggested storage time	Methods of analysis
pH	Polyethylene	÷	Short	Clesceri et al. (1989) p. 4-95B
Alkalinity	Polyethylene	÷	24 h	Clesceri et al. (1989) p. 2-35B
Conductivity	Polyethylene	÷	Short	Clesceri et al. (1989) p. 2-59B
Colour	Polyethylene	÷	Short	Clesceri et al. (1989) p. 2-2B
Turbidity	Polyethylene	÷	24 h	Clesceri et al. (1989) p. 2-13B
Suspended solids, 105°C	Polyethylene	÷	24 h	Clesceri et al. (1989) p. 2-75D
Suspended solids, fixed	Polyethylene	÷	24 h	Clesceri et al. (1989) p. 2-75D
Total solids, 105°C	Polyethylene	÷	1 Week	Clesceri et al. (1989) p. 2-72B
Total solids, fixed	Polyethylene	÷	1 Week	Clesceri et al. (1989) p. 2-77E
Oxygen	Glass, 130 ml	$Mn(II) + I^- + N_3^-$	Preservation at sampling	Clesceri et al. (1989) p. 4-152C
Sulphide	Glass, 130 ml	$ZnAc + NaOH$		Clesceri et al. (1989) p. 4-195D
Calcium	Polyethylene			Clesceri et al. (1989) p. 3-85D
Total hardness	Polyethylene			Clesceri et al. (1989) p. 2-53C
Iron	Polyethylene	1 ml 4M H_2SO_4/ 100 ml		Dougan & Wilson (1973)
Manganese	Polyethylene	1 ml 4M H_2SO_4/ 100 ml		Henriksen (1966)
Chloride	Polyethylene			Clesceri et al. (1989) p. 4-71D
Fluoride	Polyethylene	÷	Short	Clesceri et al. (1989) p. 4-87C
Sulphate	Polyethylene			Clesceri et al. (1989) p. 4-207E

÷, Do not preserve.

The sulphide analysis cannot be performed immediately because of interferences. Therefore, these interferences must be dealt with before analysis. It is suggested that the sulphide be removed as hydrogen sulphide by acidifying the sample and by collecting the hydrogen sulphide in an alkaline phosphate buffer (Segato, 1990, unpublished).

The colorimetric methods for calcium, total hardness, iron and manganese cannot always be used because of interfering substances. General experience suggests atomic absorption spectroscopic methods.

The remarks about ion chromatographic methods and precision and accuracy given for nutrients are valid for these parameters as well.

ORGANIC SUM PARAMETERS

Organic sum parameters cover, in this context, chemical oxygen demand with sodium dichromate (COD), biochemical oxygen demand (BOD), non-purgable organic carbon (NPOC), purgable organic carbon (POC), adsorbable organic halogens (AOX), purgable organic halogens (POX) and extractable organic halogens (EOX). Some of these, i.e. POC and EOX, are very seldom used in leachate analyses and will thus not be discussed further. It should be noted that 'purgable' and 'volatile' in this context are synonymous expressions.

Sample containers, preservation, storage and suggestions on methods of analysis are given in Table 2 for COD, BOD, NPOC, AOX and POX.

TABLE 2. Sample Treatment and Analysis for Organic Sum Parameters

Parameter	Preferred container	Preservation	Max. suggested storage time	Methods of analysis
COD	Glass	H_2SO_4 to pH < 2	24 h	Clesceri et al. (1989) 5220[a] ISO 6060 (1989)
BOD	Glass	÷	As short as possible, max. 24 h	Clesceri et al. (1989) 5310[a]
NPOC	Glass	Mineral acid to pH < 2	2 Weeks	Clesceri et al. (1989) 5310[a]
AOX	Glass	H_3PO_4 or HNO_3 to pH < 2	3–14 Days	Clesceri et al. (1989) 5320[a] ISO 9562 (1989)
POX	Glass	÷	As short as possible	ISO 9562 (1989)

[a] Method no.
÷, Do not preserve.

The analysis for COD includes the determination of reduced inorganic species such as ferrous ions and sulphide. High concentrations of chloride (>1000–2000 mg/litre) also interfere. The contribution from reduced inorganic species is generally recognized as a legitimate part of COD. Chloride may to some extent be masked by the addition of mercuric ions. In most leachates the interference from chloride may be overcome by dilution and by masking since the content of COD generally is sufficiently high to allow for appropriate dilution. In samples from the late methanogenic stage it may be necessary, however, to take special care that undue interference from chloride is not encountered.

The complex nature of leachate has a negative effect on the quality of analyses for COD compared to what is seen in wastewater. This is exemplified in Table 3 showing data from intercomparison exercises on COD in two wastewater (Reference Laboratory, 1980) and two leachate (Reference Laboratory, 1985) samples, respectively. It is evident that the precision for high concentrations of COD is good for both leachate and wastewater. However, the precision for the leachate sample at low concentration is considerably reduced, with a coefficient of variation both within and between laboratories about a factor of two higher than for wastewater of approximately the same concentration. For BOD, extreme pH values and low concentration of phosphorus in the leachate may influence the results of the BOD test. The composition of the standard dilution water prescribed in, for example, Clesceri *et al.* (1989) does, however, seem to compensate sufficiently for this. In some leachates from the methanogenic phase a tendency towards inhibition of the BOD test has been observed (Cossu *et al.*, submitted).

TABLE 3. Accuracy and Precision of Leachate Analyses Compared to Wastewater Analyses (after Reference Laboratory, 1985)

	Waste water		Leachate	
	High	Low	High	Low
Mean (mg O$_2$/litre)	239	37·0	1430	28·0
Number of laboratories	37	27	43	45
Standard deviation within laboratory (mg O$_2$/litre)	7·5	3·5	30	4·7
Coefficient of variation within laboratory (%)	3·0	9·0	2·1	17
Standard deviation between laboratories (mg O$_2$/litre)	24	6·8	110	9·8
Coefficient of variation between laboratories (%)	10	18	7·7	35

For AOX and POX only, data for the analysis of leachates from hazardous waste landfills have been published (Riggin et al., 1984; Först et al., 1989a). These studies show a precision and accuracy of the AOX method comparable to the analysis of wastewater (Riggin et al., 1984). For POX determinations a good agreement has been found between specific analyses and the POX determination (Först et al., 1989a). Precautions should, however, be taken to ensure that no interference from chloride is encountered, as concentrations higher than 1000 mg/litre, which may often be found in leachate, are known to give a positive bias in the AOX determination.

SPECIFIC ORGANIC COMPOUNDS

One of the first problems encountered in the analysis of leachate for organic constituents is the selection of which parameters to include in the analytical programme. If no former investigation at a given site suggests which groups of organic contaminants to expect in the leachate, a complete gas chromatography–mass spectrometric (GC–MS) screening of one or two leachate samples may be considered. An example of the use of GC–MS screening in leachate analysis is given by Schultz & Kjeldsen (1986).

As a GC–MS screening is relatively expensive, the samples to be analysed must be selected with care, e.g. by screening all available samples from drains and wells by organic sum parameters (NVOC and AOX). On the basis of the results from the GC–MS screening, it is possible to establish an analytical programme for use for mapping and monitoring. The analytical scheme for a GC–MS screening analysis is shown in Fig. 1. In all instances it is recommended that the laboratory to perform the analyses is consulted at the planning stage. In this way it can be ensured that the sampling programme, sample handling and analysis are an integrated part of the investigation.

The analysis of two groups of specific organic compounds are discussed below: volatile (or purgable) compounds and extractable compounds.

The considerations concerning sampling, preservation and storage of samples do not differ from the precautions to be taken when sampling wastewater and groundwater. Sample bottles should be made of glass with either ground glass stoppers or Teflon lined caps. The bottles should

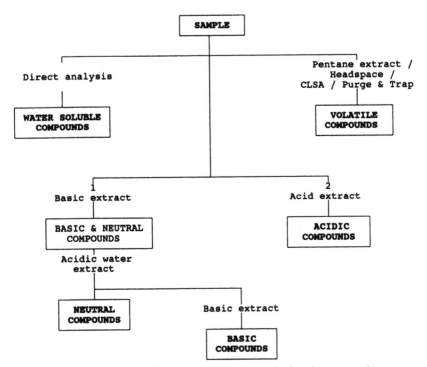

Figure 1. Diagram for total GC–MS screening of a leachate sample.

be specially cleaned to be free of any organic contaminants. The sample volume required depends on the numbers of parameters to be determined and the analytical methods employed. In general an analysis for volatile compounds requires from less than 50 ml to 1 litre while analysis for extractables in most cases requires 1 litre or more. If the available sample volume is sufficient, collection of 2–4 litres is recommended in order to enable duplicate analyses and/or analyses of spiked samples.

As to sampling, preservation and storage of samples it is recommended that samples to be analysed for volatile compounds are handled as little as possible from sampling to analysis. This implies that samples should be collected in the bottles from which analyses are to be made (headspace vials, extraction bottles, etc.). Except for headspace vials, sample bottles should be filled to the top without any headspace. Samples to be analysed for volatile compounds are not preserved, but are kept cool (+4°C), in the dark, and are analysed within 24 h after sampling.

Samples to be analysed for extractable compounds are collected in completely-filled bottles and are preserved by acidification to pH≤2 (sulphuric or nitric acid). Samples are kept cool (+4°C), in the dark, and it is preferable that the analytical work starts as soon as possible, and at most within one week of sampling.

Volatile Compounds

Volatile compounds can be analysed by either headspace, stripping ('purge and trap' or closed loop stripping analysis—CLSA) or extraction methods coupled with gas chromatographic detection. Interferences from the often complex sample matrix are seldom problematic, but it is recommended that either standard addition calibration is used or the recovery factors are determined for the specific sample matrix.

Först et al. (1989a,b) analysed leachates from both hazardous and domestic wastes for volatile aromatics and volatile chlorinated compounds by headspace-gas chromatography using a two-level standard addition calibration. They reported ≤20 000 μg/litre chlorinated compounds and 5–13 000 μg/litre aromatics in leachates from hazardous wastes and ≤14 μg/litre chlorinated compounds and 1–500 μg/litre aromatics in leachates from domestic wastes. Triplicate analyses at these concentration levels showed relative standard deviations of less than 10%.

Extractable Compounds

The analyses for extractable compounds are preceded by one or more pH-adjusted extractions of the sample by an organic solvent (often dichloromethane). The extracts are then analysed by gas chromatography (GC) or liquid chromatography (HPLC); for some groups of components this is done after an appropriate derivatization.

The complex sample matrix in many cases will result in poor recoveries and/or the co-extraction of components that can interfere with the chromatographic determination of specific compounds. The latter problem can be avoided in most cases by appropriate clean-up of the extract.

Foerst et al. (1982) applied EPA method 610 to the analysis of polynuclear aromatic hydrocarbons (PAHs) in sanitary landfill leachate. They reported poor recoveries (10–50%) for four-, five- and six-ring

PAHs. Clean-up of the extracts on a silica column removed all interferences by the HPLC determination of the PAHs.

Buisson *et al.* (1984) determined chlorinated phenols, as pentafluorobenzoyl derivatives, in sludge, wastewater and leachate. Fivefold analyses of leachate samples spiked with eight chlorophenols at the 100 ng/litre level resulted in mean recoveries of 65% and relative standard deviations of 16%, which is acceptable.

Tschochner *et al.* (1989) determined chlorophenols as their acetyl derivatives in extracts from leachate and other types of environmental samples. They reported that clean-up on a small chromatographic column packed with sodium sulphate and alumina removed all interfering substances from the sample.

All the examples given above deal with organic compounds which can be extracted or purged/stripped from the samples. As much as 90% of the organic content in leachates from domestic landfills is made up of polar, non-extractable compounds (Østfeldt & Schultz, 1991). Even though a minor part of the non-extractable organic matter in a leachate sample may be either volatile organic acids and/or water miscible solvents (methanol, ethanol, etc.), the main body is often not identified. Very little information addressing this is published in the literature, but some attempts to characterize the non-extractable organic matter in connection with leachate treatment have been made. Slater *et al.* (1985) used an ultrafiltration technique and Millot *et al.* (1987) used gel permeation chromatography (GPC) to characterize the apparent molecular weight distribution of untreated and treated leachates in order to assess treatability and to monitor the effect of treatment. They found that biological treatment removed the low molecular weight compounds almost quantitatively, while high molecular weight compounds could be removed by physical-chemical treatment (see Chapter 2.6).

Different laboratories are currently investigating methods for the specific analysis of the vast number of non-extractable organic compounds present in leachates, and there is little doubt that, in the future, methods such as direct high performance liquid chromatography coupled with mass spectrometry (HPLC–MS) and supercritical fluid chromatography will play a role in leachate analysis.

AN ILLUSTRATIVE EXAMPLE

Table 4 shows an example of an analysis of leachate from a landfill with mixed wastes which are mostly slags and fly ash. As mentioned earlier

TABLE 4. Analysis of Leachate from a Mixed Waste Landfill

Parameter	Unit	Result
Temperature	°C	10·3
pH		6·9
Conductivity	mS/m	1040
Alkalinity	meq/litre	11
Chloride	mg/litre Cl^-	3040
Sulphate	mg/litre SO_4^{2-}	15
Nitrate	mg/litre NO_3^-	0·22
Ammonia	mg/litre NH_4^+	4·4
Sodium	mg/litre Na^+	1445
Potassium	mg/litre K^+	247
Total hardness	°dH	112
Calcium	mg/litre Ca^{2+}	675
Iron	mg/litre Fe	65
Non-purgable organic carbon, NPOC	mg/litre C	23·3
COD	mg/litre O_2	140
Adsorbable organic halogens, AOX	µg/litre Cl	85
Purgable organic halogens, POX	µg/litre Cl	2·0
1,1,1-Trichloroethane	µg/litre	1·7
Trichloroethylene	µg/litre	1·0
Phenol	µg/litre	0·1
Cresols	µg/litre	650
Xylenols	µg/litre	44

some parameters present analytical problems and the problems in this specific case will be commented on below.

Earlier analyses of leachate from this landfill have shown that the ammonia analysis according to the standard method (phenate method) has a positive bias, among other reasons because of interferences from iron. Therefore, the analysis in this case was performed according to a gas diffusion method with standard addition. (Principle: ammonia is volatilized from an alkaline solution and diffused through a Teflon membrane. In the resulting solution ammonia can be measured without interferences.) This method is not recommended as a standard method, as it requires special development efforts in each case, but is mentioned to illustrate that in some cases innovative methods must be applied.

The concordance between the COD- and the NPOC-analysis is not very good. Normally the COD/NPOC-ratio is 3–4, while the ratio in this leachate is 6. Assuming that organic substances in the sample are analysed quantitatively and without interferences, the theoretical COD/NPOC ratio cannot exceed 5·3. The value measured thus shows that positive interferences in the COD determination or incomplete recovery in the NPOC measurement are encountered. In performing the COD-analysis, masking interference from chloride using mercury ions was attempted, while the ferrous iron content of the sample is included in the measurement. Both interferences give high COD-results. During the NPOC analysis particulate matter caused greater uncertainty than normally. Furthermore POC was not measured. However, previous analyses of NPOC and POC have shown that the value of POC corresponds to only a few percent of the NPOC-value. The high COD/NPOC ratio may thus also be influenced by low NPOC results due to incomplete oxidation of, for example, fine particulate matter.

As can be seen from Table 4, good accord is found between specific analyses for chlorinated compounds and non-specific measurements of purgable/volatile organic halogens (POX). AOX is, however, much larger. This has not been explained, but is often seen in leachate both from domestic waste and from hazardous waste.

BOD-analyses were not performed on this sample. This is due to the fact that the sample shows inhibitory effects to microorganisms. The only way to avoid interference is by dilution, and as the BOD-content is low, this will reduce the BOD-content of the diluted sample below the detection limit of the method. This problem often arises when performing leachate analyses on samples with a low content of organic matter.

CONCLUSIONS

The overall picture of analysis of leachates is that most standard methods give good reproducible results, but special attention must be given to interferences and sample matrix effects.

For the specific organic analyses, it must be kept in mind that even though conventional analyses can identify most of the persistent pollutants, as much as 90% of the organic matter in a leachate sample is very water soluble and therefore not identified by most conventional analyses.

In order to select the most relevant parameters and the pertinent sample handling procedures, the paramount importance of close contact with the analytical laboratory from the early planning stages must be stressed.

REFERENCES

Buisson, R. S. K., Kirk, P. W. W. & Lester, J. N. (1984). Determination of chlorinated phenols in water, wastewater, and wastewater sludge by capillary GC/ECD. *Journal of Chromatographic Science*, **22**, 339–42.

Clesceri, L. S., Greenberg, A. E. & Trussel, R. R. (eds) (1989). *Standard Methods for the Examination of Water and Wastewater*, 17th edn. American Public Health Association, Washington, DC.

Cossu, R., Serra, R. & Cannas, P. (submitted). BOD measurement in sanitary landfill leachate. *Waste Management and Research*.

DeWalle, F. B., Zeisig, T. Y. & Chian, E. S. K. (1979). Analytical methods evaluation for applicability in leachate analysis. In: *Municipal Solid Waste: Land Disposal. Proceedings 5th Annual Research Symposium, Orlando, Florida*, ed. M. P. Wanielista & J. S. Taylor. EPA/600/9-79/023A, US Environmental Protection Agency Municipal Environmental Research Laboratory, Cincinnatti, OH, pp. 176–85.

Dougan, W. K. & Wilson, A. L. (1973). Absorptiometric determination of iron with T.P.T.Z. *Water Treatment and Examination*, **22**, 100.

Foerst, D. L., Froning, B. A. & Bellar, T. A. (1982). Application of EPA method 610 to the analysis of polynuclear aromatic hydrocarbons in leachate samples. Project report EPA-600/4-82-041, Environmental Monitoring and Support Lab., Cincinnati, OH.

Först, C., Stieglitz, L., Roth, W. & Kuhnmünch, S. (1989a). Application of headspace analysis and AOX-measurement to leachate from hazardous waste landfills. *Chemosphere*, **18**(9/10), 1943–54.

Först, C., Stieglitz, L., Roth, W. & Kuhnmünch, S. (1989b). Determination of volatile organic pollutants in leachate from different landfills. *Vom Wasser*, **72**, 295–305.

Henriksen, A. (1966). An automated, modified formaldoxime method for determining low concentrations of manganese in water containing iron. *Analyst*, **91**, 647.

ISO 6060 (1989). Water Quality—Determination of the Chemical Oxygen Demand. International Organization for Standardization, Geneva, Switzerland.

ISO 9562 (1989). Water Quality—Determination of Adsorbable Organic Halogens (AOX). International Organization for Standardization, Geneva, Switzerland.

Millot, N., Granet, C., Wicker, A., Faup, G. M. & Navarro, A. (1987). Application of G.P.C. processing system to landfill leachates. *Water Research*, **21**(6), 709–15.

Reference Laboratory for Chemical Water Analysis (Denmark) (1980). Intercomparison exercise 13—5 days biochemical oxygen demand, chemical oxygen demand with sodium dichromate, turbidity. Report from the Water Quality Institute, Hørsholm, Denmark. (In Danish)

Reference Laboratory for Chemical Water Analysis (Denmark) (1985). Intercomparison exercise 30—sodium, potassium, chloride, sulphate, total solids and COD in leachate. Report from the Water Quality Institute, Hørsholm, Denmark. (In Danish)

Riggin, R. M., Lucas, S. V., Jungclaus, G. A. & Billets, S. (1984). Measurement of organic halide content of aqueous and solid waste samples. *J. Testing Evaluation*, **12**, 91.

Schultz, B. & Kjeldsen, P. (1986). Screening of organic matter in leachates from sanitary landfills using gas chromatography combined with mass spectrometry. *Water Research*, **20**(8), 965–70.

Slater, C. S., Uchrin, C. G. & Ahlert, R. C. (1985). Ultrafiltration processes for the characterization and separation of landfill leachates. *J. Environmental Science and Health* **A20**(1), 97–111.

Tschochner, F., Pilz-Mittenburg, W., Benz, T., Brunner, H., Jäger, W. & Hagenmaier, H. (1989). Determination of chlorophenols in aqueous, solid and gas samples by GC/ECD and GC/MS. *Zeitschrift für Wasser-Abwasser-Forschung*, **22**, 267–71.

Østfeldt, P. & Schultz, B. G. (1991). Investigations at the Vejen municipal landfill, organic groundwater quality. Lossepladsprojektet, Report P1, Lyngby, Denmark. (In Danish)

3. LEACHATE TREATMENT

3.1 Biological Processes

HANS-JÜRGEN EHRIG

Universität Wuppertal, Abfall- und Siedlungswasserwirtschaft, Pauluskirchstrasse 7, D-5600 Wuppertal, Germany

&

RAINER STEGMANN

Institute of Waste Management, Technical University of Hamburg-Harburg, Harburger Schlossstrasse 37, D-2100 Hamburg 90, Germany

INTRODUCTION

Leachate treatment might be a very complex process if low discharge values have to be obtained (see also Chapter 3.15). Using only the biological process the organic-degradable components can be reduced to very low values and almost complete nitrification and denitrification can be achieved. Biological processes are in most cases less expensive than chemical/physical processes. In addition, except for new biomass, no other solids are produced. Biologically treated leachate still has relatively high concentrations of COD and AOX (as chlorinated hydrocarbons) that can be further reduced by other methods.

Leachates from young and old landfills differ significantly in their BOD_5- and COD-concentrations (see also Chapters 2.4 and 2.5). Biological treatment of 'old' leachate focuses therefore mainly on the reduction of nitrogen.

Biological leachate treatment has proven to be a successful method if it is practised either separately or together with sewage (see also Chapter 3.14). The design and operation of leachate treatment plants cannot be based on the criteria of sewage treatment plants. One reason for this situation is the different composition of the leachate which results in a different degradation behaviour. In addition temperatures in leachate treatment plants might be very low in winter due to the relatively long detention times in lagoons and activated sludge plants.

Biological systems can be distinguished in anaerobic and aerobic treatment processes that are realized by means of different techniques. The following treatment methods are described in this overview; they are described in more detail in the following chapters of this book:

* anaerobic biological treatment

    —parts of the landfill body used as reactor (Chapter 2.7)
    —anaerobic sludge bed reactor (UASB) (Chapters 3.5 and 3.8)

* aerobic biological treatment

    —aerated lagoons (Chapter 3.2)
    —activated sludge process (Chapters 3.7, 3.9, 3.10)
    —rotating biological contactor (RBC) (Chapter 3.3)
    —trickling filter (Chapter 3.4)

BASIC DATA FOR THE DESIGN OF BIOLOGICAL LEACHATE TREATMENT PLANTS IN GERMANY

Although the relation between leachate production rates, leachate quality and the 'age' of a landfill can be predicted to a certain degree, it is necessary to design the treatment plant at the same time as the landfill is designed. For this reason the BOD_5-load of the leachate has to be predicted somehow. In addition the development of the landfill size relative to time has to be known, since leachate production rates are dependent upon the surface area of the landfill; this is somewhat of a problem since it is very often difficult to forecast the amount of waste that will be landfilled in the future at a specific site.

In the following, only the leachate production rates, as well as BOD_5 and NH_4-N concentrations, will be estimated. It is important to mention that values reported here are extracted from a great many data measured under controlled conditions at different landfill sites in Germany (the landfills were sealed from the surrounding strata). For this reason the design values are only valid in Germany and in those countries where waste composition, climate and landfill operation are comparable. The data used in this chapter show mean values; dependent upon the specific situation, leachate concentrations and production rates from actual landfills may differ to a high degree from these values. For this reason leachate treatment plants should be designed in such a way that fluctuations of organic loads can be handled successfully. The design criteria given in this chapter tend to be on the safe side.

The following range of mean leachate production rates may be used for the climatic conditions of Germany, where mean annual precipitation rates of about 750 mm can be expected:

- highly compacted 'young' landfills: 15–25% of the annual precipitation rate
- highly compacted 'old' landfills: 25–50% of the annual precipitation rate.

It should be mentioned that these values are not valid for landfills where silty or clay-like cover material is used. Landfills can be categorized as 'old' when they have been out of operation for more than ten years.

High leachate production rates can be expected during winter and spring seasons, when an increase of more than 100% can be expected. This is to a certain degree caused by melting snow and low evaporation rates. On the other hand during summer and autumn seasons correspondingly low leachate production rates can be expected. After long dry periods leachate production rates from large landfills will not decrease to zero. Since biological leachate treatment plants are designed for low sludge and space loadings, there is sufficient buffering capacity available to use the above-mentioned mean annual leachate production rates as a design criterion. This may not be the case in areas with different climatic conditions (i.e. much snow, rainy seasons).

It is well known that leachate BOD_5-concentrations decrease with the 'age' of a landfill (see also Chapter 2.4). As a consequence no mean BOD_5-value but a BOD_5-curve dependent upon the 'age' of a landfill should be used for the design of leachate treatment plants (see Chapter 2.4). This curve does not describe the actual leachate concentrations, but these data seem to be practical for design purposes.

In addition to these data, the ammonia concentration of the leachate has to be respected for the design of the aeration system, since nitrification and denitrification may be treatment goals in a biological treatment step. In regard to the abovementioned research activities, mean ammonia nitrogen concentrations of 1200 ppm may be used for design purposes in Germany. Again this ammonia concentration is on the safe side, since far lower NH_4-N concentrations can be expected in the first three years of operation of highly-compacted landfills.

As already mentioned the actual BOD_5-loading of a leachate treatment plant is dependent upon the leachate production rate and the BOD_5-concentration; for this reason, the BOD_5-load is directly dependent upon the size of the landfill, where the 'age' of a landfill has to be respected. Keeping this in mind the influence of the development of the landfill size with time becomes obvious.

ANAEROBIC TREATMENT

During the period of high organic concentrations in the leachate (see also Chapter 2.4), an anaerobic treatment step might be a most effective way of reducing compounds like organic acids by a high degree. The main advantage of the anaerobic treatment process is the low energy requirement, since no oxygen has to be supplied.

If anaerobic processes are not enhanced in the landfill body, separate anaerobic treatment steps should be considered (see also Chapters 3.4 and 3.5). The anaerobic treatment can only be operated temporarily, as long as the organic degradable leachate concentrations are high (up to 4–7 years) for each operational part of the landfill.

After some years of operation at the latest every landfill body works as an anaerobic reactor. Using enhancement methods not discussed here, complete anaerobic degradation could be initiated much earlier. Degradation of the BOD and COD of leachate in a 2-m layer of a pilot-scale landfill dependent upon volumetric loading rate and temperature is shown in Fig. 1. But in full-scale landfills, temperatures are normally higher with smaller variation. These data show that this operational technique should be used when new landfills are installed (see also Chapter 2.7). The separate anaerobic treatment step can be avoided by these measures.

Figure 2 shows some results from different treatment experiments with anaerobic reactors (anaerobic filters and upflow anaerobic sludge bed reactors—UASB) at 33°C from laboratory-scale plants. Operation of anaerobic filters was negatively influenced by iron and calcium precipitation that resulted in clogging problems. The free volume of the sand drain at the bottom was filled up to 60% by the precipitate after a COD-reduction of 2000–3000 kg/m^3. At UASB reactors the inorganic content of sludge increases dramatically with time and reduces elimination rates.

Anaerobic leachate treatment is an effective process but the remaining BOD$_5$- and COD-effluents are still high with COD-values of 1000–

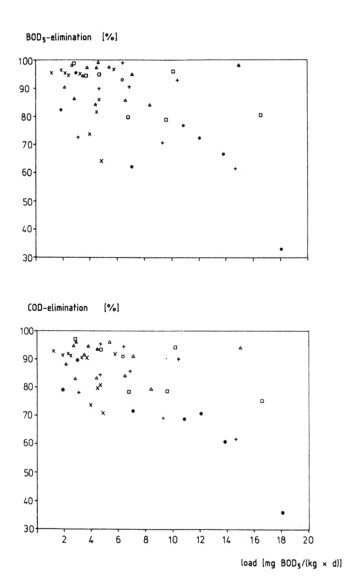

Figure 1. Anaerobic degradation of leachate in a 2-m refuse layer of a pilot-scale landfill (diameter 5 m) (Ehrig, 1985a). (Temperatures [°C]: ● = <7; × = 7–10; □ = 10–13; ▲ = 13–16; + = 16–19; ○ = >19.)

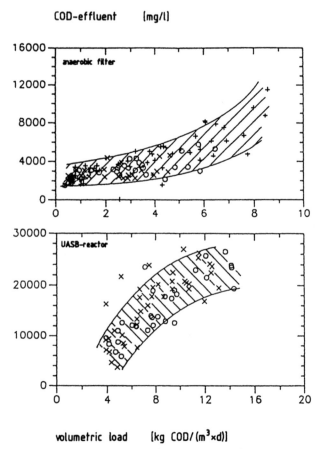

Figure 2. COD-effluents versus loading; results from anaerobic laboratory-scale experiments (Mennerich, 1988).

4000 mg/litre and a BOD_5–COD-relationship $\geq 0\cdot 3$. After the anaerobic treatment step the leachate has to be treated to final effluent standards by means of aerobic and probably chemical-physical processes.

AERATED LAGOONS

Since leachate production pollution rates may be relatively low, aerated lagoons have been used frequently for leachate treatment. The basic idea

TABLE 1. Leachate Treatment in Aerated Lagoons (Robinson & Davies, 1985)

	A		B		C	
	infl.	*effl.*	*infl.*	*effl.*	*infl.*	*effl.*
Mean flow rate (m³/day)	33		104		84	
Mean lagoon temperature (°C)		—		10		4
pH	5·8	8·0	5·9	8·0	6·3	7·7
COD (mg/litre)	5 583	164	5 400	215	9 750	210
BOD₅ (mg/litre)	3 694	24	3 500	9	7 000	37
Ammonium (mg N/litre)	129	12	158	3·9	175	0·9
Fe (mg/litre)	254	3·8	90	1·9	295	9·9
Zn (mg/litre)	6·3	0·2	5·6	0·2	11·5	0·5

A, July 1983–December 1984
B, October/November 1983
C, February 1984

is that the detention time of the leachate in the lagoon is long enough so that as many bacteria can develop per time as the number of species that are transported out of the lagoon with the effluent. Long detention times are necessary in order to degrade also the organic fraction which is not easily degradable and because of low temperatures. The maintenance and operational costs are relatively low.

Results from an aerated lagoon with a volume of 100 m³ are presented

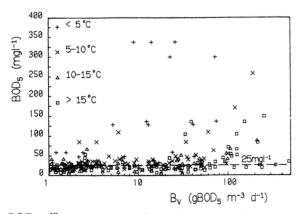

Figure 3. BOD₅-effluents versus volumetric loading (B$_v$) from a full-scale aerated lagoon (Ehrig, 1985*b*).

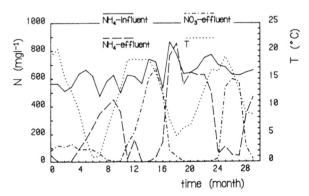

Figure 4. Temperature effects on nitrification at a full-scale aerated lagoon (Ehrig, 1985*b*).

in Table 1. Effluent values of organics and ammonia are very low. Our investigations on a technical-scale plant did not show such high treatment efficiency. Figure 3 presents BOD_5-effluent values depending on loading rate and water temperature. Also with loading rates ≤ 0.01 kg BOD_5/m^3 day the resulting effluent values in most cases are only ≤ 40 mg BOD_5/litre. When water temperatures were $<5°C$ the effluent values increased sharply. Figure 4 presents the effect of water temperature on the nitrification rate of this aerated lagoon.

ACTIVATED SLUDGE PROCESS

The detention time in activated sludge plants can be considerably shorter than in aerated lagoons. The reason for this phenomenon is that the sludge content (amount of bacteria) can be controlled to a certain degree in the activated sludge process and is 3–5 times higher than in aerated lagoons. This is achieved by installing a settling tank behind the aeration tank and recirculating the sludge back into the activated sludge tank. A certain amount of sludge (produced bacteria mass) has to be removed from the system.

Results from a full-scale treatment plant are presented in Table 2. Figure 5 shows results from highly-loaded pilot-scale activated sludge plants. Our observations on laboratory and full-scale plants show BOD_5-effluent values ≤ 25 mg/litre at F/M ratios ≤ 0.05 kg BOD_5/kg MLSS day (Fig. 6).

TABLE 2. Results from Leachate Treatment in Full-Scale Activated Sludge Plants

	Influent	Effluent
Cossu (1981)		
Volumetric load > 1·0 kg BOD$_5$/m^3 day		
Phase I		
BOD$_5$ (mg/litre)	5 294	254
COD (mg/litre)	12 359	1 566
Phase II		
BOD$_5$ (mg/litre)	5 015	231
COD (mg/litre)	11 216	1 067
Klingl (1981)		
Volumetric load 0·21 kg BOD$_5$/m^3 day		
F/M-ratio 0·017 kg BOD$_5$/kg MLSS day		
BOD$_5$ (mg/litre)	5 162	24
COD (mg/litre)	9 785	347

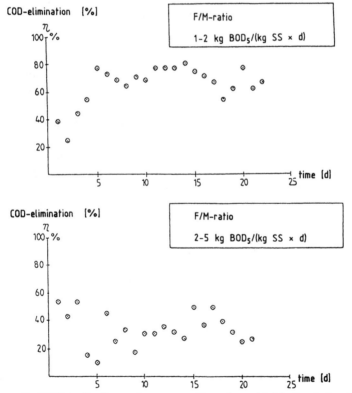

Figure 5. COD reduction rates versus time from highly-loaded activated sludge pilot plants (Damiecki et al., 1988).

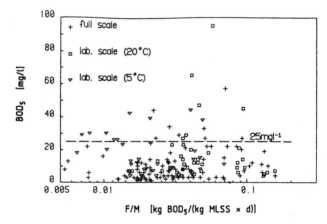

Figure 6. BOD$_5$-effluent concentrations versus F/M-ratio from laboratory-scale and full-scale activated sludge plants (Ehrig, 1985b); F/M = food/microorganism ratio, MLSS = mixed liquor suspended solids.

In addition to BOD$_5$-reduction the nitrification of ammonia is a very important process in activated sludge plants. Nitrogen elimination becomes more and more important with the age of a landfill resulting in a reduction of BOD in the landfill body.

The pH-values of this kind of leachate from landfills in the methanogenic stage are in the range 8·0–8·3. During aeration the pH increases in many cases up to pH 9. Under these circumstances the equilibrium shifts from ammonium to free ammonia; that has an inhibiting effect on nitrifying bacteria. If on the other hand the ammonium is converted to nitrate the pH decreases as a result of alkalinity destruction. As a conclusion, very careful operation and pH-control is necessary to achieve low ammonium effluent concentrations. Figure 7 shows the effluent values of ammonium, nitrate and nitrite in relation to the nitrogen-F/M-ratio from laboratory-scale experiments.

Overloading the plant in regard to nitrogen results in an incomplete nitrification step; as a consequence nitrite will accumulate. If in this situation the nitrogen load is reduced the nitrite inhibits the nitrifying bacteria so that complete nitrification will not take place for a longer period of time. In general complete nitrification can be observed at N-loading rates lower than 0·03 kg N/kg MLSS day. Knox (1985) reported ammonium removal rates of 0·1 kg N/kg MLVSS day at 20°C for a pilot plant (MLVSS = mixed liquor volatile suspended solids). With

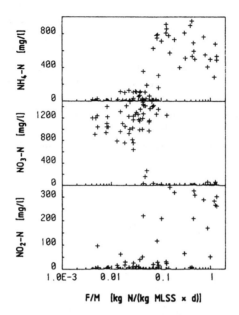

Figure 7. NH$_4$-N, NO$_3$-N, and NO$_2$-N effluent concentrations versus nitrogen-F/M-ratio; data from laboratory-scale activated sludge plants (Ehrig, 1985b).

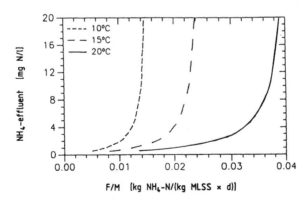

Figure 8. NH$_4$-effluent concentrations versus nitrogen-F/M-ratio and temperature; data from laboratory-scale activated sludge plants (Mennerich, 1988).

approximately 50% organics in the MLSS the loading rate is nearly 0·05 kg N/kg MLVSS day. Knox (1985) also reports a sharp decrease in removal rates with decreasing temperatures (at 5°C loading is 0·025 kg N/kg MLVSS day). Ammonium effluent values dependent upon the F/M-ratio and temperature are presented in Fig. 8. To prevent low temperatures it is necessary to cover the aeration tank and to use the heat from air blowers or other heating methods.

In order to reduce the high nitrate content in the leachate effluent and to stabilize pH-conditions in activated sludge plants a denitrification step is necessary.

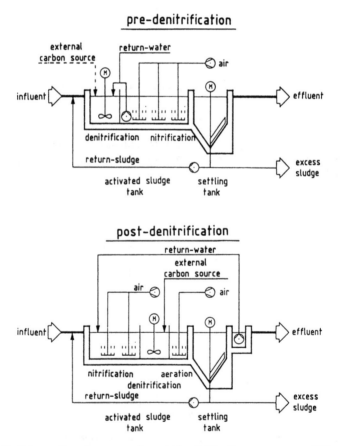

Figure 9. Schematic of pre- and post-denitrification processes (Albers & Mennerich, 1987).

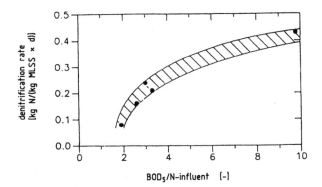

Figure 10. Denitrification rate versus BOD_5/N-ratio in the influents; data from laboratory-scale plants (Mennerich, 1988).

Pre-denitrification (Fig. 9) is more effective if small amounts of leachate have to be treated. The denitrification rate depends very much on the amount of water and sludge that is recirculated. In order to reach nitrogen effluent values lower than 5–10% of influent concentrations, extremely high recirculation rates are necessary. Using the post-denitrification process, very low nitrate effluent values can be reached. In this case the degradable organics in the leachate cannot be used as a carbon source and a separate pH-control device in the nitrification tank might be necessary. For the pre-denitrification process the denitrification rate dependent upon the BOD_5/N-ratio in the influent is presented in Fig. 10.

TRICKLING FILTERS AND ROTATING BIOLOGICAL CONTACTORS (RBC)

This process differs from the activated sludge process in so far as the bacteria are attached to the rotating contactors or the media in the trickling filter. The air supply takes place naturally, i.e. the rotating contactor is partly in the air and partly in the water, while in trickling filters air vents in general from the bottom to the top. These methods of aeration consume low amounts of energy. When treating highly organically-polluted leachate, clogging by inorganic precipitates and/or produced biomass could lead to the conclusion that these processes may not be appropriate under these conditions. But in many cases nitrification processes are more effective in fixed film reactors, so that trickling filters and RBC may be used for the treatment of leachate from landfills in the methanogenic stage.

Knox (1985) presents results from trickling filter pilot-scale test plants. The leachate was of low organic concentration (BOD_5: 80–250 mg/litre; COD: 850–1350 mg/litre) and relatively high ammonium concentrations (NH_4: 200–600 mg N/litre). At low temperatures the BOD_5-effluent values exceeded 25 mg/litre. The influence of temperature on nitrification is shown in Fig. 11. Temperature effects are lower when RBCs are used, since they are normally covered or in-house. The nitrogen removal rates relative to area loading (g N/m² day) in laboratory-scale plants are presented in Fig. 12. The results show that about 95% of the ammonium concentrations are oxidized also when high loading rates are present (>10 g N/m² day). Problems occur when the loading exceeds 2 g N/m² day because the ammonium oxidation is not complete and increasing NO_2-concentrations are produced (Fig. 13). In order to avoid toxic nitrite concentrations the nitrogen loading should therefore not exceed 2 g N/m² day. Results from full-scale plants are presented in Table 3.

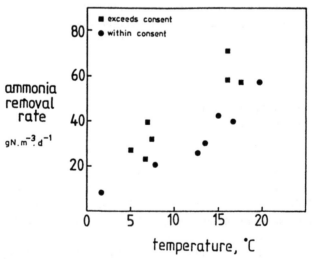

Figure 11. Ammonia removal versus temperature; data from pilot-scale trickling filters (Knox, 1985).

Biological Processes

199

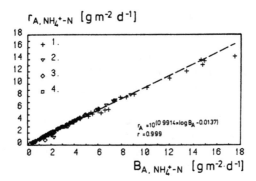

Figure 12. Ammonium removal (r_A) versus area load (B_A); results from laboratory scale RBCs (Ehrig, 1983).

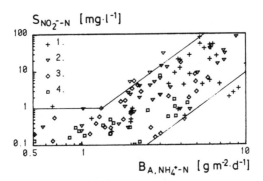

Figure 13. Nitrite-effluent concentrations ($S_{NO_2^--N}$) versus ammonium area load (B_A); results from laboratory-scale RBCs (Ehrig, 1983).

TABLE 3. Characteristic Data of Two Full-Scale RBCs ((a) Fleischer & Wittek, 1985; (b) Knox, 1985)

	a	b
Area (m²)	28 500	30 000
Specific area (m²/m³)	220	180
Settling tank	Yes	Yes
Area load (g N/m² day)	1–2	0·65–2·85
Temperature (°C)	—	8–18
Mean NH₄-effluent (mg N/litre)	0·28	2·96
Mean NO₂-effluent (mg N/litre)	0·02	<1

SUMMARY AND CONCLUSIONS

Biological treatment processes are very effective methods of reducing significantly biodegradable organics as BOD_5 and the main part of COD. Also from leachates with low organic concentrations and BOD_5–COD ratio < 0·2, the COD reduction by means of biological treatment may be up to 50%. This method is also an effective way to oxidize ammonium to nitrate by nitrification and to denitrify the nitrates to gaseous nitrogen. The decreasing elimination rates during periods of low water temperatures especially for nitrification processes have to be respected. The following effluent values can be reached by means of biological processes:

$BOD_5 \le 25$ mg/litre

$COD \le 400-1500$ mg/litre

$NH_4 \le 20-50$ mg N/litre (activated sludge)

≤ 10 mg N/litre (RBC or two step biological treatment)

$AOX \le 10-20\%$ elimination

There is still the question of what kind of substances create the final COD that is not accessible to further biological reduction. In the authors' opinion these compounds are humic- and fulvic-like substances similar to those produced in nature during the decomposition of highly-concentrated organics.

Biological treatment methods are generally less costly than chemical/physical processes and the organics are converted to a high degree to CO_2 and water; except for bacterial sludge production no other sludges or concentrates are produced. For this reason, also, when physical-chemical processes are required to further reduce the effluent concentrations, a first stage biological step should be considered.

The design parameters are different from those used for the design of sewage treatment plants. In most cases nutrients have to be added, since there is a phosphorus deficit in the leachate.

Regarding operation, foam build-up may take place during certain periods of time and has to be faced; otherwise sludge may leave the plant with the foam.

Since relatively long detention times have to be provided for the treatment of leachate, cold temperatures in the activated sludge plants and lagoons may affect the treatment efficiency significantly. Especially in activated sludge plants, insulation and heating (i.e. by means of air supply blowers or by heating using landfill gas) may be necessary.

Installations like pumps, aeration facilities, pipes, etc., have to be built and/or operated in such a way that the precipitates from leachate do not cause any problems. Submersible pumps may be coated with precipitates so that sufficient cooling no longer takes place. In addition clogging of the air supply system occurs.

The sludge concentrations in activated leachate treatment plants are, in general, higher than in sewage treatment plants; the organic content of the sludge may be lower. Heavy metal accumulation in the sludge has been observed quite often; this may cause problems in regard to the final storage of the sludge.

The design of leachate treatment plants is not easy since it is difficult to predict the leachate production and quality sufficiently accurately with time. There is a significant reduction in BOD_5- and COD-concentrations when the landfill has reached the methanogenic stage. After this change in leachate quality, biological treatment has mainly to focus on nitrogen removal; the design criteria change and the ammonium sludge loading and/or ammonia volume loading have to be respected. In this phase anaerobic treatment is no longer valid.

REFERENCES

Albers, H. & Mennerich, A. (1987,). Chemisch-physikalische Nachreinigung von Deponiesickerwasser in Minden-Heisterholz (Chemical-physical treatment of leachate at Minden-Heisterholz). *Müll und Abfall* **8**, 326.

Cossu, R. (1981). Biological treatment of leachate in a full scale plant. 5th European Sewage and Refuse Symposium, Munich, Germany, p. 605.

Damiecki, R. & Lanzrath, B. (1988). Ergebnisse aus der Untersuchung verschiedener Sickerwasseraufbereitungsverfahren; Pilotmassstab (Results from investigations on different leachate treatment processes; pilot scale). Zentrum für Abfallforschung, TU Braunschweig, Germany, **3**, 339.

Ehrig, H.-J. (1983). Biological oxidation of sanitary landfill leachate with high ammonium concentrations. International Seminar on Rotating Biological Discs, 6–8.10.1983, Fellbach, West Germany.

Ehrig, H.-J. (1985a). Untersuchungen zur Behandlung von Sickerwässern in Festbettanlagen (Investigations on leachate treatment in fixed film reactors). *Veröffentlichungen des Institutes für Stadtbauwesen TU Braunschweig*, **39**, 288–303.

Ehrig, H.-J. (1985b). Laboratory and full scale experiments on physical-chemical treatment of sanitary landfill leachate. Proc. Conference New Directions and Research in Waste Treatment and Residual Management, University of British Columbia, Vancouver, Canada, p. 200.

202 Ehrig, Stegmann

Fleischer, B. & Wittek, J. (1985). Erfahrungsbericht über die Sickerwasserbehandlung mit Tauchtropfkörpern (Report on experiences of leachate treatment in trickling filters). *Veröffentlichungen des Institutes für Stadtbauwesen TU Braunschweig*, **39**, 401–10.

Klingl, H. (1981). Erfahrungen über die Reinigung von Müllsickerwässern an einer Deponiekläranlage im bayerischen Raum (Experiences on leachate treatment in a full scale plant in Bavaria). *Münchner Beiträge zur Abwasser-, Fischerei- und Flußbiologie*, **33**, 321.

Knox, K. (1985). Leachate treatment with nitrification of ammonia. *Water Research*, **19**, 895–904.

Mennerich, A. (1988). Beitrag zur anaerob–aeroben Behandlung von Sickerwässern aus Hausmülldeponien (Anaerobic–aerobic leachate treatment). *Veröffentlichungen Institut für Siedlungswasserwirtschaft TU Braunschweig*, **44**.

Robinson, H. & Davies, J. (1985). Automatic answer to leachate treatment. *Surveyor*, 14 March.

3.2 Aerated Lagoons

HOWARD ROBINSON

Aspinwall & Company Ltd, Walford Manor, Baschurch, Shrewsbury, UK, SY4 2HH

INTRODUCTION

The development of simple, economic, robust, automatic and reliable means by which leachate can be treated on a landfill site assumes major importance if, by this means, leachate can be discharged to disposal routes which would otherwise be unacceptable, whether these routes are to domestic sewage treatment works, land irrigation schemes (including grass plots and reed or peat beds), or direct to surface watercourses.

This chapter presents innovative work by the author in the field of on-site treatment of landfill leachate in the UK during a five-year period, describing three particular schemes, through which technology has been developed and progressed until full-scale treatment of leachate can now be provided with confidence, at relatively low cost, for any modern landfill site.

BACKGROUND

Until recently, although the treatability of leachate has been demonstrated many times on a laboratory scale (e.g. Robinson & Maris, 1983, 1985), the only full-scale plants in operation were in Germany (Stegmann & Ehrig, 1980), Canada and the USA, where sewage treatment technology had been modified for leachates by building plants that were innovative, but relatively expensive and sophisticated. What have been required, and are appropriate for the UK, are demonstrations on a full-scale that treatment of leachate can be accomplished simply and reliably and at reasonable cost, in robust plants on landfill sites. To date, about six such plants are operating, the first since 1983, and several more are in various stages of planning, design and negotiation. These plants have minimal impact on the environment, (for example, no smells at all are produced), and have maintained high quality effluents for extended periods, including during adverse winter conditions.

The three plants which are the first running are on the sites listed below, and have been designed and constructed in an ordered sequence as they have been required.

(1) Bryn Posteg Landfill, Powys—operational since July 1983
(2) Compton Bassett Landfill, Wiltshire—operational since July 1985
(3) Whiteriver Landfill, County Louth—operational since February 1986

Detailed monitoring of all three plants has been undertaken, on behalf of the site operators, and has been partially supported at two sites by research contracts from the United Kingdom Department of the Environment.

THE LEACHATE TREATMENT PLANTS

Bryn Posteg Landfill Site.   This site in central Wales includes the first purpose-built, automated plant in the UK to treat large quantities of strong leachate. The scheme is centred on a large, HDPE-lined lagoon, with a long mean period of retention. Aeration is provided by high-efficiency floating aerators, allowing up to $150\,m^3$ of leachate per day to be treated to a high standard, with minimal attention from on-site staff. Effluent has been discharged automatically to a local, rural sewage treatment works since July 1983 (which would not have been able to accept untreated leachate) and substantial savings in trade effluent charges have been made. It has been possible to monitor the performance of the plant closely for four years now, including during periods of severe weather (see Figs 1 and 2) and results have been published widely (Robinson, 1985a,b, 1987a; Robinson & Davis, 1985; Robinson & Grantham, 1988). Table 1 gives typical results for operation at high rates

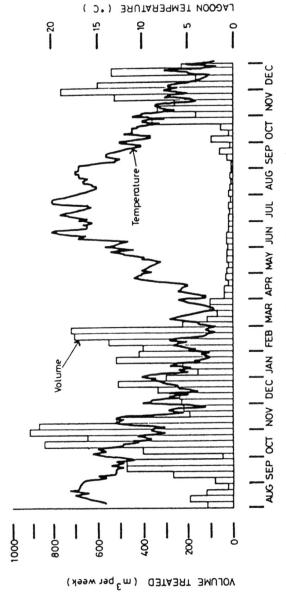

Figure 1. Bryn Posteg Treatment Plant: treatment conditions for a 17-month period.

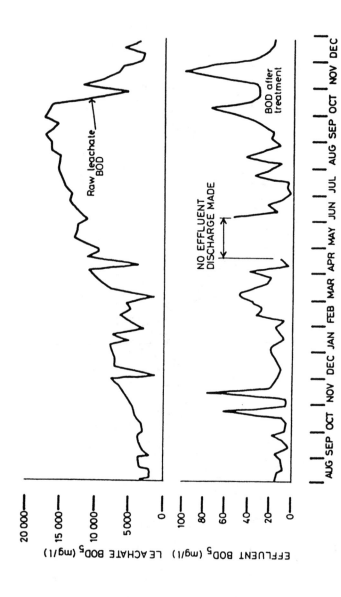

Figure 2. Bryn Posteg: results for a 17-month period.

TABLE 1. Performance of Bryn Posteg Leachate Treatment Plant (during one month in winter)

Parameter	Concentration (mg/litre except pH)	
	Leachate	*Effluent*
pH-value (typical)	6·3	7·7
COD	9 750	210
BOD$_5$	7 000	37
Ammonia nitrogen (as N)	175	0·9
SS	160	45
Chloride	1 380	810
Fe	295	9·9
Zn	11·5	0·5

Mean treatment rate of 84 m^3 per day; mean lagoon temperature 4°C; cost of electricity and nutrients for operation about 25p per m^3; trade effluent charge savings about £1.40 per m^3.

during winter months. The capital cost of this plant was about £60 000 when it was built in 1982–83.

Compton Bassett Landfill Site. In Wiltshire, UK, another plant has been operating since July 1985 at a site with a much more complex leachate management problem. Leachate is generated from several sources, including an old landfill containing well-stabilised leachate (typically BOD 1000 mg/litre, ammonia 700 mg/litre) (see Table 2),

TABLE 2. Compton Bassett Landfill: Comparison of Leachate Quality in Cell 4 before (September 1985) and after (September 1986) Establishment of Methanogenic Conditions

Determinand	Cell 4	Cell 4
Date sampled	17.9.85	19.9.86
pH value	6·3	7·3
COD	23 200	2 000
BOD	18 000	580
Ammoniacal N	770	840
Chloride	1 580	2 020
Calcium	1 480	143
Iron	811	23·9
Zinc	1·5	2·4

(Results in mg/litre except pH value).

more recent tipping cells containing less stabilised leachate (e.g. BOD 5000 mg/litre) and ammonia (500 mg/litre), through to strong fresh leachate in other cells (typically up to BOD 20 000 mg/litre, ammonia 700 mg/litre).

The design of the plant was based on that used at Bryn Posteg, but modified in view of the experiences there, and of the more difficult nature of the site at Compton Bassett. The plant is operated automatically by means of two simple microprocessors, one to control the treatment scheme, the other to control pumping of leachates from the various areas, allowing these to be 'blended' for optimum treatment.

Effluent is discharged to a very small rural sewage treatment works, and hence to a small stream, so much tighter consent conditions have been imposed than at Bryn Posteg. The most important constraint is an ammonia limit of 75 mg/litre; since ammonia is removed by incorporation in biomass produced during removal of COD and BOD, and since during the early years in the site's life most leachate comes from stabilised old wastes (unlike normal landfills), it has been necessary to overcome a shortfall of organic matter in the leachate. This has been done by the novel technique of accepting into the lagoon a liquid waste stream from a local jam factory. This liquor had caused previous disposal difficulties, but is ideal for this purpose, with a BOD value typically as high as 200 000 mg/litre, and negligible concentrations of ammonia.

This liquor is dosed automatically from a site holding tank, and provides a simple solution to the problems posed by the relatively high concentrations of ammonia currently requiring to be treated. Continuing development of the site over the next 30–40 years, with sealing and capping of the earlier cells to minimise leachate generation from them,

TABLE 3. Typical Effluent Quality at Compton Bassett (10-month period)

Determinand	May	Dec.	March	Consent conditions
COD	380	250	225	1 500
BOD	12·0	13·0	11·0	500
SS	50	44	8·0	500
Ammonia-N	1·2	1·0	3·1	75
Chloride	800	930	1 050	—
Iron	1·8	0·7	0·2	—
Zinc	0·3	0·3	0·2	—

(Results in mg/litre).

will assist in the longer term in blending of leachate to obtain optimum ratios of BOD : N for ammonia removal.

Table 3 gives typical results for operation of the plant to date. More detailed results have been published elsewhere (Robinson, 1986, 1987*a,b,c*). Because of the increased size of the plant, and greater complexity of the pumping and control scheme, costs were much greater at about £120 000 for the plant itself, but this has been considered to be good value compared to the enormous quantity of airspace (more than 3 million m^3 at least) which will be made available for infilling as a result of the protection which the plant provides for the local hydrogeological environment.

Whiteriver Landfill Site. This site is operated by Louth County Council in the Republic of Ireland, receiving smaller quantities of wastes than either of the two previous sites (from 80 to 110 tonnes per day).

The leachate treatment plant which has been installed at this site is important, however, because it demonstrates that increasing levels of knowledge and technology need not result in increasing levels of cost being incurred. This plant began to operate during February 1986, and has consistently produced a high quality effluent. Table 4 gives some typical recent results from operation of the plant in winter.

The capital cost of the treatment plant was about £48 000 (sterling) in 1986, and *maximum* running costs amount to about £12 per day.

TABLE 4. Typical Performance of Whiteriver Leachate Treatment Plant, Results for December 1986

Determinand	*Leachate*	*Effluent*
pH-value	6·8	8·9
COD	1 733	291
BOD	980	39
Ammoniacal-N	104	0·19
Chloride	328	489

(Results in mg/litre, except pH-value).

CONCLUSIONS

The ability to treat leachate to a high standard on a landfill site using a simple, automatic, robust and reliable plant, without excessive costs being incurred, has become a very important task.

Three full-scale treatment plants have been operated for up to four years, and detailed monitoring data and experience obtained. These results have been published in detail, widely in the technical literature; more plants will be built in subsequent years.

While not always necessary at all sites, the author considers that the ability to treat leachate reliably and to high standards on landfills will be vital for the protection of both surface and ground water adjacent to landfills.

REFERENCES

Robinson, H. D. (1985a). Treatment of domestic waste leachate in a full-scale automated plant. *Proceedings of the International Conference on New Directions and Research in Waste Treatment and Residuals Management*, University of British Columbia, Vancouver, Canada, pp. 166–84.

Robinson, H. D. (1985b). The treatment of high-strength landfill leachates on on-site plants. Proceedings of the IWM Open Meeting, *Leachate Treatment at Landfill Sites* held at the Council Chamber, Basildon District Council, UK, 14 Nov.

Robinson, H. D. (1986). Design and operation of leachate control measures at Compton Bassett Landfill Site, Wiltshire. Paper presented to an Open Meeting of the South-West Centre of the IWM, held at the Cadbury Country Club, Yatton, Bristol, UK, 25 July.

Robinson, H. D. (1987a). Wastes management—utilisation and disposal. Paper presented on behalf of the Institute of Wastes Management to the Public Works and Municipal Services Congress combined with the International Pollution Abatement Fair, National Exhibition Centre, Birmingham, UK, 9 April 1987, Paper 15(2).

Robinson, H. D. (1987b). Design and operation of leachate control measures at Compton Bassett Landfill Site, Wiltshire, UK. *Waste Management and Research*, **5**, 107–22.

Robinson, H. D. (1987c). A jammy solution to a leachate problem. *Surveyor*, **168** (4948), 21–3.

Robinson, H. D. & Davies, J. N. (1985). Automatic answer to leachate treatment. *Surveyor* (14 March 1985) **165**(4835), 7–8, 26.

Robinson, H. D. & Grantham, G. (1988). The treatment of landfill leachates in on-site aerated lagoon plants: experience in Britain and Ireland. *Water Research*, **22**, 733–47.

Robinson, H. D. & Maris, P. J. (1983). The treatment of leachates from domestic wastes in landfills. Aerobic biological treatment of a medium-strength leachate. *Water Research*, **17**(11), 1537–48.

Robinson, H. D. & Maris, P. J. (1985). The treatment of leachates from domestic waste in landfill sites. *Journal of the Water Pollution Control Federation*, **57**(1), 30–8.

Stegman, R. & Ehrig, H.-J. (1980). Operation and design of biological leachate treatment plants. *Progress in Water Technology*, **12**, 919–47.

3.3 Rotating Biological Contactors

KEITH KNOX

Knox Associates, 21 Ravensdale Drive, Wollaton, Nottingham, UK, NG8 2SL

INTRODUCTION AND BACKGROUND

Pitsea is a landfill situated close to the Thames Estuary on 284 hectares of reclaimed marshland. It is described in more detail elsewhere (Knox 1983, 1987). In 1983 a modification to the site licence required a leachate removal system to be installed.

The leachate within the refuse mass at Pitsea is typical of that found at sites in the methanogenic stage and is characterised by high ammonia and relatively low BOD concentrations. Leachate drains laterally into a perimeter collection ditch and is then pumped into a large holding lagoon within the refuse. Major quality changes occur within this system which, in effect, forms a vital first stage in the leachate treatment process for the site.

A detailed assessment of leachate quality in the lagoon and of bench-scale treatability trials (Knox, 1983) indicated that the residual ammonia in it could be fully nitrified by aerobic biological treatment. Outdoor pilot-scale trials were therefore undertaken, using activated sludge and trickling filter plants, to establish design information (Knox, 1985a). The studies lasted several years and allowed detailed information on leachate quality variations and temperature variations to be obtained. Nitrification kinetics were found to be similar to those in sewage treatment, with no inhibition in the leachate environment. However, with the activated sludge process, operational problems were encountered with solids control; clarification was poor and it was concluded that attached-growth systems might be more suitable for such leachates.

Some research was undertaken with a pilot-scale (5 m²) RBC unit (Knox, 1985b). The results from this work and from literature sources (Lue-Hing *et al.*, 1974; Ehrig, 1985) were used to generate design data for a full-scale RBC plant.

At the same time, hydrogeological studies were being undertaken. These indicated that removal of ~150 000 m³/a of leachate would be required to maintain it at a suitable level within the site.

Detailed cost comparisons between the activated sludge and RBC options showed their capital costs were comparable, but the running costs were much lower for the RBC, mainly because of lower power requirements. In addition it was expected that fewer operational problems would be encountered and that in the long run an RBC would require less frequent attention. The capital costs for a trickling filter are much greater than for the other two options.

During the design process it became clear that it would be cost-effective to use landfill gas to heat the leachate in Winter. It would thereby be possible to reduce the overall size of plant required by forty per cent. In addition, the maintenance of more even flows throughout the year would make flow control, nutrient dosing, and pH control simpler. A gas-heated RBC was therefore chosen as the treatment option for Pitsea.

In this chapter the performance during the first two years of operation is reviewed.

DESIGN BASIS OF TREATMENT PLANT

Influent Quality

Typical characteristics of the lagoon leachate during the period of the pilot-scale treatment studies are shown in Table 1.

The values of key parameters used in designing the plant are summarized in Table 2, with the effluent quality objectives shown.

Reactor Loading Rates

At a flow rate of 150 000 m³/a (411 m³/day) the plant was expected to have to remove, on average, 52.5×10^3 kg of NH_3-N per year. The

TABLE 1. Physical Characteristics, Sanitary Analyses and Major Constituents Analyses of Pitsea Leachate

Physical characteristics	
Colour	Peaty brown
Odour	Faint, slightly ammoniacal

Sanitary analyses	
pH	8·0–8·5
Total organic carbon, TOC	200–650
Chemical oxygen demand, COD	850–1 350
Biochemical oxygen demand, BOD (ATU)	80–250
Ammonia nitrogen (as N)	200–600
Organic nitrogen (as N)	5–20
Oxidised nitrogen (as N)	0·1–10
Alkalinity (as $CaCO_3$)	2 000–2 500
Phosphate (as P)	0·3–1·3
Total suspended solids (105°C)	100–200
Volatile suspended solids (550°C)	50–100
Fatty acids, C_1–C_6 (as C)	20

Major constituent analyses	
Conductivity	13 700
Total dissolved solids	8 600
Total hardness (as $CaCO_3$)	1 100
Na	2 185
K	888
Mg	214
Ca	88
Cl	3 400
SO_4	340

Conductivity in μmho/cm; all other analyses except pH in mg/litre. Fatty acids basd on borehole and perimeter ditch analysis.

TABLE 2. Key Parameters used in Designing the Plant

	Range	Yearly average	Discharge limit	Removal required (%)
NH_3-N	80–600	350	10	96–98·5
BOD	50–300	100	40	60–90
SS	50–200	100	—	—
VSS	25–100	50	—	—

All parameters in mg/litre.

TABLE 3. Loading Criteria used for the Three Leachate Treatment Plant Options

Temp-erature (°C)	RBC (gN/m² day)	Activated sludge (gN/m³ day)	Trickling filter (gN/m³ day)
5	1·4	120	10
5–10	1·9	160	20
10–15	3·2	280	30
15–20	4·6	400	50
20	5·6	480	60

loading criteria used for the three options, derived mainly from pilot-plant data, are shown in Table 3.

The gas heating system was designed to maintain rotor temperatures at around 20°C. The RBC plant was therefore designed to provide a surface area of 30 000 m² to give an average design loading of 4·8 gN/m² day.

Plant Description

The layout of the plant is shown in Fig. 1.

Leachate is pumped to the plant via a pumping station located at the central lagoon roughly 1 km away. Two 33 m³/h (nominal) centrifugal pumps are installed. These are electrically driven and are controlled by level controls in the leachate balancing tank.

The main plant area contains:

Leachate balancing tank. 100 m³ vertical cylindrical tank, 6-m diameter by 4-m high for intermediate storage of leachate prior to feeding by gravity to the rotors. The flow rate from the tank can be controlled either manually or by a temperature-controlled motorised valve actuator. The tank is also used to mount the heat exchange coils for the heating system.

Heating system. An immersion tube system consisting of two 100-mm diameter stainless steel exchangers is installed. The units, manufactured by Lanemark Thermal Systems Ltd, are each equipped with a burner rated at 1·58 GJ/h, capable of a heat output of 250 kW. They are supplied from a 1-ha gas field with 20 operational wells, producing ~170 m³/h of gas.

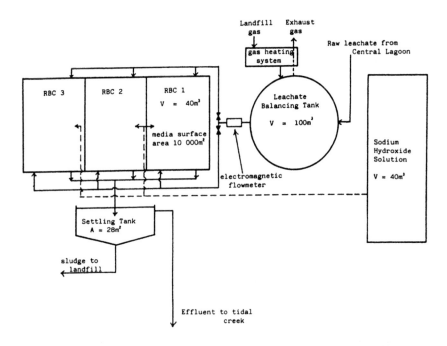

Figure 1. Schematic layout and photograph of Pitsea leachate treatment plant.

TABLE 4. Pitsea Leachate Treatment Plant: Design Summary

Reactor loadings	
Total ammonia mass loading	52.5×10^3 kgN/year
Total media surface area	30 000 m^2
Ammonia loading rate	4·8 gN/m^2 day
Speed of rotation	1·5 rpm
Reactor sizing	
Number of tanks	3
Liquid volume per tank	40 m^3
Hydraulic retention time	
@ 100 m^3/day	1·2 days
@ 411 m^3/day	7 h
@ 1000 m^3/day	2·9 h
Rotor diameter	3·66 m
Settling tank	
Diameter	6 m
Liquid depth	2·3 m
Liquid volume	65 m^3
Surface area	28·3 m^2
Surface loading rate at 411 m^3/day	14·5 m/day
Weir loading rate at 411 m^3/day	24 m^3/m day
Alkali dosing	
Type of alkali	NaOH
Storage tank	40 m^3
Concentration of dosing solution	20% v/w
Anticipated requirement	0·5 kg NaOH/m^3
Phosphate dosing	
Requirement	4 g/m^3 as P
(added as triple superphosphate fertiliser direct to rotor tanks)	
Gas heating system	
Heat exchange/balancing tank	100 m^3
Heat exchange system: two 4-in i.d. stainless steel W-shaped heat exchangers, each with burner	
Burner ratings (each)	1·58 GJ/h
Gas supply	
Gas field	1 ha
Number of wells	24
Total gas flow rate (max)	171 m^3/h
Power consumption (expected)	
Rotors	7·5 kW
Pumps, etc.	10 kW

Rotating biological contactors. The plant consists of three units of 10 000 m³ each. Each shaft is 6·6-m long, with a 3·6 m media diameter. The media consist of black, corrugated polypropylene sheets, welded into modules, having a specific surface area of 180 m²/m³.

Each rotor is mounted in a reinforced-concrete tank with GRP cover. The speed of rotation was originally supplied at 1·3 rpm but this has since been upgraded to 1·5 rpm. The degree of submergence is fixed and is ~40%. The pipework to the rotors is designed so that they can be used in parallel, for single stage operation, or in series for a two-stage operation.

Settling tank. After leaving the RBC units the leachate is fed to a 6-m diameter cylindrical steel settling tank, with bridge and scraper. Collected sludge is periodically pumped for on-site disposal. The clarified effluent passes by pipeline to discharge into a tidal creek which is a tributary of the Thames Estuary.

A single-storey building, 8 m × 6 m, houses pumps, spares, instrumentation and a small laboratory facility. The design summary is given in Table 4.

RESULTS AND EXPERIENCE

General Operational Aspects

The plant was constructed during 1985 and commissioning was started at the end of August that year. Visible biological growth became evident on the media within a few weeks, and continuous discharge of treated effluent was begun in October.

The gas heating system became available for commissioning in late January 1986, but made little impact during the plant's first winter.

Flow rates and loadings were increased during Summer 1986 to ~350 m³/day, but became physically restricted by a build-up of calcium carbonate between the feed tank and rotors. This was rectified during the late Autumn of 1986. Some benefit was gained from the heating system during the 1986/87 Winter; however, performance was not consistent enough to allow sustained increases in flow, until the ambient temperatures rose during Summer 1987. In recent months the flow rate has

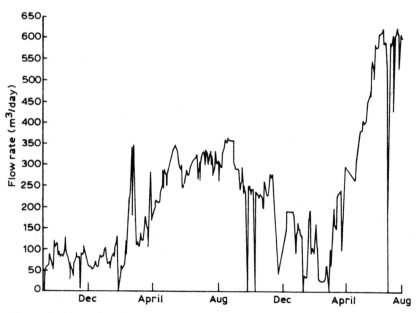

Figure 2. Pitsea leachate treatment plant: flow rate.

consistently been held at around 600 m³/day, which is the maximum which can be supplied to the plant with the existing feed pumps. Leachate flow rate is shown in Fig. 2.

Several modifications have been made:

- The electromagnetic flowmeter was repositioned early in 1987 to the influent side of the plant (see Fig. 1). In its original position downstream of the settling tank, entrapment of air caused erratic and erroneously high flow rates to be recorded at higher flow rates.
- The gearing on the rotors has been modified to increase their speed to 1·5 rpm, from 1·3 rpm as originally supplied.
- Automatic re-starting facilities have been incorporated into the electrical controls for the feed pumps, rotors and scraper. Considerable problems had been encountered because of frequent short-duration cuts in the power supply to the site. These had led to drying out of the rotors and to poor quality effluents on occasion.
- The caustic soda dosing system has been repiped so that it doses into each rotor, rather than into the leachate balancing tank. The pH shift so caused had resulted in precipitation of calcium carbonate, leading to severe scaling in the pipework feeding the rotors.

Process Performance

Ammonia and BOD removal. Influent and effluent quality for ammoniacal nitrogen and BOD are shown in Figs 3 and 4. (All BOD results are for a 5-day test, with allyl thiorea added.)

The effects of having to adjust flows to accommodate the uneven performance of the heating system whilst maintaining effluent quality can be seen in the flow rates during both of the Winter periods shown. During the Summer of 1987, however, flows consistently above the design requirement were maintained.

The influent has been considerably weaker than expected, although it has continued to show significant seasonal variations. NH_3-N concentration has been around half of the design average for much of the time since commissioning; BOD concentrations have rarely exceeded 30 mg/litre in the influent.

Effluent quality has generally been excellent, with NH_3-N and BOD concentrations consistently below the discharge consent limits. Mean and 95 percentile values to date are presented in Table 5.

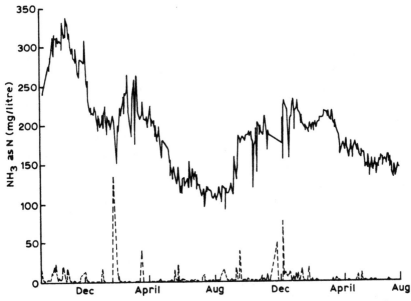

Figure 3. Influent (——) and effluent (– – –) NH_3-N-concentrations from the RBC-plant.

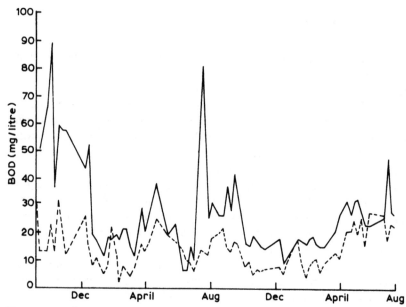

Figure 4. Influent (——) and effluent (- - -) BOD$_5$-concentrations of the RBC plant.

On occasions when the NH$_3$-N concentration has exceeded 10 mg/litre, the cause has nearly always been either overnight heating failure, or electrical failure. When both flow and temperature have been maintained consistently, the effluent NH$_3$-N concentration is usually less than 1 mg/litre.

Comprehensive data on the relationship between ammonia removal rates and temperature are not yet available. Current loadings of around 3 gN/m^2 day are almost completely removed, at rotor temperatures of 18°C. It is not possible to examine the effects of increasing these until the influent concentration increases, because the flow rate is at the maximum

TABLE 5. Mean and 95-Percentile Effluent Concentrations of the RBC-Plant

	Mean	*95 Percentile*
NH$_3$-N (mg/litre)	2·96	14
BOD (mg/litre)	14	27

TABLE 6. Influent and Effluent Quality on 11 August 1987 at a Loading of 3·1 gN/m² day

Determinand	Influent	Effluent
pH	8·21	8·04
NH_3-N (mg/litre)	153	0·1
NO_x-N (mg/litre)	34	105
NO_2-N (mg/litre)	11	0·1
TOC (mg/litre)	303	264
BOD (mg/litre)	27	23
Alkalinity (as $CaCO_3$) (mg/litre)	1 444	629
Phosphate (as P) (mg/litre)	0·9	0·6
Suspended solids (mg/litre)	135	107
Volatile suspended solids (mg/litre)	28	31

obtainable. An example of analysis of influent and effluent at 3·1 gN/m² day is given in Table 6.

There is little indication that the process has become oxygen-limited at such loadings; oxygen diffusion is generally regarded as the rate-limiting factor for nitrification in RBCs (Harremoes, 1983). Dissolved oxygen (DO) concentrations at the influent and effluent ends of the three rotors for 14.8.87 are shown in Table 7 when loadings were still around 3·1 gN/m² day.

The high DO concentrations at the effluent ends indicate that there is still scope for further increases in loading.

There has been little or no occurrence of elevated nitrite levels, even on occasions when ammonia removal has been poor. This contrasts with experiences during pilot-plant operation and with the observations of other workers (Ehrig, 1985). Slight elevations have occasionally been observed, but NO_2-N concentration has never exceeded 7 mg/litre, as shown in Fig. 5.

As noted in earlier papers (Knox, 1983, 1985a) most of the TOC

TABLE 7. Dissolved Oxygen Concentrations at the Influent and Effluent Ends of the Three Rotors

	Rotor 1	Rotor 2	Rotor 3
Influent end (mg/litre)	0·09	0·06	0·00
Effluent end (mg/litre)	3·47	4·75	5·49

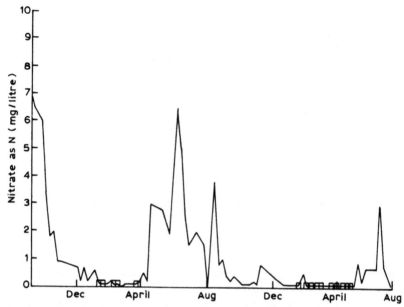

Figure 5. Effluent NO$_2$-concentration from the RBC-plant.

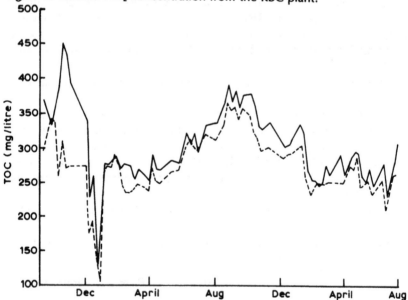

Figure 6. Influent (——) and effluent (– – –) TOC-concentrations from the RBC-plant.

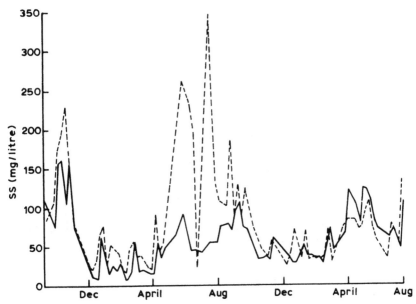

Figure 7. Influent (– – –) and effluent (——) suspended solids concentrations from the RBC-plant.

content consists of non-degradable, biologically-inactive compounds. The analyses in Table 6 indicate a small amount of removal of TOC and this is confirmed as fairly typical by Fig. 6.

Suspended solids removal and sludge production. SS and VSS concentrations for the influent and effluent are shown in Figs 7 and 8. Influent and effluent concentrations are strongly correlated for both parameters; the extent of removal is generally minor, except that some of the more extreme high values in the influent are greatly reduced in the effluent.

These results correspond with pilot-scale observations and suggest that the leachate feed contains poorly settling solids, which pass largely unaffected through the plant. It is unlikely that the performance of the settling tank can be blamed for this; the applied surface loading rates are not high, attention has been paid to weir levelling and sludge recovery rates correspond well with the calculated rates of biological solids production.

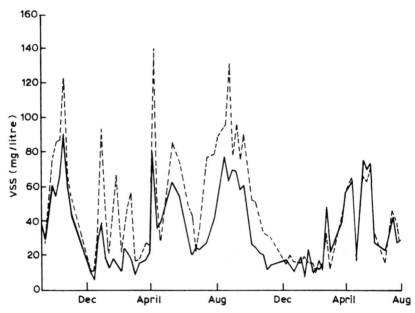

Figure 8. Influent (– – –) and effluent (——) volatile suspended solids concentrations from the RBC-plant.

Measurement of recovered sludge concentration and quantities pumped, over a three-month period, indicated the figures for sludge production shown in Table 8.

These values are similar to those obtained in other nitrification systems. It is interesting to note (Figs 4 and 8) that there is some correlation between effluent BOD and effluent VSS. This may have implications for other landfills where the discharge limit on BOD may be tighter than the 40 mg/litre which applies at Pitsea. If the presence of high concentrations of non-settling VSS is a general characteristic of methanogenic leachates rather than being specific to Pitsea, additional means of solids removal may be required.

TABLE 8. Sludge Concentrations and Yield over a Period of 3 Months

Typical solids content	2–4%
SS yield per kg NH_3-N removed	0·168 kg
VSS yield per kg NH_3-N removed	0·110 kg
Typical ratio VSS/SS	66%

The physical appearance of the biofilm has varied a lot. During pilot-plant operation, with smooth PVC discs, the biofilm was even, smooth, light-brown in colour and no more than 2 mm in thickness. On the full-scale plant, with black polypropylene media, the film has occasionally developed a very granular appearance, during which times it has been far more easily shed from the media surface. At other times, it has developed a much thicker (2–3 mm) gelatinous coating, very light brown in colour, whilst a more reddish-brown, granular-looking film develops elsewhere on the discs.

No microbiological investigations have been undertaken to investigate these changes. However, there have been no noticeable changes in process parameters or performance which correlate with any of them and their significance remains unclear.

Nutrient requirements. The requirement for caustic soda has been intermittent. For more than half of the period of operation to date, it has not been necessary to add alkali in order to maintain the pH above 7·5. Whilst the influent ammonia concentration has been well below expectation, influent alkalinity has not decreased in the same proportion so that the leachate has had sufficient buffering capacity.

Phosphorus dosing was discontinued in late June 1987 for a trial period. No effect upon process performance has been noted and after two months, the phosphorus content of the surplus sludge has remained unchanged; it varies from sample to sample but averages around 3% of volatile solids. This is close to the literature expectation for nitrifying sludge. The phosphorus concentration in the influent varies but is typically in the range 0·3–1·3 mg/litre. This is evidently sufficient for the present influent ammonia concentrations.

Heavy metals. Typical concentrations of heavy metals in the influent and effluent are shown in Table 9. They are not regarded as a cause for concern, either with the process performance or with effluent quality.

Gas Heating System

Figure 9 shows the temperatures maintained in the rotors, compared with ambient air temperature. The data plotted are 24-hour averages. The objective of maintaining high rotor temperatures in the winter has not yet been achieved. There are several reasons for this. Numerous problems have arisen with the burners and their control systems, often

TABLE 9. Metals and Heavy Metals in Pitsea LTP Influent
and Effluent, 10.6.87 (mg/litre)

Metal	Influent	Effluent
Calcium	144	145
Magnesium	178	177
Iron	1·19	0·99
Manganese	0·39	0·28
Zinc	0·25	0·27
Copper	0·18	0·18
Nickel	0·17	0·13
Chromium	0·06	0·06
Lead	<0·1	<0·1
Cadmium	<0·02	<0·02
Mercury	<0·01	<0·01

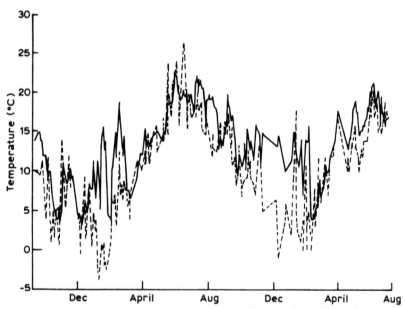

Figure 9. Air (– – –) and rotor (——) temperature data (24-hour mean values).

resulting in them cutting out or failing to start. Corrosion of one of the heat transfer tubes has led to this having to be replaced. A further factor has been that the same gas field has also been used to supply a small power generator. It is likely that further gas wells will be needed in order to ensure an adequate supply of gas to both systems.

Mode of Operation

For most of the time the plant has been operated in a single-stage mode. However, it has proven beneficial to be able to operate in a two-stage mode. During the Spring of 1987 it was operated in such a way, on a 10-day cycle using two rotors as stage 1, and the third as stage 2. This proved very successful, allowing a full development of biofilm throughout the length of all three rotors; it also allowed flow rates to be increased with less risk of producing a poor quality effluent. The change back to single-stage operation was made because greater headlosses occurred in two-stage mode, and the flow rate became restricted by plant hydraulics rather than by the process capability.

CONCLUSIONS

The effluent quality achieved and the loading rates applied confirm the suitability of RBCs for the treatment of methanogenic leachates, at least to the stage of full nitrification. The mass loadings applied during the period reviewed in this chapter had not yet reached the design loadings which had been successfully treated in pilot-scale trials. Review of subsequent performance will indicate whether higher mass loadings can successfully be treated at full-scale.

Within two years of start-up, the plant was able to discharge flows in excess of the design requirement. This was due to the fact that the leachate was weaker than anticipated, possibly because of continuing improvements in the efficiency of the ditch/lagoon system as a pre-treatment process. It nevertheless illustrates a common problem in designing leachate treatment facilities: it is difficult to quantify the problem with the degree of precision which is possible in other effluent treatment problems. Even in the unusual position of having extensive monitoring data, which most sites still do not have, the leachate strength and composition may be different from that expected at the design stage.

In addition there are uncertainties in estimates of the volume of leachate requiring treatment. The degree to which plants should be oversized to allow for such uncertainties depends to an extent on the consequences of failure to treat surplus leachate and the timescale for upgrading facilities.

This chapter does not discuss manning requirements: during the period of review, the plant had a full-time operator. It is anticipated, however, that having established the operating procedures and performance characteristics, it should be possible to reduce the degree of routine supervision to roughly four man-hours per week, with a monthly requirement for more lengthy maintenance work.

REFERENCES

Ehrig, H. J. (1985). Biological treatment of sanitary landfill leachate with special aspects on the high ammonia concentration. Paper presented at Conference: Waste Treatment and Residuals Management, University of British Columbia, Canada, June 1985: pp. 232–48.

Harremoes, P. (1983). The application of biofilm kinetics to rotating biological contactors. EWPCA–IAWPRC Seminar: Rotating Biological Discs, Fellback, Germany, 6–8 October, pp. 19–39.

Knox, K. (1983). Treatability studies on leachate from co-disposal landfill. *Env. Poll.*, (B) **5**, 157–74.

Knox, K. (1985a). Leachate treatment with nitrification of ammonia. *Water Research*, **19**(7), 895–904.

Knox, K. (1985b). Control and treatment of leachate from methanogenic landfill sites. IWM Symposium, Basildon, UK; November 1985.

Knox, K. (1987). Practice and trends in landfill in the UK. ISWA Symposium Process, Technology and Environmental Impact of Sanitary Landfill, Cagliari, Italy, October 1987.

Lue-Hing, C., Obayashi, A. W., Zenz, D. R., Washington, B. & Sawyer, B. (1974). Nitrification of a high ammonia content sludge supernatant by use of rotating discs. In *29th Purdue Indiana Waste Conference*. Ann Arbor Science, Ann Arbor, MI, pp. 245–254.

3.4 Aerobic and Anaerobic Fixed Film Biological Reactors

GLENN P. VICEVIC, PETER J. TOP & ROBERT G. W. LAUGHLIN

Ortech International, 2395 Speakman Drive, Mississauga, Ontario, Canada L5K 1B3

INTRODUCTION

Landfill leachate treatment has been investigated at both the laboratory and pilot scale, employing a number of different technologies. Anaerobic digestion, aerobic oxidation, adsorption and physical/chemical treatment, as well as such exotic technologies as electron beam irradiation and reverse osmosis/ultrafiltration, have all been tested on various leachates. Biological treatment processes have been the most widely applied. The authors conducted two concurrent pilot-scale leachate treatability studies at the Britannia Sanitary Landfill, Mississauga, Ontario, Canada, as the result of two individual and distinct projects. An anaerobic, upflow fixed-film filter and an aerobic rotating biological contactor (RBC) were operated for an extended test period. Although the two separate projects had different ultimate objectives, the fact that they both involved the treatment of the same leachate provided the opportunity to compare these technologies under identical field conditions.

The Britannia Sanitary Landfill site services the municipal solid waste and non-hazardous solid industrial waste needs of the cities of Brampton and Mississauga. It commenced operations in January 1980 and occupies 83 ha, 65 of which are used for landfilling. The site encompasses eight 8-ha separate landfill cells which are fitted with leachate collection underdrains, 5 m below original grade. MSW is dumped, compacted and then covered with 15 cm of clean fill each day.

Leachate from the individual cell underdrains discharges to a perimeter collection system which also receives sewage from the site office buildings. The leachate flows by gravity to a pumping station located at the south end of the site, where it is sent through a force main to the municipal sanitary sewer system.

PILOT-PLANT SPECIFICATIONS

Ancillary Equipment

The pilot plant was constructed beside a leachate pumping station. An insulated steel building, 3.7 m $\times$ 4.9 m $\times$ 3.7 m high, was erected on a reinforced concrete slab immediately west of the pumping station. An underground leachate supply/discharge piping network was installed between the building and a service manhole. A positive displacement pump with an explosion-proof electric motor was installed in the manhole. The leachate pumped from the manhole was stored in the pilot plant in a 1-m^3 polyethylene surface equalization tank. After mechanical problems with the original tank were experienced, this tank was later replaced with a 0.75-m^3 fibreglass reinforced plastic (FRP) tank. When the leachate throughput became very high (>1000 litres/day), both of the tanks were connected in series to provide sufficient storage capacity for continuous operation of the pilot plant. A pair of peristaltic pumps transferred the leachate from the equalization tank to each of the treatment reactors. Control of the organic loading to the units was facilitated by adjusting the flow rates of these pumps.

The Anaerobic Upflow Fixed-Film Reactor

The anaerobic filter constructed for the Brittania leachate study was custom designed for the project. An engineering drawing of the digester is presented in Fig. 1. The reactor measured 2.5 m $\times$ 1 m diameter with the bottom 0.25 m a cone, reducing to a 2.5-cm threaded fitting. The tank was constructed of FRP, while the top of the digester was sealed by a set of PVC flanges with a rubber gasket. Numerous sampling/feed ports were installed, as well as the requisite sensor ports. The total available wetted volume of the digester was 1970 litres, of which 75% was occupied by modular PVC media in a crossflow configuration. In order

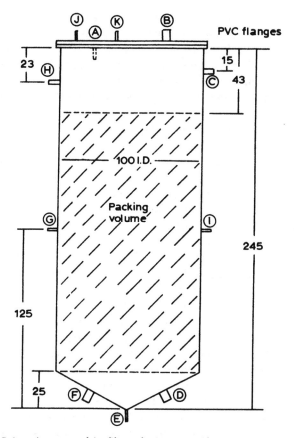

Figure 1. Brittania anaerobic filter design specifications. A, pH probe; B, pressure regulator; C effluent; D, sludge drain; E, influent; F, sludge drain; G, temperature sensor; H, recycle; I, sample port; J, pressure sensor; K, biogas effluent. (Measures are given in cm.)

to fit these modules in the reactor, they were cut in semicircular segments, which were placed at the bottom of the reactor, just above the top of the conical base. The media were secured by compression and forced into place. The next layer was made up of the scrap material from the original cutting, and, finally, it was topped off by two additional semicircular pieces. The media provided $> 90 \, m^2/m^3$ and offered a void space of $>95\%$. This meant that the actual wetted volume of the reactor was 1887 litres, with a total surface area of $170 \, m^2$.

The leachate was pumped from the equalization tanks through a limestone column and into the leachate recirculation pump suction, as illustrated in Fig. 2. Here, leachate was mixed with the recycle from the digester effluent. This was done to augment the buffering capacity of the feed and to assist in maintaining the elevated temperature in the reactor.

The recirculation pump, operating at about 2 litres/min, was a positive displacement progressive cavity model with a variable speed controller. A nutrient solution, di-ammonium phosphate, was fed to the digester by a peristaltic pump, also in the recirculation pump suction line, based on the formula from Parkin (Parkin & Owen, 1986). The recirculation pump, in turn, transferred the nutrient and leachate feed through a 12 m, 1·3-cm tube, 1·3-cm pipe, double pipe heat exchanger. The heating fluid was hot water, supplied by a standard domestic electrical water heater (136 litre, 1·25 kW), and recirculated through the water heaters by a centrifugal pump. Nominally, the hot water entered the heat exchanger at 40°C and exited at 38°C, heating the digester to 34–37°C.

The reactor was assembled with two temperature probes, one just below the inlet fitting (TI-2) and the other at the middle of the tank (TI-1). A double junction pH probe was threaded from the top flange and was submerged in the digester liquid by 15 cm. The reactor freeboard was approximately 15 cm from the flange. A pressure gauge (PI-1) and a pressure relief valve (PRV-1) were installed, together with two plastic impingers to remove any moisture from the biogas. These impingers connected directly to a cumulative, liquid displacement flowmeter (FI-1). To ensure that the anaerobic filter was isolated from atmospheric oxygen, a 15-cm water seal was maintained on the leachate effluent line. Finally, the entire reactor tank was wrapped with several inches of fiberglass/aluminum foil insulation.

The Aerobic RBC Fixed-Film Reactor

From the equalization tanks, leachate was supplied to the RBC by a peristaltic pump. In order to provide the requisite nutrient supply to the reactor, a solution of di-ammonium phosphate was added to the feed line as illustrated in Fig. 3. Nutrient requirements were calculated, based on Grady & Lim's formula (Grady & Lim, 1980). The RBC was provided by CMS Rotordisk Inc., and was their smallest commercially available unit. The four-stage reactor had a total disk area of 46·5 m^2 and a hydraulic retention capacity of 0·31 m^3. This unit size has been shown,

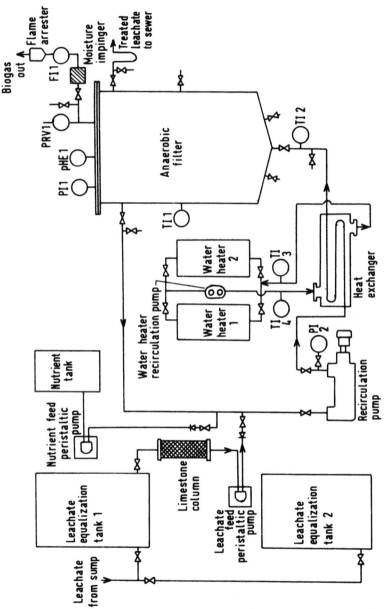

Figure 2. Britannia Landfill leachate anaerobic treatment plant.

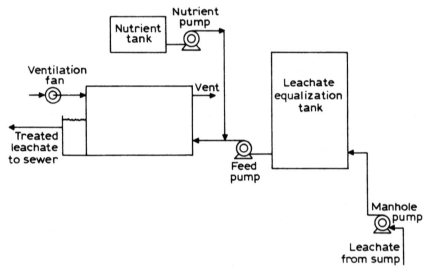

Figure 3. Brittania aerobic leachate treatment plant.

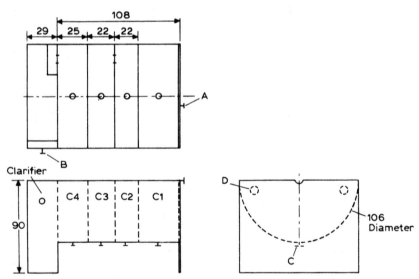

Figure 4. CMS Rotordisk pilot-scale rotating biological contactor. A, influent port; B, effluent port; C, compartment sampling port; D, inter-compartment flow passage. (Measures are given in cm.)

in previous studies, to be adequate for obtaining scale-up data. A schematic drawing of the RBC is presented in Fig. 4.

The disks were rotated by an electric motor at a speed of 4–6 rpm. Rotation of the disks provides oxygen for the biomass and good mixing in the individual stages of the reactor. The staged configuration approaches the kinetics of four constantly stirred tank reactors (CSTRs) in series. A small fan was installed to ensure sufficient air flow to the biomass. Immediately adjacent to the fourth stage, a small, 255 litre clarifier was connected in line to help manage sludge production and observe the treated effluent.

RESULTS

Table 1 presents the average leachate composition data over the length of the two studies.

TABLE 1. Britannia Sanitary Landfill Leachate Average Composition

Parameter	Concentration (mg/litre)
COD	9 254
BOD	5 340
SS	483
VSS	328
pH	6·86 pH units
Total ammonia (as N)	196
Alkalinity (as $CaCO_3$)	3 455
TOX (as Cl)	1·05
Chloride (as Cl)	1 078
Sulphate	115
VFA	4 581
Zn	3·6
P	3·6
Fe	109
Ca	700
Mg	339
Pb	0·07
Ni	0·83
Cr	0·14
Al	1·09
Cu	0·07
Mn	7·6
Cd	0·02

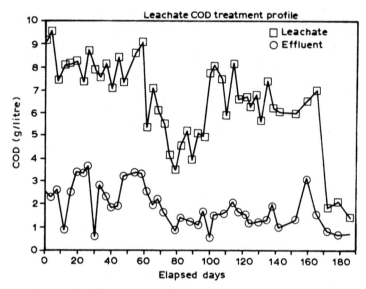

Figure 5. Anaerobic filter: COD in leachate and effluent.

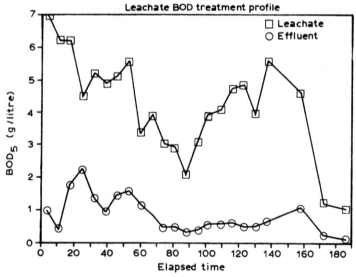

Figure 6. Anaerobic filter: BOD_5 in leachate and effluent.

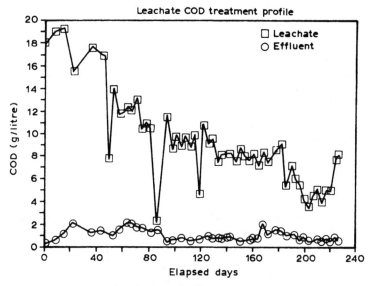

Figure 7. RBC: COD in leachate and effluent.

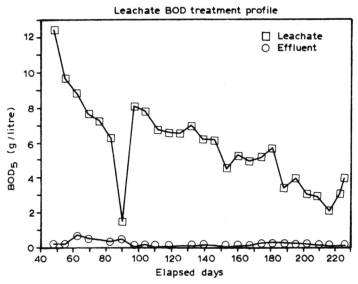

Figure 8. RBC: BOD$_5$ in leachate and effluent.

TABLE 2. Average Performance Data

Parameter	Anaerobic filter	RBC
Organic loading (kg COD/m³ day)	1·5	2·8
Hydraulic retention time (days)	5·8	2·9
COD removal (%)	72	86
BOD removal (%)	80	95
Specific biogas production (litres biogas/kg COD removed)	490	N.A.
Methane content (% v/v)	70	N.A.

N.A., Not applicable.

The COD and BOD removal profiles for both the reactors are presented in Figs 5–8. Although both systems were operated from the same leachate source, the RBC data begin approximately 120 days prior to the anaerobic filter data. From the four graphs, it can be seen that the leachate feed concentration was extremely variable throughout the study. COD levels as high as 19 000 mg/litre and as low as 1500 mg/litre were observed. Despite the variability in feed concentration, both systems demonstrated effective removal rates. Over the study period, the anaerobic filter averaged COD and BOD removal efficiencies of 72% and 80%, respectively, while the RBC averaged 86% and 95%. It should be noted that, throughout the study, the leachate throughput was consistently increased and, consequently, these results do not reflect steady-state conditions. The various other performance data averages are presented in Table 2, while the best operating conditions are presented in Table 3.

TABLE 3. Best Performance Data

Parameter	Anaerobic filter	RBC
Organic loading (kg COD/m³ day)	2·9	3·4
Hydraulic retention time (days)	2·3	1·8
COD removal (%)	71	84
BOD removal (%)	83	94
Specific biogas production (litres biogas/kg COD removed)	480	N.A.
Methane content (% v/v)	74	N.A.

N.A., Not applicable.

TABLE 4. Metals Removal Efficiency

	(% Removal)	
Element	Anaerobic filter	RBC
Co	74	54
Zn	61	92
B	36	11
Bi	16	N.D.
Si	N.D.	20
Fe	73	93
Mn	71	82
Ca	56	92
Mg	36	19
Cu	N.D.	67
Al	N.D.	90
V	24	15
Mo	43	49
Pb	N.D.	35
Ni	69	46
Cr	N.D.	70

N.D., Not detected.

Biological reactors have the capability to remove some metals from wastewaters. The Britannia leachate contained a variety of metals (see Table 1). Table 4 compares the metal removal efficiencies for the RBC and anaerobic filter throughout the project. The mechanism by which this removal takes place is known as bioadsorption and has been documented (Tzesos, 1983). Anaerobic systems also remove metals by the reduction of sulfate to insoluble metal sulfides. Samples of sludge from both reactors showed high concentrations of these metals.

Sludge production from the RBC was measured by recording sludge level accumulation in the clarifier over a known period of time. The rates varied throughout the measurement period, but ranged from 0·03 to 0·003 kg sludge/litre leachate throughput. Suspended solids (SS) removal for the RBC averaged 32% and total organohalides (TOX) removal was measured at 37%. In addition, despite sulphate levels as high as 600 mg/litre and dissolved oxygen levels of 1–2 mg/litre during the study, no evidence of nuisance organisms, such as *Beggiatoa* and *Thiotrix*, were observed.

The anaerobic filter exhibited negligible SS removal during the study. The volatile suspended solids (VSS) to SS ratio for the leachate and the effluent was constant. Samples from the bottom of the reactor did not contain heavy sludge as expected, nor were any plugging problems encountered. It was, therefore, difficult to quantify sludge production during the study.

Volatile fatty acids (VFAs) were measured throughout the anaerobic treatment study by HPLC. Measured as the sum of the acetic, propionic and *n*-butyric acid concentrations, VFA levels were extremely variable throughout the study. VFA concentration in the leachate averaged 4581 mg/litre, with a standard deviation of 2210, and effluent levels averaged 1295 mg/litre, with a standard deviation of 1106. As well, the alkalinity of the leachate remained in the 3500 mg/litre range, even following treatment. TOX removal efficiency was slightly higher than that observed by the RBC at 41%.

DISCUSSION

The aerobic RBC and the anaerobic filter both demonstrated the ability to remove organics effectively from landfill leachate at organic loadings up to 3·4 and 2·9 kg COD/m^3 day, respectively. The RBC performed at a higher COD/BOD treatment efficiency throughout the programme. It should be noted that, during the anaerobic study, it was intended to increase the loading to the filter beyond 2·9 kg COD/m^3 day, but, when this increase was due, the leachate feed became extremely dilute. The limited pumping capacity of the pilot plant would not allow any further increase in loading. It is expected that the RBC, and especially the anaerobic filter, could be pushed to higher loadings without significant loss in efficiency.

It was noticed upon decommissioning the pilot plant that some of the media in the anaerobic filter had shifted, leaving a large gap. This effectively promoted a significant amount of short-circuiting in the reactor. A thin biofilm was observed over >90% of the packing. In some areas, a brittle crust had built up between the biofilm and media. Subsequent analysis found this crust to be composed of 30% calcium, 9% iron and 7% Mg. Although this crust layer did not appear to inhibit biological activity, a reduction of active surface area and media mechanical support difficulties could be encountered as a result of long-term operation.

The methane content of the biogas averaged 70% (v/v) and the specific biogas production rate approached the stoichiometric value of 500 litres of 70% biogas/kg COD destroyed (Grady & Lim, 1980). Anaerobic filter VFA concentrations did not appear to correlate with performance variability or fluctuations in the unit. The filter demonstrated the ability to operate at VFA levels from 35 mg/litre to 3800 mg/litre, although some literature sources suggest 500 mg/litre (Anderson *et al.*, 1982), and 2000 mg/litre and 6000 mg/litre (Parkin & Owen, 1986) as upper limits.

The observation of considerable metal removal and moderate TOX removal is important in the light of the leachate composition. Metals and halogenated organics, the results of household hazardous products, batteries, etc., are normally detected in leachates from landfill sites. The fact that both the aerobic RBC and anaerobic filter have the capability of partially removing particular constituents, along with the organic content, augments their usefulness as leachate treatment systems. The aerobic unit which produces more sludge than its anaerobic counterpart exhibits a higher degree of bioadsorption. It is also apparent that, since these metals do concentrate in the organic sludge, there remains the question of its environmentally appropriate disposal. It is the authors' belief that managing this small volume of sludge is much easier than resolving the problem of the uncontrolled release of contaminants into the environment. The variability of the sludge production in the RBC and the moderate SS removals can be at least partly attributed to the variability of the leachate and the clearly underdesigned clarifier.

Perhaps the most striking result of the two studies was the resilience of both reactors to consistently maintain high treatment efficiencies. Whether the organic loading increased significantly or the leachate feed ceased due to mechanical problems, within 24 hours both units were operating normally once again.

With respect to the reactor's performance compared to other work, it can be seen that the study's results confirm the site-specific nature of leachate. Henry *et al.* (1983) found that a 24-litre anaerobic filter, operated at 25–34°C, removed 70% of the COD at a hydraulic retention time (HRT) of 5 days; the quantity of methane produced was 400–500 litres biogas/kg COD destroyed with a concentration of 80%. In a later study Henry *et al.* (1987) operated 3-litre packed bed anaerobic reactors at ambient temperature with an 8-day HRT and 1·45 kg COD/m^3 day, while achieving 95% COD removal at 580 litres biogas/kg COD destroyed and 74% methane. Young & Maris (1987) tested a 12·7-litre upflow anaerobic sludge blanket (UASB) reactor on high strength landfill leachate at 29°C, HRT of 1·9 days and an organic loading of 11·6 kg COD/m^3 day. Reported COD removal was 83%, with

499 litres of biogas/kg COD destroyed and 75·5% methane. Wright & Austin (1988) report that, at a full-scale leachate treatment plant in Halifax, Canada, two 135-m³ hybrid anaerobic filter/UASB reactors operating at 35°C are expected to treat 7 kg COD/m³ day with 95% COD removal efficiency and 500 litres biogas/kg COD destroyed at 72% methane. Finally, Coulter (1984) describes a leachate study employing an RBC where 86% COD removal was achieved at 9·6 kg COD/1000-m² media.

The results of our two studies yielded overall performance which fall within previously reported work. The fact that the other researchers were able to operate at higher loadings suggests the Brittannia reactors could have been optimized further, had the pilot plant been capable of delivering more organic loading to the units.

CONCLUSIONS

The anaerobic filter and the aerobic RBC both demonstrated the ability to treat variable strength landfill leachate with high organic removal efficiency, in addition to metals and total organohalide (TOX) removal. The anaerobic reactor produced biogas composed of 70% (v/v) methane at approximately the stoichiometric rate. The RBC did not exhibit any evidence of the growth of nuisance organisms that could contribute to reduced reactor efficiency. Both systems were shown to be extremely resilient to varying organic loadings and leachate strengths over the 200-day study. The two studies have shown that the anaerobic filter and the aerobic RBC are effective landfill leachate treatment technologies.

REFERENCES

Anderson, G. K., Donnelly, T. & McKeown, K. J. (1982). Identification and control of inhibition in the anaerobic treatment of industrial wastewaters. *Process Biochemistry*, **17**(4), 28–32.
Coulter, R. G. (1984). Pilot plant study of leachate treatability by rotating biological contactors. 2nd International Conference on Fixed-Film Biological Processes, Arlington, VA, 10–12 July, pp. 777–813.
Grady, C. P. L. & Lim, H. C. (1980). Biological wastewater treatment: theory and applications. *Pollution Engineering and Technology*, (12), 298–9.

Henry, J. G., Prasad, D., Scarcello, J. & Hilgerdenoar, M. (1983). Treatment of landfill leachate by anaerobic filter, Part 2: Pilot studies. *Water Poll. Res. J. Canada*, **18**, 45–6.

Henry, J. G., Prasad, D. & Young, H. (1987). Removal of organics from leachates by anaerobic filter. *Water Res.*, **21**(11), 1395–9.

Parkin, G. F. & Owen, W. F. (1986). Fundamentals of anaerobic digestion of wastewater sludge. *Journal of Environmental Engineering*, **112**(5), 867–921.

Tzesos, M. (1983). The role of chitin in uranium adsorption by *R. arrhizus*. *Biotechnology and Bioengineering*, **25**, 2025–40.

Wright, P. J. & Austin, T. P. (1988). Nova Scotia landfill leachate treatment facility is first of its kind in Canada. *Env. Sci. and Eng.*, **1**(5), 12–14.

Young C. P. & Maris, P. J. (1987). Optimization of on-site treatment of leachate by a two phase biological process. In *Proceedings of Safe Waste 87 Conference*, Cambridge, UK. Industrial Seminars, Tunbridge Wells, UK, pp. 334–9.

3.5 Anaerobic Lagoons and UASB Reactors: Laboratory Experiments

NICK C. BLAKEY,[a] RAFFAELLO COSSU,[b] PETER J. MARIS[a]* & FRANK E. MOSEY[a]

[a] Water Research Centre plc, Medmenham, PO Box 16, Marlow, Bucks, UK, SL7 2HD
[b] Institute of Hydraulics, University of Cagliari, Piazza d'Armi, I-09123 Cagliari, Italy

INTRODUCTION

Interest in the digestion of solid and liquid wastes has risen significantly over the past 10 years. Research has progressed over a very short period of time from laboratory and pilot-scale studies to the implementation of full-scale plants to treat high organic content wastes. These innovations have included developments directed towards rapid anaerobic reactors to treat soluble wastes. Background detail and requirements for the digestion of leachate liquors which are produced following the disposal of solid waste to landfill are described in this chapter. Leachate from putrescible solid wastes is a high organic content liquid and many workers throughout the world have carried out research into the treatment of this material. The primary objective has been to remove or reduce the pollution content of the leachate and then to reduce the potential environmental impact around landfill sites. However, the use of efficient reactors has suggested the possibility of energy export from the anaerobic process. Export would be in the form of biogas, with methane as the major component. This biofuel can be used primarily to maintain temperature levels in the reactor, but under some conditions gas production will be greater than that required for the plant and export may be possible.

* Present address: Monitor Environmental Services, 3 Dewpond Close, Stevenage, Herts, UK, SG1 3BL.

ANAEROBIC FERMENTATION OF MUNICIPAL SOLID WASTES

The putrescible components of municipal solid waste are ideally suited to support microbial growth and a proportion of this material will already be heavily enriched with bacteria. Decomposition and fermentation of this material may begin before it reaches the landfill site.

The main products of this fermentation are a range of short-chain fatty acids known as 'volatile fatty acids', VFAs or simply 'volatile acids'. They comprise between 70 and 90 per cent of the gross polluting load of the liquors leaching from young landfill sites, but they only become volatile (and malodourous) under acidic conditions. Also present are smaller quantities of methyl, ethyl, propyl and butyl alcohols which react with these acids to produce highly volatile esters. Unlike the VFAs, these compounds are volatile under neutral conditions and are responsible for the characteristic ripe, fruity odour of decomposing wastes and of the liquors leaching from them. Many other trace organic compounds contribute to the overall odour and in particular thiols are often responsible for the sulphurous overtone in smell connected with recently landfilled wastes.

As time passes within the landfill, the initial rapid fermentation of putrescible material to volatile acids is replaced by syntrophic and methanogenic fermentations leading eventually to the production of methane and carbon dioxide gases. Degradation of the volatile acids dramatically decreases the gross polluting strength of landfill leachates, but the methane generated may provide a hazard at the landfill and in its vicinity.

ANAEROBIC DIGESTION OF LEACHATE

The anaerobic fermentation which occurs naturally within landfills can be speeded up and engineered in a leachate treatment plant to provide high-rate partial treatment of strong leachates.

The anaerobic fermentation of volatile fatty acids to carbon dioxide and methane is carried out by a close-knit symbiotic consortium of syntrophic and methanogenic bacteria. These bacteria are fastidiously anaerobic, complex in their symbiotic relationships and largely self-regulating. They also have the ability to grow together to form fast-settling flocs or pellets or as a biofilm attached to an inert support surface. Their major industrial application has always been the stabilisa-

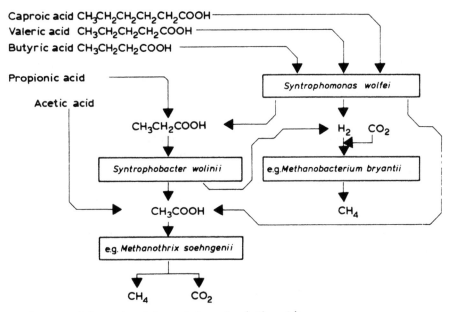

Figure 1. Schematic of degradation of volatile acids.

tion and deodourisation of the highly putrescible sludges arising from the treatment of sewage but more recently the fermentation has been extensively adapted to provide high-rate partial treatment of industrial wastewaters mostly from the pulp, paper and sugar-beet industries.

The schematic diagram in Fig. 1 shows how syntrophic and methanogenic bacteria cooperate together in landfills and wastewater treatment plants to convert complex mixtures of volatile acids to acetic and propionic acids, which are then fermented to carbon dioxide and methane gases.

Syntrophomonas wolfei is a versatile anaerobic bacterium that converts the range of volatile acids,

octanoic acid ($C_7H_{15}COOH$)
heptanoic acid ($C_6H_{13}COOH$)
caproic acid ($C_5H_{11}COOH$)
valeric acid (C_4H_9COOH)
butyric acid (C_3H_7COOH)

to acetic and propionic acids. This generates large quantities of hydrogen gas which would normally stop the reaction. *Syntrophomonas wolfei* relies on the presence of methanogenic bacteria such as *Methanobacterium bryantii* to act as hydrogen-scavengers in order to grow and ferment these acids.

The propionic acid produced by *Syntrophomonas wolfei*, or directly by other anaerobic bacteria, is converted to acetic acid by *Syntrophobacter wolinii*, a slow-growing bacterium which specialises in the reaction

$$C_2H_5COOH + 2H_2O \rightarrow CH_3COOH + CO_2 + 3H_2$$

Again the reaction is thermodynamically complex and the syntrophic bacterium relies on close association with a methanogen which acts as its hydrogen-scavenger.

The resultant acetic acid can then act as a direct substrate for the acetoclastic methanogens: *Methanosarcina barkeri*, *Methanosarcina mazei* and *Methanothrix soehngenii*.

Destruction of the acetic acid to form carbon dioxide and methane

$$CH_3COOH \rightarrow CO_2 + CH_4$$

provides most of the purification and, by reducing acidity, helps to provide the neutral pH conditions favoured for the growth of other bacteria.

Adapting high-rate partial treatment techniques to leachate digestion has proven successful, although a single method has not been universally adopted. Many individual process streams have been proposed, ranging from completely-mixed or plug flow in single or two-stage reactors with suspended or attached growth systems, to anaerobic lagoons and in-situ treatment using leachate collection and recycle through the landfill (Barber & Maris, 1984; Maris *et al.*, 1985; Blakey & Maris, 1987).

The Water Research Centre has investigated many treatment methods over the last 15 years. More recently research has concentrated on the Upflow Anaerobic Sludge Blanket (UASB) Reactor, but that does not imply that other techniques or processes are inferior or less efficient. Indeed, the trials described below show that low technology processes such as anaerobic lagooning can be applied as a further option in leachate management.

SIMULATION OF ANAEROBIC LAGOONS

Experimental

The tests summarised below were designed to investigate the effects of temperature, nutrient supplementation and microbial inoculation upon

TABLE 1. Summary of the Anaerobic Reactor Experiments Including Nutrient Supplement and Microbial Inoculation (Leachate Volume: 10 Litres)

Temperature (°C)	Reactor	Additions
25	A1	Raw leachate (RL)
	A2	RL + phosphorus (P)
	A3	RL + P + digested sewage sludge, (dS)
10	B1	RL
	B2	RL + P
	B3	RL + P + dS
4	C1	RL

the quality of domestic waste landfill leachate, when temporarily stored under anaerobic conditions. Each experiment was carried out in duplicate in a 10-litre reactor under controlled temperature conditions; 25°C, 10°C and 4°C, respectively (Table 1).

The leachate was obtained from a landfill site in southeast England which was sited in a disused portion of a sand/gravel pit. The site is lined with a polythene membrane and has received both domestic and commercial wastes since 1978. The leachate was pumped from a central sump, fed by French drains installed above the liner.

The bacterial culture used in the tests was a sample of digested sewage sludge obtained from a small experimental digester operated on mixed primary plus secondary sludges. Full details of the physical and chemical composition of both the digested sewage sludge and the leachate are shown in Table 2.

Where phosphorus nutrient addition was required a working solution of phosphoric acid (10% w/v as P) was prepared and used to raise the concentration of phosphorus to approximately 100 mg/litre, and to lower the BOD:P ratio to below 100:1 in the leachate. A litre of digested sewage sludge was also added to Reactors A3 and B3.

Leachate was withdrawn from the reactors at fortnightly intervals. Following settlement, the supernatant was bottled for routine analysis. When the experiments were completed, detailed analysis of mixed liquor, settled supernatant and sludge solids was carried out for comparison with the original feedstock.

TABLE 2. Composition of Landfill Leachate Used in All Reactors and Also of the Digested Sewage Sludge Used in Those Reactors Receiving Microbial Inoculation (Concentrations in mg/litre except pH)

Determinand	Leachate		Digested sewage sludge	
	Total	Dissolved	Total	Supernatant
pH	7·30	—	6·75	—
COD	13 700	13 000	74 600	—
BOD	9 580	9 170	—	—
TOC	4 500	4 450	—	—
MLSS	18 370	—	48 750	—
MLVSS	6 350	—	37 100	—
Kjeldahl-N	755	695	—	—
Ammoniacal-N	690	685	—	210
Nitrate-N	39·8	36·5	—	<0·05
Nitrite-N	0·2	<0·05	—	<0·05
Total phosphate	25·2	0·62	—	445
Dissolved orthophosphate	—	0·38	—	445
Total alkalinity (as $CaCO_3$)	6 450	—	—	1 055
Chloride	1 750	1 750	—	—
Sulphate	140	—	—	—
Sulphide	5·1	—	0·74	—
Volatile acids (as C)				
Acetic	1 290	—	—	470
Propionic	1 770	—	—	370
i-Butyric	200	—	—	20
n-Butyric	270	—	—	150
i-Valeric	230	—	—	30
n-Valeric	460	—	—	50
i-Caproic	10	—	—	<5
n-Caproic	170	—	—	20
Total (TVA)	4 400	—	—	1 110
Metals				
Sodium	1 210	1 210	—	9
Magnesium	340	300	—	33
Potassium	1 500	1 450	—	75
Calcium	1 000	720	—	565
Chromium	0·09	0·04	—	0·12
Manganese	5·2	4·0	—	1·47
Iron	120	90	—	1·85
Nickel	0·20	0·10	—	0·30
Copper	0·04	0·04	—	0·07
Zinc	4·90	0·36	—	0·09
Cadmium	0·04	0·01	—	0·01
Lead	<0·1	<0·1	—	<0·1

MLSS, mixed liquor suspended solids.
MLVSS, mixed liquor volatile suspended solids.

Organics Removal

The relationship between COD removal and time, at different temperatures and under different operating conditions, is shown in Fig. 2.

The digested sewage sludge addition was found to have a beneficial effect on removal efficiency in those reactors incubated at 25°C; after 28 days 65% removal of COD was achieved compared with 46–50% in the unseeded reactors. After 42 days this had increased to 87% compared with 68–75% in the unseeded reactors. No significant differences in removal efficiency were observed in any of the reactors after 70 days, where a removal efficiency of approximately 90% was achieved. In those reactors incubated at 10°C the addition of sewage sludge appeared to have no significant effect. This was thought to be due in part to a reduction in bacterial metabolism brought about by the thermal shock of rapidly reducing the leachate feedstock temperature from environmental to experimental conditions. In addition, metabolic inhibition effects would be more evident at this low experimental temperature where the presence of an additional organic substrate may have had a synergic effect.

In the 10°C reactors a 50% COD removal was achieved after about 70 days. COD removal in excess of 80% was observed after approximately

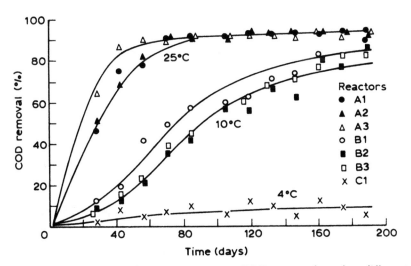

Figure 2. The effect of temperature on COD removal under different operating conditions (see Table 1 for details of reactor additions).

170 days. Unlike the tests carried out at 25°C, a metabolic lag phase of up to 40 days was also indicated at this temperature. It is possible that the effect could have been obscured at 25°C because the reaction rate was more rapid and the sampling frequency low.

At 4°C, no significant bacterial activity was observed in the reactors. A maximum COD removal of no more than 12% was recorded.

Stabilisation of the leachate was considered complete at 25°C after approximately 80 days with residual values of 90–1000 mgCOD/litre and 350–400 mgTOC/litre. Total volatile acids (TVA) were completely absent after 58 days. At 10°C stabilisation was achieved more slowly taking approximately 160–170 days. Residual COD and TOC were higher than that at 25°C, although TVA again were almost completely absent. Stabilisation was not achieved at 4°C where the organic substrates in the supernatant liquor changed marginally.

The volatile acids in the leachate feedstock at the start of the experiment were shown to be in the ratio 4:3:3, representing propionic, acetic and all the other volatile acids (butyric, valeric and caproic) respectively. During the anaerobic degradation of the leachate a variation in this ratio was observed which is illustrated in Fig. 3.

Above a concentration of 3500 mgTVA/litre there was little variation

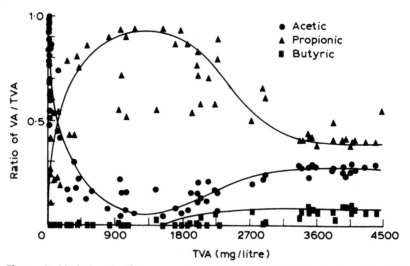

Figure 3. Variation in the proportions of the individual volatile acids (VA) in the reactors during the removal of total volatile acids (TVA) (as C).

in the initial volatile acid concentrations. However, a rapid increase in the percentage proportion of propionic acid was observed as the TVA concentration dropped from 3500 to 1300 mg/litre, accounting for some 90% of the TVA present. This corresponded with a decrease in acetic acid to less than 10% of TVA and the virtual disappearance of the other volatile acids determined in the leachate feedstock.

Hoeks & Borst (1982) observed similar behaviour in the metabolic conversion of volatile acids under anaerobic conditions. The observations can be explained by considering two distinct reaction pathways; those acids with an even carbon chain length are cleaved, two carbon atoms at a time, to yield the methanogenic substrate in the form of the acetate ion. Acetate is then rapidly converted to methane and carbon dioxide (Fig. 1). Where an odd number of carbon atoms exist the last fragment following the cleavage process consists of the propionate ion which is converted to acetate by a slower carboxylation process. This explains the apparent increase in the proportion of propionic acid at concentrations below 1000 mgC/litre. The eventual conversion of propionate to acetate can be clearly seen in Fig. 3 just prior to the total removal of volatile acids from solution.

The theoretical oxygen demand, due to the presence of volatile acids in the leachate, can be calculated by assuming a stoichiometric oxygen requirement sufficient to oxidise the individual acids to carbon dioxide and water (Table 3). The theoretical oxygen demand of 13 541 mg/litre is

TABLE 3. Calculation of the Theoretical Oxygen Demand of the Individual Acids Present in the Leachate Feedstock. (Assumes a Stoichiometric Relationship Between Oxygen and Volatile Acid (VA) Sufficient for Oxidation to CO_2 and H_2O)

Volatile acid (VA)	Molecular weight	Oxygen requirement		TVA conc. in leachate		Theoretical oxygen demand
		(moles/ mole VA)	(mg/ mg VA)	(mgC/ litre)	(mgVA/ litre)	
Acetic	60·05	2·0	1·066	1 290	3 224	3 437
Propionic	74·08	3·5	1·512	1 770	3 642	5 507
Butyric	88·12	5·0	1·816	470	863	1 567
Valeric	102·13	6·5	2·037	690	1 174	2 391
Caproic	116·16	8·0	2·204	180	290	639
			Totals	4 400	9 193	13 541

only marginally smaller than the COD measured in the leachate feedstock at the start of the experiment (13 700 mg/litre). This calculation shows that the COD of the leachate is almost entirely composed of the volatile acids indicating that the hydrolytic and acetogenic (acid fermentation) stages in leachate production occurred almost entirely within the confines of the landfill and not in the reactors.

Metal Removal

The biological reduction of sulphate to hydrogen sulphide under anaerobic conditions provides an efficient precipitant for most heavy metals present in landfill leachate. Calcium behaves in a similar way by precipitating as carbonate and phosphate salts which accumulate in the sludge as the biological reduction of organic substrate progresses. Experience has shown that most heavy metals present in landfill leachate are almost completely precipitated in the presence of sulphide where anaerobic conditions persist. However, the efficiency of removal in these circumstances is regulated by the settleability of solids.

This process alone does not account for the apparent relationship between temperature and metal removal. Indeed it could be argued that the absence of soluble organics has the dominant role in metal removal (Table 4). Under anaerobic conditions the chemistry of leachate is highly complex with pH–Eh mediated reactions resulting not only in sulphide production and precipitation of calcium and other metal salts, but complexation reactions involving soluble organic material. Work carried out by Lawson et al. (1984) showed that with long sludge ages, and with persistently high concentrations of soluble organic ligand, chelation and suspension of metal in solution may result.

Application of Removal Kinetics

COD removal was found to follow a first order rate equation such that removal rate was independent of the initial COD concentration. The rate can be expressed by the following relationship:

$$-d[COD]/dt = K[COD] \qquad (1)$$

TABLE 4. Metal Removal in the Reactors under Different Temperature Conditions

	Leach-ate feed-stock (mg/litre)	Filtered effluent (Day 200)[a] (mg/litre)			Removal (%)		
		4°C	10°C	25°C	4°C	10°C	25°C
Calcium	720	645	230	9·5	10·4	68·1	86·8
Iron	90	43	3·6	2·0	52·2	96·0	97·8
Manganese	4·0	2·4	0·53	0·25	40·0	86·8	93·8
Zinc	0·46	0·13	0·08	0·08	63·9	77·8	77·8
Nickel	0·10	<0·10	0·15	<0·10	—	zero	—
Copper	0·04	<0·02	<0·02	<0·02	—	—	—
Cadmium	0·01	<0·01	<0·01	<0·01	—	—	—
pH	7·3	7·3	7·6	7·6	—	—	—
COD (settled sample)	13 700	11 200	1 500	710	18·2	89·1	94·8

[a] Sample filtered through a Whatman GF/C filter paper.

where

$$[COD] = \text{initial COD concentration (mg/litre)}$$
$$t = \text{retention time (days)}$$
$$K = \text{reaction constant (day}^{-1})$$

Integrating eqn (1) between the initial time (t_o) and time t yields

$$\ln[COD]_t/[COD]_o = -Kt \qquad (2)$$

First order rate constants for the removal of COD in the reactors at the three different experimental temperatures have been obtained from the slope of the plot $\ln[COD]_t/[COD]_o$ against time (Fig. 4). Initial removal data up to 81 days and 150 days were used for the reactors at 25°C and 10°C, respectively.

The effect of temperature on the reaction constant can be determined from an Arrhenius type relationship where

$$K_\tau = K_{25}\theta^{\tau-25} \qquad (3)$$

where K_τ and K_{25} are the reaction constants at temperatures τ and 25°C respectively, and θ is a constant dependent on both temperature and bacterial activity. Taking the experimental values of K determined above

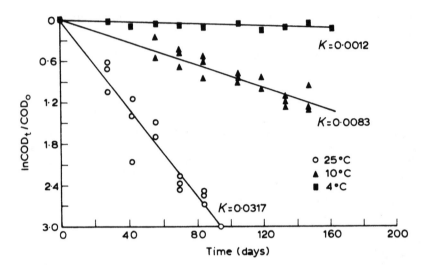

Figure 4. First order rate constants for the removal of COD in reactors at different temperatures (see eqn (2)).

at 25°C and 10°C a value for θ in eqn (3) can be calculated ($\theta = 1\cdot093$). For temperatures lower than 10°C the value of $K_{4°C}$ can be substituted ($\theta = 1\cdot17$). These values agree well with those reported in the literature where the range for θ, under different biological conditions, is $1\cdot0$–$1\cdot2$ (Vismara, 1982).

Using the relationships described above, an expression can be derived linking the mean operating temperature and the target percentage COD removal (η_{COD}) to the volume of a suitable anaerobic lagoon

$$\eta_{COD} = [COD]_o - [COD]_t/[COD]_o \qquad (4)$$

Substituting eqn (4) into eqn (2) gives

$$\eta_{COD} = 1 - e^{-Kt} \qquad (5)$$

If Q equals the daily leachate flow (m³/day) and V is the volume of the anaerobic lagoon, then rearranging eqn (5) and substituting V/Q for t we obtain the expression

$$V = -Q \ln(1 - \eta_{COD})/K \qquad (6)$$

This approach can only provide an approximate guide to lagoon size, since performance should be measured by a continuous-flow pilot plant

TABLE 5. Estimates of Lagoon Volume for a 30 m³/day Leachate Flow under Different Temperature Conditions for a Range of Treatment Efficiencies (eqn (6))

Target COD removal (%)	*Lagoon volume (m³ × 10³)*		
	4°C	*10°C*	*25°C*
10	2·63	0·38	0·01
25	7·19	1·04	0·27
50	17·3	2·51	0·66
75	34·7	5·01	1·31
90	57·6	8·32	2·18

rather than the batch experiments described. Suitable caution in the use of the data should therefore be observed.

Mean daily flow from a hypothetical landfill receiving a mean annual rainfall of about 700 mm/year has been estimated to be approximately 3 m³/ha day (Ehrig, 1983). Extrapolating this figure to a 10-ha site, daily flow would be around 30 m³. Incorporating this figure into eqn (6) yields a relationship between leachate temperature and lagoon volume for a range of treatment efficiencies (Table 5).

THE UPFLOW ANAEROBIC SLUDGE BLANKET (UASB) REACTOR

Background

The UASB-reactor has achieved widespread acceptance in Europe as a high-rate partial treatment process for high organic strength wastewaters. Landfill leachate is a particularly suitable substrate for such a process, with the leachate from recently emplaced wastes being of high organic strength and low suspended solids. In addition, the UASB lends itself to a design where liquid, gas and solid phases can be separated within the one vessel. In the past, however, the use of this treatment technique was thought to be inappropriate because the process was too slow (long retention times), often unreliable and sensitive to loading shock and toxic constituents. However, many of these considerations have proved unfounded and anaerobic reactors have performed with high efficiency.

The potential of a UASB reactor can be illustrated by data from a small pilot-scale plant operated by WRC using a medium to high strength (>10 000 mg COD/litre) leachate.

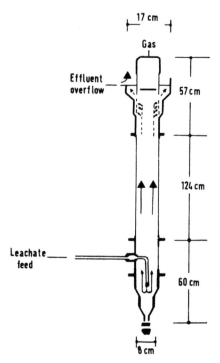

Figure 5. Laboratory-scale anaerobic digestion plant (UASB-reactor).

Reactor Performance Data

The laboratory-scale anaerobic reactor used for this study had a volume of 12·7 litres and is shown in Fig. 5. It was constructed from 75-mm diameter Excelon tubing and unplasticised PVC pipe fittings, and heated by a Grant FH15 flow heater which supplied hot (40°C) water to a heating coil wrapped around the main body of the reactor. A gas collector at the top of the reactor was connected via butyl rubber tubing to a small wet-type gas meter to measure the volumes of gas produced. Inserted in the gas line between the gas collector and the gas meter was a tee-piece fitted with a rubber septum to allow gas samples to be collected in 50-ml syringes for analysis of the gas composition.

The reactor was inoculated with a highly active enriched culture of methanogenic and acetogenic bacteria. The culture comprised 0·5–2·0-mm pellets of bacteria, which permitted a very rapid start-up. However, the pelletised nature of the culture was gradually lost during the course of the trial, to be replaced by a dense, fast-settling, metal-rich, bacterial sludge derived from the leachate. Leachate fed to the reactor was routinely dosed with phosphate (45 mg P/litre of feed) to correct any nutrient deficiency. During the initial period of operation, the reactor operated at a loading rate of 5–10 kg COD/m³ day to allow the bacteria to acclimatise to their new growth medium.

During the experiment the process was stressed by deliberately increasing the flowrate through the reactor. At the same time the strength of the leachate drawn from the landfill increased naturally from around 10 000 mg COD/litre to around 35 000 mg COD/litre. This caused the space loading of the reactor to increase from 5·5 kg COD/m³ day to 33 kg COD/m³ day.

After four months' operation the nature of the leachate changed when high concentrations of inorganic sludge were drawn from the sump at the landfill site. These inorganic sludges were deliberately fed to the reactor to test their effect on the process. They proved to be non-toxic to the methanogenic bacteria but their presence in the blanket caused severe blockages at the bottom of the reactor. Removal of a large proportion of the bacterial blanket to clear these blockages caused a temporary drop in the performance efficiency of the fermentation and, for a short period, virtually untreated leachate broke through into the digester effluent. Normal performance was restored within a few weeks, as new bacteria grew to replace losses.

The average performance data for an UASB reactor are summarised in Tables 6 and 7. The unit was operated at an average loading rate of about 11 kg COD/m³ day, with an average hydraulic retention time of about 1·8 days and a fermentation temperature of 29°C. The average removals of COD, BOD, total organic carbon (TOC) and suspended solids (SS) were 82%, 85%, 84% and 90%, respectively. During this period, the

TABLE 6. Removal Efficiency Data for the UASB Reactor

	Leachate feed (mg/litre)		UASB effluent (mg/litre)		Removal (%)	
	Range	*Average*	*Range*	*Average*	*Range*	*Average*
COD	11 450–33 440	22 070	886–9 320	3 864	65·7–92·2	82·5
BOD	7 910–24 130	14 760	329–6 290	2 262	69·2–96·1	84·7
TOC	4 010–11 390	7 425	294–3 040	1 200	66·2–92·7	83·8
SS	860–12 930	3 360	149–846	322	54·5–97·2	90·4

TABLE 7. Performance Data for the UASB Reactor

	Range	Average
Temperature (°C)	27·5–32·0	29·1
Flow (litres/day)	2·9–12·3	6·9
HRT (days)	1·0–3·2	1·8
Loading rate (kg COD/m³ day)	3·6–19·7	10·5
Gas production (litres/day)	23·1–84·5	51·6
Gas composition		
Methane (%)	69·5–81·0	75·9
Carbon dioxide (%)	19·0–30·5	24·1
Gas yield (ml/g COD removed)	382–610	496
Methane yield (ml/g COD removed)	310–494	377

composition of the gas varied from 70 to 80% methane (30 to 20% carbon dioxide) accompanied by traces of hydrogen (24–59 ppm).

The yield of gas from the reactors was difficult to calculate accurately for individual periods because of the violent fluctuations in strength and flow of the leachate but the overall average figures of 496 ml gas per gram of COD removed and 377 ml methane per gram of COD removed are reasonable estimates of yield.

Anaerobic digestion naturally generates hydrogen sulphide from the biological reduction of sulphate present in the leachate. This provides an unusually efficient precipitant for most of the toxic metals, which accumulate as inert solids within the sludge blanket. Calcium behaves similarly, being precipitated as carbonate and phosphate salts. This surplus sludge was periodically removed from the reactor. Toxic metals

TABLE 8. Metals in the Anaerobic (UASB) Digester Sludge Including Suspended Solids

Determinand	Concentration (mg/litre)		Percentage removal
	Leachate feed	Digester effluent	
Iron	890	47·7	94·6
Manganese	22·6	0·79	96·5
Zinc	14·4	0·21	98·5
Nickel	0·33	0·084	74·5
Copper	0·18	0·006	96·5
Cadmium	0·029	0·001	96·7
Suspended solids	7 207	96·1	96·7

were almost completely precipitated by the fermentation. Average removals of metals and suspended solids during this trial are shown in Table 8.

This work showed the potential of the process, and provided justification for larger scale trials to assess the process as an alternative treatment method for medium- to high-strength leachates, produced during the early life of a landfill (acetogenic phase).

SUMMARY AND CONCLUSIONS

Data from the anaerobic lagooning experiments have provided an encouraging insight into an alternative approach to landfill leachate management, particularly where a balancing lagoon may be required to mediate both flow and strength of leachate to a secondary treatment system. However, full-scale application of the process may be limited by the long retention time required to reach significant COD removal, particularly at the less favourable temperatures. This in turn may cause problems with on-site odours from hydrogen sulphide production and it is therefore unlikely that the procedure would find application in close proximity to housing.

More specifically, phosphorus enrichment was found to be unnecessary. The data also revealed that pH control with lime addition was not required although some 'fine-tuning' of pH levels may be required to maximise organic removal at low temperatures. Digested sewage sludge seeding appeared to be beneficial during the initial stages of the process by apparently decreasing the time period of biomass acclimatisation. The cost of lime and phosphorus addition would be high with very little benefit, whereas the sewage sludge addition, at no material cost, may have some use in the process.

Potential advantages of such a system when adopted alongside alternative leachate management options would include:

- Minimising the potential blockage/clogging of spray equipment with ferric hydroxide precipitate and/or bacteria growth during leachate recirculation. Odour problems associated with high volatile acid concentrations may also be reduced.
- Reduction in heavy metal toxicity to vegetation where spray irrigation to land was adopted. Oxygen deficiency at root level caused by high organic load would be alleviated.

• Lower disposal/treatment tariffs would apply at the sewage works. Lower levels of heavy metal discharges would also result.
• Where on-site aerobic treatment follows anaerobic lagooning, the lower organic and heavy metal content of the leachate would result in a lower sludge volume. Electricity costs in supplying air may consequently be reduced.

The performance data from the UASB-reactor show that it was operated at an average loading rate of about 12 kg COD/m^3 day, an average hydraulic retention time of about 1·75 days and at a fermentation temperature of 29°C for a total period of about seven months. The average removals of COD, BOD_5, TOC and SS were 82%, 85%, 84% and 90%, respectively. During this period, the composition of the gas varied from 70 to 80 per cent methane (30 to 20 per cent carbon dioxide) accompanied by traces of hydrogen (24–59 ppm). Hydrogen sulphide was only rarely detected in the gas.

The yield of gas from the reactors was difficult to calculate accurately but the overall average figures of 496 ml gas per gram of COD removed and 377 ml methane per gram of COD removed should be reasonably reliable.

The anaerobic digestion process naturally generates hydrogen sulphide from the biological reduction of sulphate present in most natural waters. This provides an unusually efficient precipitant for most of the 'toxic metals'. Most of these 'toxic metals' are removed from solution and suspension during passage of the leachate through the reactor and accumulate as inert solids within the sludge blanket. Calcium behaves similarly, becoming precipitated as carbonate and phosphate salts which accumulate within the blanket to produce a fast-settling surplus sludge which must be periodically removed from the reactor. It also produces some scaling of the pipework.

'Toxic metals' are almost completely precipitated by the fermentation and the efficiency of their capture by the anaerobic plant appears to depend only upon the efficiency of capture of suspended solids generally.

Anaerobic digestion techniques are more difficult to assess at full-scale, with very little published data on which to form an opinion. However, application of this technique looks most encouraging especially where medium- to high-strength leachates are anticipated. The experimental trials and theoretical calculations suggest that the process can be self-sufficient in energy requirements, with digester temperatures being maintained by gas production from the reactor. At lower leachate strengths the energy requirement for maintaining reactor temperatures could be provided by using the gas produced within the landfilled waste.

REFERENCES

Barber, C. & Maris, P. J. (1984). Recirculation of leachate as a landfill management option: benefits and operational problems. *Quarterly Journal Engineering Geology*, **17**(1), 19–30.

Blakey, N. C. & Maris, P. J. (1987). On-site leachate management—anaerobic processes. Proceedings ISWA Sanitary Landfill Symposium, Cagliari, Italy, 19–23 October.

Ehrig, H. J. (1983). Quality and quantity of sanitary landfill leachate. *Waste Management and Research*, **1**, 53–68.

Hoeks, J. & Borst, R. J. (1982). Anaerobic digestion of free volatile fatty acids in soils below waste tips. *Water, Air and Soil Pollution*, **17**, 165–73.

Lawson, P. S., Sherritt, R. M. & Lester, J. N. (1984). Factors affecting the removal of metals during activated sludge wastewater treatment. 1. The role of soluble ligands. *Arch. Environ. Contam. Toxicol.*, **13**(4), 383–90.

Maris, P. J., Harrington, D. W. & Mosey, F. E. (1985). Treatment of landfill leachate; management options. *Water Pollution Research J. Canada*, **20**(3), 25–42.

Vismara, R. (1982). *Depurazione Biologica*. Hoepli, Milan, Italy.

3.6 Physico-chemical Treatment of Leachate

RAFFAELLO COSSU

Institute of Hydraulics, University of Cagliari, Piazza d'Armi, I-09123 Cagliari, Italy

ROBERTO SERRA & ALDO MUNTONI

CISA, Environmental Sanitary Engineering Centre, Via Marengo 34, I-09123 Cagliari, Italy

INTRODUCTION

In the past, physico-chemical treatment of leachate was considered mainly on the basis of its efficiency as a single treatment, and then justified only for 'old leachate' (see Table 1). Currently these processes are considered as the only ones also able to treat 'young leachate' in order that the limits for discharge, which are becoming stricter and stricter, can be met (see also Chapter 3.15).

Criteria reported in Table 1 are valid if only the parameters COD, BOD and TOC are considered, but need to be revised if other parameters (particularly micro-pollutants) are considered, such as AOX (organic halogen compounds), heavy metals and ammoniacal compounds.

Among the physico-chemical processes used both for old leachate with low content of biodegradable compounds and for pre-treated young leachate, activated carbon adsorption, reverse osmosis, precipitation/flocculation and chemical oxidation are the more frequently studied and applied.

These treatment methods are examined herein, and some of the most interesting results for single stage or combined treatment in full- or pilot-scale tests are mentioned.

TABLE 1. Selection Criteria for Removal of Organics from Municipal Waste Leachate (Chian & De Walle, 1977)

COD (mg/litre)	>10 000	500–10 000	<500
COD/TOC	2·7	2·0–2·7	2·0
BOD/COD	0·5	0·1–0·5	0·1
Age of fill	Young	Medium	Old
Biological treatability	Good	Fair	Poor
Chemical precipitation	Fair	Poor	Poor
Ozone	Fair–Poor	Fair	Fair
Reverse osmosis	Fair	Good	Good
Activated carbon	Fair–Poor	Good–Fair	Good
Ion exchange resin	Poor	Good–Fair	Fair

FLOCCULATION/PRECIPITATION

Process Description

The aim of flocculation is to create flocs of particles that settle rapidly. Usually, flocculation follows coagulation in order to remove colloidal particles. These particles have dimensions in the range 1 nm–1 μm, and are characterized by a large specific surface. As a consequence, they are sensitive to surface forces.

During coagulation, colloidal particles are destabilized in order to enhance their agglomeration into larger particles and their removal by gravity. Destabilization is obtained by means of chemical reagents (coagulants) which are able to minimize repulsive forces through neutralization of electrical charges present in colloidal particles; this occurs by means of bonding or adsorption mechanisms.

All these considerations concern hydrophobic colloidal particles, for which stabilization derives from negative electrical charges. The more usual coagulants are Al and Fe(III) salts, which are characterized by multivalent ions with opposite charges. These salts have an acid behaviour and, consequently, change the physico-chemical characteristics (pH, alkalinity) of wastewater. Their efficiency depends on the alkalinity of wastewater.

Polymeric organic compounds (polyelectrolytes) are also frequently used as coagulants due to their capacity for charge neutralization (cationic polyelectrolytes) and to increase bridging between particles (Weber, 1972).

The agglomeration of destabilized colloidal particles is enhanced by controlled stirring, and is further facilitated by addition of specific chemicals ('flocculating agents'). Among these, activated silica or clay (inorganic flocculants) and polyacetate (organic flocculants) are mainly utilized.

Moreover, Al and Fe salts also behave as flocculants, since their low solubility allows precipitation with floc agglomeration and concomitant capture of colloidal particles by electrostatic action or adsorption.

Coagulation/flocculation is able to reduce colloidal suspension which is partially responsible for turbidity and colour. Also, dissolved organic substances, principally those with higher dimension (about 1 nm), are involved in the flocculation process, because they are adsorbed in the flocs and successively removed through gravity settling. Sometimes the term 'precipitation' is used to describe the phase that immediately follows flocculation, and, also, to the formation of insoluble compounds obtained by adding reagents which shift the chemical equilibrium towards the insoluble form of the compound or the elements which have to be removed. Precipitation is mainly applied to metals removal (particularly heavy metals), with metal hydroxide or metal sulphide formation, or phosphorus removal by formation of insoluble compounds with cationic metals, including Al or Fe coagulants.

Treatment Experiences

Several experimental studies utilizing coagulation/flocculation for the removal of organic substances from raw leachate have been conducted, principally in the 1970s. Salts of Al and Fe in combination with lime were mainly utilized as precipitation agents. Results were unfavourable, as COD removal efficiency lower than 40% was observed (Thornton & Blanc, 1973; Ho et al., 1974; Spencer & Farquhar, 1975; Chian & De Walle, 1976; Bjorkman & Mavinic, 1977). The reason for these low efficiencies can be attributed to the inability of the process to remove substances other than molecules of large dimensions and high molecular weight. Chian & De Walle (1976) showed the possibility of removing only the organic fraction with molecular mass greater than 50 000. Therefore, higher treatment efficiency is possible only for 'old leachate' (low BOD/COD ratio) or for biologically pre-treated leachate. In fact 'young leachate' is characterized by high levels of volatile fatty acids, i.e. small dimensions and slightly precipitable molecules, so the removal involves only a minor fraction of the organic compounds in raw leachate.

More favourable results were obtained for suspended solids and colour removal (Thornton & Blanc, 1973; Ho *et al.*, 1974; Keeman *et al.*, 1983). A removal efficiency of 75% for suspended solids and 50–70% for certain heavy metals was obtained (Keeman *et al.*, 1983) with high dosage of lime. These results were predictable, based on the limits and possibilities of precipitation, and demonstrate that this process is not able to satisfy discharge limits usually adopted for raw leachate treatment effluents.

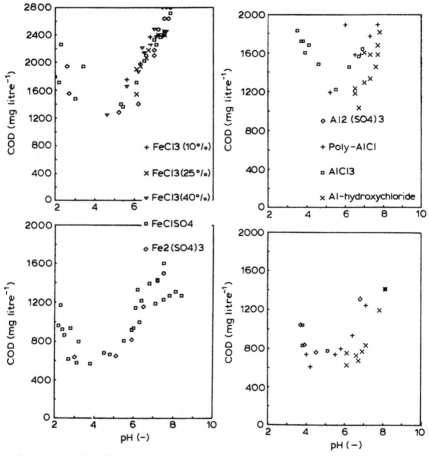

Figure 1. COD effluent versus pH for flocculation with iron salts (left) and with aluminium salts (right) using two different leachates (Ehrig, 1986).

More recently, tests have been carried out on the utilization of flocculation for biologically treated leachate. In a large experimental study (Ehrig, 1986), 50% COD removal efficiency for leachate with low BOD content (<25 mg/litre) and low BOD_5/COD ratio was obtained, with Fe and Al salts, for which the most suitable pH values were, respectively, 4·5–4·8 and 5–5·5 (Fig. 1). No particular difference in COD removal efficiency was observed with the two salts, or with different stirring methods. A minimum dosage between 250 and 500 g Fe^{3+} (or Al^{3+}) per m^3 of leachate was observed (Fig. 2). Dosage of Fe salts was very close to the theoretical dosage (0·37 g Fe^{3+}/g $CaCO_3$) for consuming all alkalinity during the process. A correlation between the COD value before and after the flocculation process was also found for biologically pre-treated leachate at an optimal pH value (Fig. 3).

Limiting factors in the application of this process are high sludge production in ratio to dosage of reagents (Fig. 4), chloride or sulphate increases and low pH in the effluent. An interesting application of the coagulation/flocculation process can be found in the Minden-Heisterholz leachate treatment plant (Albers & Krückberg, 1988). The coagulant ($FeClSO_4$) was used in dosages ranging from 600 to 1200 g Fe^{3+}/m^3 depending on pH. The maximum removal efficiency was observed for pH 5, and results were in accordance with theoretical dosage for iron (0·37 g Fe^{3+}/g $CaCO_3$).

For the flocculation phase a 75% COD removal of the influent COD (400–500 mg/litre) and a 65% removal of the influent AOX (350 μg/litre)

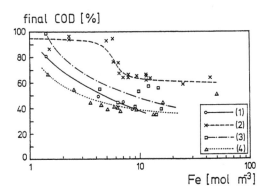

Figure 2. Correlation between iron dosage and final COD for flocculation tests (four different leachates) at optimum pH values (Ehrig, 1986). 1 mol = 56 g Fe.

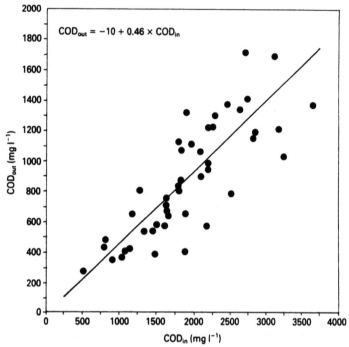

Figure 3. Correlation between effluent COD and influent COD obtained from flocculation tests (Ehrig, 1989c).

were obtained. However, it was emphasized that these efficiencies were maintained only if the effluent concentrations from the adsorption section, situated before the flocculation section, were maintained high enough.

Although these results are of interest with regard to the possible application of flocculation in combination with other processes, some researchers cite numerous disadvantages such as the increase of salt content (Doedens & Theilen, 1989) and the low efficiency of ammoniacal compound removal.

MEMBRANE SEPARATION PROCESSES

Membrane Process Principle

These processes exploit the semi-impermeable properties of certain membranes, which are characterized by water and certain solute permeability but impermeability to other solutes and to particles.

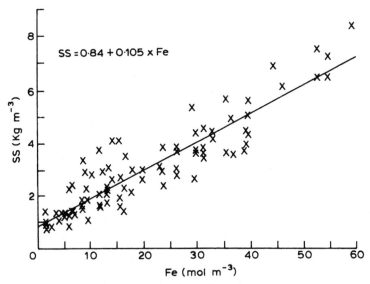

Figure 4. Sludge production in leachate flocculation process (Ehrig, 1986).

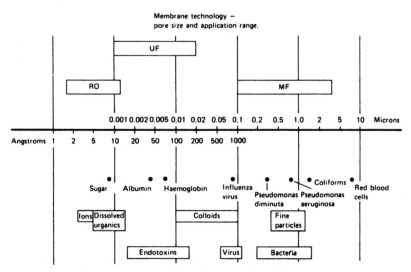

Figure 5. Application range of membrane technology (Jordain, 1987). RO, reverse osmosis; UF, ultrafiltration; MF, microfiltration.

Essentially two processes are involved: ultrafiltration and reverse osmosis. Sometimes microfiltration is considered as a membrane process, although it is not based on a semi-impermeable membrane.

The application range of membrane technology, according to the size of compounds to be removed, is summarized in Fig. 5. Usually the application is based on the molecular weight of the compound to be removed. Ultrafiltration is able to remove compounds in the molecular weight range 10 000–100 000, whereas reverse osmosis is applicable to compounds of lower dimension (Degrémont, 1979; Jordain, 1987).

This difference in dimension and weight for removal of compounds is the reason for the different importance placed on the osmotic phenomenon in the two processes. In fact, since both methods modify chemical potential by separation of dissolved compounds, a flux of solvent towards a solution with higher salt concentration is produced. At equilibrium, the pressure, measured by the different water levels on the two sides of the membrane, equalizes the osmotic pressure of the salt solution (Fig. 6).

The following equation can be used to calculate osmotic pressure

$$\pi = \Delta C \,.\, R \,.\, T \tag{1}$$

Therefore, it can be deduced that for constant temperature (T) the osmotic pressure is directly proportional to the difference of molar concentrations (ΔC) of solutions across the membrane; consequently, for equal mass concentration, higher molar mass compounds involve lower osmotic pressure.

Since reverse osmosis and ultrafiltration involve increasing molar mass compounds, the membranes need decreasing operational pressure. In fact, the aim of the process is to guarantee an efficient solvent flux (permeate) towards the solution, characterized by lower pH, by overcoming the osmotic pressure by means of external pressure (Fig. 6). Operating pressures for reverse osmosis are in the 10 bars range, whereas for ultrafiltration, lower pressure is required.

Numerous experimental studies (mentioned in Pohland, 1987) demonstrated that differences in ultrafiltration and reverse osmosis processes involve flow mechanisms inside the membrane. In the ultrafiltration process, a porous mechanism, through which solute molecules are separated when exceeding a 'critical pore' size, is involved. Different mechanisms have been proposed for reverse osmosis; a diffusion mechanism through nonporous homogeneous media (Lonsdale *et al.*, 1971) or other mechanisms which hypothesize the presence of water with a regular structure in the empty spaces between polymeric chains. This

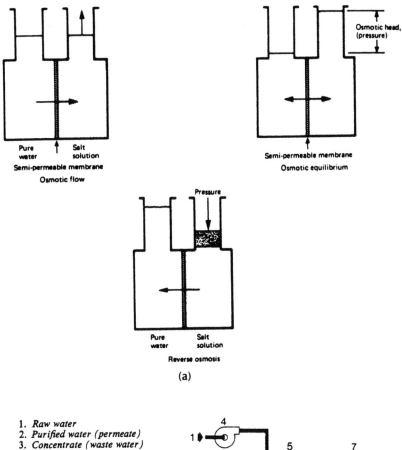

(a)

1. *Raw water*
2. *Purified water (permeate)*
3. *Concentrate (waste water)*
4. *High-pressure pump*
5. *Reverse osmosis module*
6. *Semi-permeable membrane*
7. *Discharge valve*

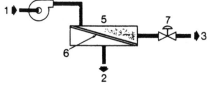

(b)

Figure 6. Principle of reverse osmosis (a) and simplified flow diagram (b).

fact does not even allow the passage of solutes with smaller dimensions than the pore size, nor of solutes which are not compatible with the regular configuration of water (Gregor & Gregor, 1978).

Referring to the scheme in Fig. 6, three parameters can be used to identify the performance of the process

Efficiency, $$R = \frac{C_f - C_p}{C_f} \cdot 100$$

Salt transport, $$ST = \frac{C_p}{C_f} \cdot 100$$

Volumetric concentration factor, $C_{fv} = \dfrac{Q_f}{Q_c} \cdot 100$

where

C_f = feed (raw wastewater) concentration
C_p = permeate concentration
Q_f = raw wastewater flow-rate
Q_c = concentrate flow-rate.

Hereafter, some details regarding reverse osmosis, mostly used for leachate treatment, will be presented.

According to the Merten model (Merten, 1966), solvent (permeate) and concentrate flow can be described (Weber & Holz, 1989) by the following equations

$$\Phi_p = \frac{K_p \cdot a}{l} \cdot (\Delta P - \pi) = A \cdot (\Delta P - \pi) \tag{2}$$

$$\Phi_s = \frac{K_s \cdot a}{l} \cdot (C_f - C_p) = B \cdot (C_f - C_p) \tag{3}$$

and

$$\Phi_s = \Phi_p \cdot C_p$$

where

K_P = permeability coefficient of membrane versus permeate
a = membrane surface area
l = membrane thickness
K_s = permeability coefficient of membrane versus salt

A = membrane permeability to permeate
B = membrane salt conductivity
ΔP = applied pressure difference across the membrane
π = osmotic pressure.

According to Van't Hoff's law,

$$\pi = R \cdot T(C_f - C_p) = b \cdot (C_f - C_p) \tag{4}$$

Equation (3) can be rearranged to

$$\Phi_p = A[\Delta P - b(C_f - C_p)] \tag{5}$$

Since usually $C_f \gg C_p$, eqns (5) and (3) can be rewritten

$$\Phi_p = A(\Delta P - bC_f) \tag{6}$$

$$\Phi_s = B \cdot C_f \tag{7}$$

Permeate flux depends on applied pressure, whereas solute flux depends only on concentration. If the process is studied with reference to the free-solute situation ($\Phi_o = A \cdot \Delta P$), the following equation (Weber & Holz, 1989) is valid

$$\frac{\Phi_p}{\Phi_o} = 1 - \frac{b}{\Delta P} \cdot C_f \tag{8}$$

None of these equations, however, are fully able to predict the type of compounds which will flow across the membrane, and they are only useful for describing macroscopic phenomena.

Generally, two types of membrane are utilized for reverse osmosis application; cellulose acetate and aromatic polyamide. The most important characteristics of these materials are reported in Table 2.

Four types of configuration are utilized to increase compactness and membrane surface area per unit volume (Pohland, 1987):

Plate module	165 m²/m³
Tubular module	335 m²/m³
Spiral wound module	1000 m²/m³
Hollow fine-fibre module	16 500 m²/m³

A scheme for a spiral wound module is shown in Fig. 7.

Cellulose acetate membranes are used in tubular, spiral wound and, more recently, hollow fibre modules. With respect to aromatic polyamidic membranes, which are mainly used in hollow fine-fibre

TABLE 2. Characteristics of Spiral-Wound Acetate and Hollow Fibre Polyamide Membranes (Modified from Degrémont, 1979)

Material	Aromatic polyamide	Cellulose acetate
Configuration	Hollow fibres	Spirally-wound hollow fibres
Physical data:		
Normal working pressure	28 bar	30–42 bar
Maximum back-pressure of treated water	3·5 bar	
Maximum operating temperature	35°C	30°C
Maximum storage temperature	40°C	30°C
Chemical characteristics:		
pH acceptable	4–11	4·5–6·5
Hydrolysis	Unaffected	Highly sensitive
Bacterial attack	Unaffected	Highly sensitive
Oxidizing agents	Highly resistant	Moderately resistant
Operating life	3–5 years	2–3 years
Salt passage (NaCl)	5–10%	5–10%

modules, the cellulose acetate membranes guarantee a higher flow-rate per surface unit (Degrémont, 1979).

Among the membrane characteristics, fouling is the most important to be considered during operation. Fouling can be caused by different factors:

—scaling (carbonate, sulphate)
—metal oxide precipitation
—coarse particles in the influent water
—biological growth

The importance of these is related to the quality of the influent water or wastewater. The tubular module is the configuration with a lesser tendency towards fouling problems and also facilitates cleansing.

Treatment Experiences

Due to the capacity of reverse osmosis to remove organic dissolved fractions with intermediate molecular weight, this method was con-

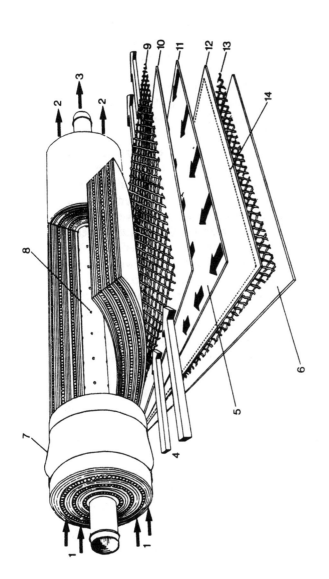

1. *Raw water*
2. *Reject*
3. *Permeate outlet*
4. *Direction of flow of raw water*
5. *Direction of flow of permeate*

6. *Protective coating*
7. *Seal between module and casing*
8. *Perforated tube for collecting permeate*
9. *Spacer*
11. *Membrane*

11. *Permeate collector*
12. *Membrane*
13. *Spacer*
14. *Line of seam connecting the two membranes*

Figure 7. Spiral-wound module scheme (Degrémont, 1979).

sidered for leachate treatment. Early studies (Chian & De Walle, 1976) using cellulose acetate and polyamidic membranes, showed high COD removal (more than 80%), although some operating problems, such as membrane fouling, were observed. Therefore subsequent studies examined the application of reverse osmosis as post-treatment after a biological stage or other physico-chemical processes, deputed to remove low molecular weight organic compounds, colloidal particles and suspended solids in order to enhance both higher efficiencies for reverse osmosis and a longer membrane life.

Recently, numerous pilot- and full-scale experiments to test reverse osmosis performance in leachate treatment have been conducted in Germany. Leachate treatment at the Venneberg-Lingen Landfill, composed of activated sludge oxidation with ammonia removal in a biological section and two-stage reverse osmosis section, provided very interesting information regarding RO performance (Weber, 1988; Weber & Holz, 1989). The experimental data indicated very high efficiency in COD and AOX removal, although it was not possible, even with two-stage RO, to obtain, with raw leachate, an ammonia concentration lower than 10 mg/litre. Reverse osmosis performances instead were positively affected by biological pre-treatment (see Chapter 3.9, Table 1).

Regarding removal of ammonia, one of the most important leachate parameters that usually exceeds the discharge limit, it was observed that a pH of 6·5 is optimal and that careful dosage of precipitation inhibitors was important to prevent sulphate scaling. Moreover, it was observed (Weber & Holz, 1989) that biological pre-treatment affects only slightly the b coefficient in Van't Hoff's law, and mainly leachate conductivity by causing a decrease (see Chapter 3.9).

A further leachate treatment plant (Logemann & Glas, 1989) showed that two-stage RO is able to guarantee more than 99% efficiency in COD, BOD and TKN removal. High removal efficiency was also observed for heavy metals ($>90\%$), with the exception of Cd and As for which approximately 70% efficiency was observed (Logemann & Glas, 1989).

Very interesting results were obtained when combining reverse osmosis with anaerobic pre-treatment (Jans *et al.*, 1987, see Chapter 3.8).

Not all studies, however, observed these positive results. The retention capacity for ammonia has been disputed, and Tables 3 and 4 (from Ehrig, 1989*a*) summarize different study results. The possibility of obtaining more than 80% efficiency in COD, BOD and AOX removal

TABLE 3. Operational Parameters and Treatment Results of Three Reverse Osmosis Plants (Ryser & Ritz, 1985; Logemann, 1987; Große, 1988) (cited in Ehrig, 1989a)

Operational parameters	Plant location		
	a	b	c
Number of stages	1	2	1
Operational pressure (kPa)	360	—	400
CF	4·9	≈4	5
Permeate flow rate (litre/m² h)	12·6	≈12	14·5
Permeate concentration			
BOD_5 (mg/litre)	112	<5	—
COD (mg/litre)	456	<10	—
NH_4 (mgN/litre)	564	<10	—
Cl (mg/litre)	1 020	<30	—
AOX (µg/litre)	<110	—	—
Elimination			
BOD_5 (%)	89	>95	98
COD (%)	89	>99	99
NH_4 (%)	58	>98	94
Cl (%)	43	>98	95
AOX (%)	93	—	—

a, Rastatt (Germany); b, VAM (The Netherlands); c, Uttigen (Switzerland).
CF, Concentration factor = (permeate flow)/(concentrate flow).

also for a one-stage RO process is evident, but it can be deduced that it is necessary to consider at least a second stage to obtain high efficiency in NH_4^+ and Cl^- removal.

Other pilot-scale tests conducted in Germany led to similar conclusions (Steensen, 1989). Results showed that an RO single-stage plant is sufficient to satisfy all the German discharge limits, except ammonia (Table 5). For ammonia removal two different solutions are suggested: nitrification before single-stage RO or a two-stage RO plant.

Semi-full-scale tests (Theilen, 1989a) conducted with different membranes and leachate (Fig. 8) showed that a two-stage RO plant is able to satisfy the threshold of 50 mg/litre whereas some question exists concerning the possibility of achieving the threshold of 10 mg/litre, required by German law for high environmental protection. This study indicates that optimal NH_4^+ removal can be obtained for pH values below neutral (Fig. 9), in accordance with other test results. Regarding

TABLE 4. Influent and Effluent Values from Pilot- (a–h) and Full-Scale (i, k) Reverse Osmosis Experiments (Marquardt et al., 1988). Table cited in Ehrig (1989a)

Parameter		Identity of experiments									
		a	b	c	d	e	f	g	h	i	k
COD	Influent	4 855	8 620	147	2 690	1 500	1 540	7 340	1 990	7 300	1 240
(mg/litre)	Effluent I	29·5	570	12	64	48	42		128	130	—
	Effluent II		<10				<5		33	15	<10
BOD$_5$	Influent	300	2 760	<50		170	147	2 700	540	2 100	<100
(mg/litre)	Effluent I	<10	51	<10		6	4	135	67	28	
	Effluent II		<20				<2	27	15	<5	<5
AOX	Influent	5·2	1·4		1·6	1·7	1·2	27	1·02	1·5	
(mg/litre)	Effluent I	0·11	0·15		<0·1	0·33	0·22	1·4	0·22	0·2	
	Effluent II		<0·05				<0·05	0·2	0·05		
NH$_4$	Influent	1 708	620	25	834	466	680	5 200	795	1 300	290
(mg/litre)	Effluent I	88	135	1	201	85	240	1 800	245	270	—
	Effluent II		<20				40	180	63	<10	<10
Cl	Influent		1 050	1 180			1 960			3 010	1 570
(mg/litre)	Effluent I		285	228			940			620	—
	Effluent II		<20				18			<20	<10

I, Effluent first stage; II, effluent second stage.
a–k, different landfills.

TABLE 5. Leachate and Permeate Quality Observed with Different Reverse Osmosis Modules, Compared with Effluent Concentration Limits (CL) in Germany (Steensen, 1989)

Module	L1	A Permeate CF 4	B Permeate CF 4.8	L2	C Per-meate >4	L3	C Per-meate 4	CL
Conductivity (μS/cm)	14 000	6 500	2 900	6 530	238	9 720	720	—
COD (mg/litre)	1 210	120	45	483	<15	5 270	39	200
BOD$_5$ (mg/litre)	30	10	<10	12	<10	2 310	14	20
NH$_4$-N (mg/litre)	240	190	100	3	0	620	53	50
NO$_3$-N (mg/litre)	732	450	250	412	19	4	0	—
Cl (mg/litre)	2 130	800	180	1 276	14	1 702	21	—
AOX (μg/litre)	1 700	280	35	1 245	5	1 855	18	500
Ni (μg/litre)	156	<10	<10	127	<30	74	<30	500
Cr (μg/litre)	94	5	<3	30	<3	95	<3	500
Pb (μg/litre)	39	—	<4	6	<1	19	2	500
Cd (μg/litre)	<1	—	<1	<1	<1	<1	<1	100
As (μg/litre)	27	<20	<20	<30	<30	<30	<30	—
Cu (μg/litre)	—	—	—	7	<5	25	<5	500

A = Tubular module, cellulose acetate membrane (6·9 m^2).
B = Plate module, composite membrane (7·6 m^2).
C = Tubular module, composite membrane (0·9 m^2).
L1 = Leachate biologically treated without nitrification.
L2 = Leachate biologically treated with nitrification.
L3 = Raw leachate.

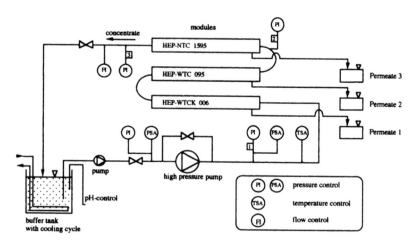

Membrane	HEP-WTCK006	HEP-WTC095	HEP-NTC1595
Material	Polyamide	Cellulose Ac.	Cellulose Ac.
pH resistance	3–10	3–7	3–7
Chlorine resistance	None	0·5 ppm	100 ppm
T_{max}	70°C	40°C	30°C
P_{max}	60 bar	40 bar	42–50 bar
Permeate flux	50 litres/h . m^2	≈40 litres/h . m^2	33 litres/h . m^2
Flow velocity	2·5–4 m/s	2–3 m/s	1·6 m/s
Salt rejection	≈98%	≈95%	≈95%

Figure 8. Pilot-plant flow sheet and membrane characteristics from German experience (Theilen, 1989a).

selection of specific processes for ammonia removal, pre-treatment through stripping or biological nitrification–denitrification has been proposed as more suitable than a two-stage RO plant (Ehrig, 1989a). In fact, apart from the operating problem with pH control (i.e. NH_4 removal efficiency is a function of pH), the presence of ammonia in the concentrate flux complicates concentrate treatment, particularly when 'evaporation treatment' is applied.

The 'concentrate' treatment is one of the most problematical aspects of reverse osmosis. This flux has usually been disposed of again in a landfill, but this practice is no longer acceptable as the concentrate can be regarded as a hazardous waste, requiring special disposal (Seyfried & Theilen, 1991). For these reasons different technologies based on

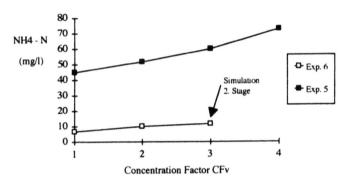

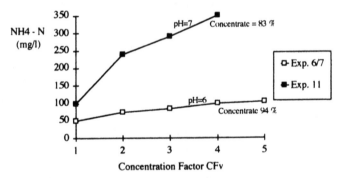

Figure 9. NH₄-N concentration versus concentration factor CF$_V$ in reverse osmosis experiments (Theilen, 1989a).

evaporation and drying processes have been developed to minimize the volume of the concentrate.

The presence of more concentrated waste also relates to other physico-chemical processes. The amount of residues changes according to the global leachate treatment process (Weber & Holz, 1991).

In conclusion the reverse osmosis process is increasingly applied to leachate treatment for its efficiency, modularity and possibility of easy automatic control. Disadvantages of the process can be summarized as follows (Seyfried & Theilen, 1991):

—the retention of small molecules (e.g. ammonia, small AOX molecules) is not fully satisfactory;

—high concentration of organics and precipitation of inorganics can cause problems of fouling, biofouling and scaling at the membrane surface;

—the energy consumption is high in connection with the high operational pressures (30–50 bar);

—the treatment of the concentrate by evaporation and drying is very expensive.

ACTIVATED CARBON ADSORPTION

Adsorption Phenomena

Adsorption phenomena involve the interphase accumulation or concentration of substances at a surface or interface (Weber, 1972). The interfaces may be related to liquid–liquid, gas–liquid, gas–solid or liquid–solid phases. The latter is of particular interest for application to wastewater treatment.

With regard to adsorption onto a solid matrix, three types of mechanism can be distinguished in which various forces predominate: electrical forces between the solution and the adsorbent; Van der Waals forces (physical adsorption); and chemical forces (chemical adsorption). The first is important in adsorption of ions, whereas the other two are concerned with adsorption of molecules. Moreover, physical adsorption is weak and predominates at lower temperatures, whereas chemical adsorption is stronger and is enhanced by higher temperatures.

Generally, all three adsorption phenomena occur simultaneously, thus hindering determination of the affinity mechanism for various types of compounds. However, specific characteristics related to organic molecules can be identified in the majority of adsorption processes. For example, Mattson *et al.* (1969) showed that the adsorptive interactions of aromatic hydroxyl compounds with activated carbon result from the formation of donor–acceptor complexes of organic molecules with carbonyl oxygen surface groups. As this process is a surface phenomenon, the adsorptive capacity of a solid depends on development of the superficial area. Several natural adsorbents are characterized by a low specific area (50–200 m^2/g) whereas activated carbon, the more widely used adsorbent, can have a specific area of 100–1500 m^2/g (Degrémont, 1979).

The quantity of substance adsorbed per unit weight of adsorbent depends on the concentration of substance present in solution at equilibrium and on temperature. At a fixed temperature, there is an increase in quantity adsorbed with increase of concentration, albeit not in direct proportion. The better known models among those proposed for interpretation of this behaviour are those of BET, Langmuir and Freundlich. The last is the most widely used, and the relation is known as Freundlich's isotherm

$$X/M = K . C^{1/n} \qquad (9)$$

where:

X = mass of substance adsorbed

M = mass of adsorbent

K = Freundlich's constant, at fixed temperature

C = concentration of substance to be adsorbed present in liquid phase at equilibrium

n = constant, at fixed temperature, of value>1

Equation (9) is normally reported graphically in a bilogarithmic diagram. A straight-line relationship leads to coefficients K and $1/n$, representing, respectively, interception and gradient of the line.

Finally, the process depends on hydrodynamic characteristics (relative speed of the two phases) and on pH. Usually acid pH facilitates adsorption onto activated carbon (Weber, 1972; Degrémont, 1979).

Activated carbon adsorption is applied to wastewater for separation of marginally biodegradable organic micropollutants and highly toxic inorganic micropollutants (heavy metals). Furthermore, it has been demonstrated (De Walle & Chian, 1974) that the organic fraction preferentially removed by activated carbon is the fulvic one with a molecular weight of 100–10 000. Organic substances characterized by a lower or higher molecular weight than this range are not efficiently retained, the first probably due to their high polarity (volatile acids, hydroxylated acids, sugars, etc.), the latter due to the large dimensions of molecules which could clog the pores and decrease adsorption capacity for other molecules (Ehrig, 1987a).

Application to Leachate Treatment

The possibility of separating organic substances with a moderately high molecular weight suggests consideration of the adsorption process as a treatment method for leachate from old landfills or for biologically pre-treated leachate. Results from studies carried out in the 1970s (Karr, 1972; Cook *et al.*, 1974; Ho *et al.*, 1974; Pohland, 1976) indicated a COD reduction efficiency of 50–60% but as high as 80% with lime pre-treatment (Cook *et al.*, 1974). However, many of these results, which were obtained on a laboratory scale, considered only performances after the first few bed volumes.

A drop in efficiency after 140 bed volumes was observed in long-term studies (Chian & De Walle, 1976) thereby indicating the need to provide high activated carbon dosage. On full-scale (Steiner *et al.*, 1977), considerable variability in a quality of leachate, reduced efficiency and excessive fouling of activated carbon was observed.

Freundlich parameters have been quantified by Pohland (1976) according to the following equation

$$X/M = 1\cdot78 \times 10^{-4} . C^{1\cdot57}$$

where C = residual COD (mg/litre).

In another study (McDougall, 1980), an isotherm for soluble organic carbon (SOC) was developed (Fig. 10).

More recently Ehrig (1989*b*) determined the following equation

$$X/M = 1\cdot32 \times 10^{-7} . C^{3\cdot72}$$

where C = residual COD (mg/litre).

The application of these results to a full-scale plant would currently be rather problematical due to the nonuniformity of observations from various experiments. In one full-scale plant, for example, a certain constancy in adsorption capacity of activated carbon was observed (Albers & Krückberg, 1988). In fact, a linear relationship between the amount of active carbon and COD removed was determined with a mean value of 3·0–3·2 mg COD removed/g activated carbon with doses ranging from 800–1200 g/m³ (see Chapter 3.7, Fig. 3).

This indicates the complexity in understanding the process when applied to leachate as biochemical processes on the carbon can interfere with the adsorption process (Albers & Krückberg, 1988).

A further critical aspect is the cost. Ehrig (1988) demonstrated how adsorption isotherms valid for leachate were too steep for low-cost treatment (Fig. 11). Also, isotherms for methanogenic phase leachates, albeit less steep with respect to those for acid phase leachates, were still characterized by a highly steep slope. Therefore, the adsorption capacity is rather low compared to those required to attain values for discharge; thus the doses required would be high. The use of other less expensive

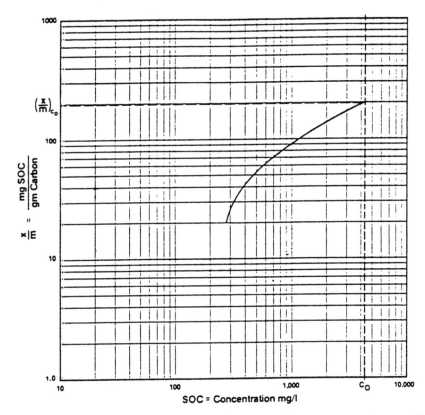

Figure 10. Adsorption isotherm experiments mentioned in McDougall (1980). SOC, soluble organic carbon.

adsorbents, such as aluminium and powdered lignite coke does not compensate for the lesser adsorption capacity (Ehrig, 1985*a*).

In spite of these limitations, it has been observed (Ehrig, 1985*b*) that the use of the activated carbon processes (powdered activated carbon in an aeration tank is preferable to granular active carbon filters), combined with flocculation/precipitation processes which reduce the presence of large molecules such as humic-like substances, is able to guarantee low levels of COD and AOX for biologically-pretreated leachates.

In order to limit treatment costs, a dosage of 1–3 kg AC/m^3 should be expected, which provides a COD reduction ranging from 100 to 400 mg/litre. These high doses imply the presence of a significant

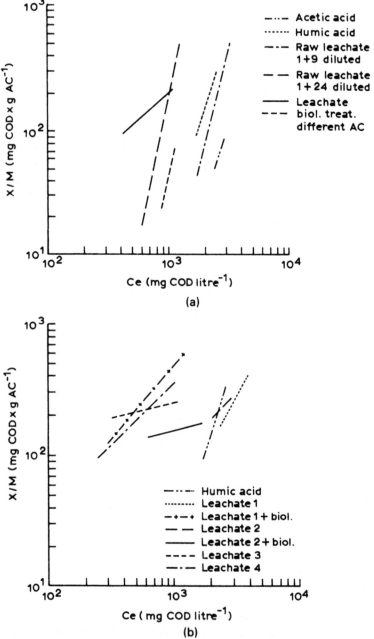

Figure 11. Activated carbon adsorption isotherm with leachate from acetic phase (a) and from methanogenic phase (b) (Ehrig, 1988).

quantity of residues which may require disposal as hazardous waste (Doedens & Theilen, 1989).

CHEMICAL OXIDATION

As mentioned previously, a typical problem for physico-chemical processes is the production of residues with concentrated pollutants. However, processes aimed at completely destroying pollutants, such as chemical oxidation, may provide a final solution to the problem. Strong oxidants, such as ozone, permanganate, chlorine compounds and hydrogen peroxide, have been considered. Most of these compounds have both a significant bacterial disinfection and oxidation capacity; therefore, they are frequently utilized in disinfection treatment of wastewater. They are less frequently used for organic compound oxidation, principally for economic reasons due to the high dosage required. However, since leachate is a wastewater often with a low flow rate, chemical oxidation for refractory organic compounds removal has been proposed.

Chemical Oxidants for Leachate Treatment

Early studies (Ho *et al.*, 1974) examined the performance of chemical oxidants such as chlorine, calcium hypochlorite, potassium permanganate and ozone. Most favourable results were obtained with calcium hypochlorite, but the COD removal efficiency was less than 50%. Moreover, all oxidants required high dosages which consequently caused numerous problems, such as increases in chloride and hardness when using, respectively, Cl_2 and $Ca(OCl)_2$.

With regard to the utilization of ozone, no residual compound problem was observed, but it was not possible to obtain more than 50% COD removal (Chian & De Walle, 1976). Ozone treatment is not effective for acid leachate with high levels of volatile acids which are resistant to ozone. High dosages and a long period of contact are necessary (Bjorkman & Mavinic, 1977).

Hydrogen peroxide (H_2O_2) was also examined principally for odour removal through hydrogen sulphide destruction. A dosage of 1·5–3·0 parts of H_2O_2 per part dissolved sulphide is required. A reduction of sulphide to 0·5 mg/litre in leachate was observed (Robinson & Maris, 1979). Utilization of H_2O_2 was also studied in combination with lagoon treatment (Fraser & Tytler, 1983) in order to increase clarification, remove iron sulphide (responsible for blackening the leachate), maintain aerobic conditions and remove suspended solids through iron precipitation. This latter phenomenon occurs by means of the following reaction (Fraser & Tytler, 1983)

$$Fe^{2+}(aq) + \text{hydrogen peroxide} \rightarrow Fe^{3+}(OH) \downarrow$$

Ferric hydroxide acts as a flocculant agent and aids in the removal of solids.

Currently chemical oxidation of leachate is still in the experimental phase. However, the most suitable oxidants appear to be ozone and hydrogen peroxide, whereas chlorine and chlorine compounds have been judged unsuitable due to the high toxicity levels of residual compounds. Concern has also been expressed for H_2O_2, although of generally less importance compared to chlorine compounds. In addition ferric hydroxide sludge, which causes an increase in the salt content of the leachate, is also of concern and is a negative aspect in applying flocculation/precipitation.

Although ozone does not cause these problems, additional experimental studies are needed before full-scale application is feasible, especially in terms of a need to increase the current low COD removal. However, the capability of oxidizing chemicals to modify the structure of complex organic substances appears to be very attractive with the possibility of converting complex compounds into simple and more easily biodegradable compounds being currently studied (Ehrig, 1987*b*). Preliminary results with ozone (Fig. 12) show a low COD reduction, but a small

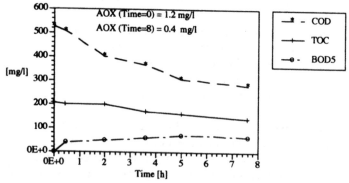

Figure 12. Organic substance parameters in leachate after ozonation (Ehrig, 1987*b*).

BOD_5 increase. AOX compounds are also detectably reduced. Collectively, these results suggest the selective use of chemical oxidation treatment if combined with biological pre- and post-treatment.

Wet Oxidation

The capacity of chemical oxidation to convert complex organic compounds is a major reason for developing wet oxidation treatment of leachate. The fundamental principle of wet oxidation is that the reactivity of oxygen for wastewater, together with the oxidation capability, increases with temperature. The process must occur in a liquid phase to enhance oxygen diffusion and its contact with the reacting compounds. Consequently, the process requires high temperature (up to 350°C) and high pressure (up to 250 bar) (Collivignarelli & Bissolotti, 1988). Application to industrial wastewater with high organic load (Baldi *et al.*, 1985) and landfill leachate has been proposed. Preliminary results (see Chapter 3.10) show the possibility of obtaining an alteration in biodegradability (BOD_{20}/COD ratio increase), and wet oxidation would seem to be suitable for leachate treatment if associated with pre- and post-treatment processes.

Obviously numerous aspects have still to be considered, such as operational problems and costs, prior to full-scale utilization. Some researchers state that wet oxidation is too expensive, especially because of energy consumption (Pfeiffer, 1989).

Catalytic Oxidation

In order to increase the efficiency of oxidation processes, the possibility for oxidation to become more reactive has been studied. Particularly ozone and hydrogen peroxide have been considered, since in the presence of intermediate radicals (i.e. hydroxyl, atomic oxygen, etc.), they become increasingly reactive.

Ozone reactivity can be improved with high pH and UV radiation. In Fig. 13 is reported the scheme of a test plant for leachate treatment where ozonation is combined with UV radiation. The results are dependent on ozone dosage, retention time in the reactor and pH value (optimal value is pH 7). With an ozone/COD ratio of between two and three an efficiency of 50% has been reached. The regulation of pH value is essential to reach a good degradation of AOX, while it has no influence in the degradation of COD.

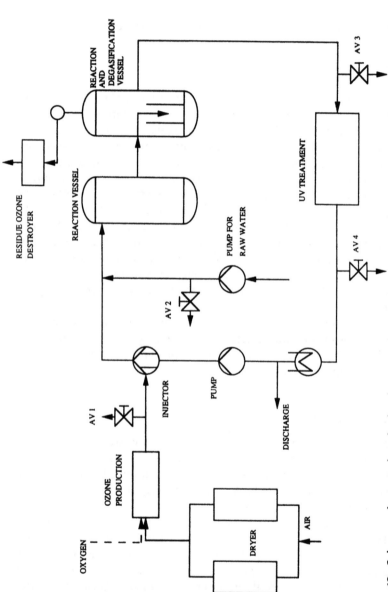

Figure 13. Scheme of a test plant for leachate treatment using ozone and UV rays. AV1–AV4, sampling points. (Kayser & Steenson, 1991, unpublished data).

Using hydrogen peroxide, improvement can be obtained with Fe^{2+} ions (Fenton's reagent) and UV radiation (Schwarzer, 1981). Good performances with raw leachate and with condensate from leachate distillation have been reported. An increase in COD removal (up to 80–95%), AOX removal (up to 85–90%) in MSW leachate (Pfeiffer, 1989) and in toxic industrial landfill leachate (Thomanetz, 1989) has been observed.

More recently, a method using a catalytic reactor for removal of highly volatile pollutants has been proposed (Gajewski *et al.*, 1989). The process consists of a stripping phase for transporting compounds in the gas phase, followed by a strong oxidation to convert compounds into CO_2, H_2O and HCl, if chloridic-organic substances are present. Ammonia changes into nitrogen (N_2) through the reaction

$$2NH_3 + \tfrac{3}{2}O_2 \rightarrow 3H_2O + N_2$$

A schematic lay-out for the ammonia removal process from leachate using this method is presented in Fig. 14. This process appears to be promising since no waste is produced and final concentrations of NH_3 are lower than 10 mg/litre, but numerous aspects have yet to be investigated.

OTHER TREATMENT

Stripping

Stripping is a physico-chemical process suitable for ammonia removal from wastewater. This method consists of increasing the pH to 10·5–11·5, thus moving the equilibrium existing in aqueous phase towards ammonia gas

$$NH_4^+ + OH^- \rightarrow NH_3 + H_2O$$

Air stripping allows NH_3 transport from the liquid to the gas phase.

Controversial results for leachate treatment were observed at the Bränåsdalen leachate treatment plant (Damhaug & Jahren, 1981), consisting of sodium hydroxide addition and multistage stripping by countercurrent aeration. High ammonia removal efficiency, approximately 90% with 100–200 mg/litre ammonia concentration in influent leachate was observed, and an air consumption of 2·5 m^3 per litre of treated leachate. Other experiments with lime addition for pH control showed low removal efficiency ($\approx 53\%$) (Kettern, 1989).

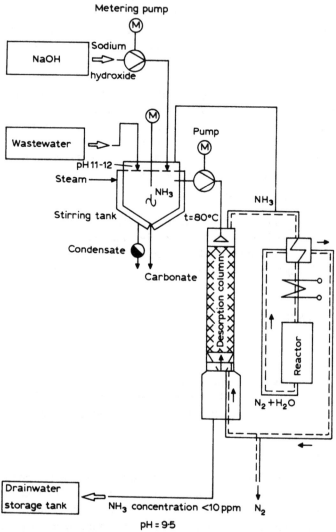

Figure 14. Catalytic oxidation process for ammonia removal in leachate (Gajewski *et al.*, 1989).

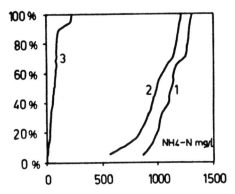

Figure 15. Cumulative frequencies of NH₄-N of raw leachate (1), effluent biological stage (2) and effluent stripping stage (3) (after Kettern, 1989).

More recent pilot-scale tests conducted in Germany at a treatment plant consisting of an activated sludge stage, followed by pH control (with lime and soda) and stripping, indicated satisfactory removal efficiencies (Fig. 15). The operating pH was fixed at 11, and the leachate/air ratio was established in the 1/3000–1/4000 range, considered optimum according to previous studies. Recently, such stripping processes have been considered a useful complement for treatment of residuals from other processes, such as concentrates from reverse osmosis.

Evaporation

The purpose of evaporation is to concentrate a polluting substance in a solid phase, and obtain a liquid stream after condensation clean enough to be discharged. To date, laboratory or pilot-scale studies have shown that it is extremely difficult to obtain a strictly solid phase and a condensate stream effectively free of polluting compounds (Ehrig, 1987a). Some results are reported in Table 6, while a scheme of full-scale plant is shown in Fig. 16. In this treatment plant, the vaporization section is preceded by stripping for ammonia removal.

In Italy, evaporation for leachate treatment was studied (Andretta *et al.*, 1986) to separate raw leachate into two equal streams. The 'concentrate' stream was recirculated into the landfill whereas the 'distillate' stream was subjected to polishing treatment (biological section with nitrification–denitrification, sterilization and adsorption section) in

TABLE 6. Results of Full-Scale and Pilot Plant Evaporation Experiments. (Data adapted from Ehrig, 1987a; Ehrig, 1989a). All values are expressed in mg/litre)

Experiments	COD		NH$_4$-N		Phenol		Hydrocarbons		TOC	
	Influent	Effluent	Influent	Effluent	Influent	Effluent	Influent	Effluent	Influent	Effluent
(Amsoneit, 1987)	4 200–5 510	227–508	475–1 400	1–219	242–420	0·63–27·4				
(Rauhenbach et al., 1986)										
Without stripping	4 000	800	2 174	1 242						
With stripping	8 160	222	1 310	20						
(Braun et al., 1985)										
Various tests on three different leachates[a]	3 000	460	220	160						
	2 900	57								
	1 150	45								
	13 900	390	1 800	1 300						
	5 300	500	850	500						
	32 600	2 500	2 100	1 000						
			1 200	480						
			3 200	680						
(Leonhard & Tiefel, 1988)	1 740–5 820	35–951	920–1 983	11–393	2·6–12·6	0·2–16·5	3–139	220	282–1 806	9–240
Acid evaporation with ammonia stripping in a full-scale plant[b]	3 870	301	1 394	81	8	4	50	22	970	67
(Leonhard, 1983)										
Acid/basic pilot-scale evaporation	7 500	940	1 445	58	9·7	0·7			2 288	210
Effluent additional stripping		250		11		0·9				11

[a] Data for ammonium refers to NH$_3$.
[b] Average value.

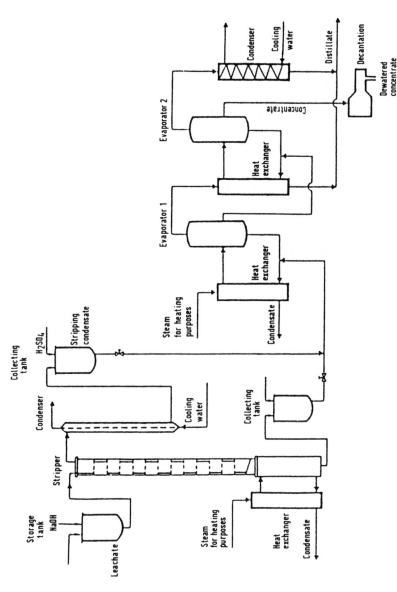

Figure 16. Leachate treatment process based on stripping and vaporization units (Leonhard & Tiefel, 1988, in Ehrig, 1989a).

order to remove organic substances and ammonia. Some problems, typical of this method, must be solved to advance development of an evaporation method. The most important problems can be summarized as follows (Ehrig, 1987a; Theilen, 1989b):

—foaming, due to high organic level;
—incrustation and corrosion with damage to the materials;
—stratified layers on the evaporation surface;
—need of further treatment for ammonia and halogenated hydrocarbon removal;
—high energy costs for raw leachate application.

The above problems advise against raw leachate treatment, whereas a development is foreseen for 'reverse osmosis concentrate' treatment (Marquardt et al., 1988; Ehrig, 1989a; Seyfried & Theilen, 1991).

Other processes are being studied, such as forced-circulation evaporation consisting of a fluidized-bed heat exchanger and a cyclone vapour section (Marquardt, 1989; Seyfried & Theilen, 1991).

Problems regarding disposal of crystals produced as the evaporation residue should be under consideration.

Landfills have been rejected due to the crystal redissolution risks. Three different methods have been proposed: upgrading and recycling of salts (principally $NaCl$, Na_2SO_4), immobilization, and disposal in underground caverns (Tiefel, 1989).

CONCLUDING REMARKS

All the examined treatment processes are affected by the presence in the leachate of organic substances, particularly simple molecules such as volatile acids, ammonia, and small AOX molecules.

Several methods are better able to separate organic substances with an extremely high molecular weight, such as humic acids (flocculation/precipitation, ultrafiltration), whereas others only concern substances with an average molecular weight such as fulvic compounds (reverse osmosis, adsorption). No method, however, can guarantee particular efficiency for compounds with a low molecular weight, both due to their polarity (adsorption) and to their dimensions (membrane processes, flocculation/precipitation). Therefore, physico-chemical treatment should be accompanied by biological treatment for removal of

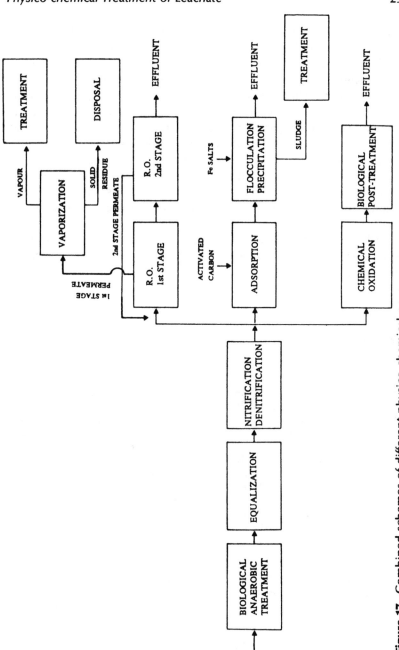

Figure 17. Combined schemes of different physico-chemical processes.

easily degradable organic substances, especially in leachates from young landfills. Thus, physico-chemical treatment deals mainly with removal of organic substances which are refractory to biostabilization, i.e. fulvic or humic compounds, halogenated compounds, some heavy metals and those compounds which would be present if specific biological treatment is not anticipated (nitrogen compounds). Biological pre-treatment is advantageous for the separation of some types of organic compounds and helps to avoid operational problems, i.e. biological fouling in reverse osmosis, elimination of interference in adsorption processes.

Nitrification pre-treatment has a positive influence on the process of flocculation/precipitation and on reverse osmosis as the decrease in alkalinity allows a reduction of the amount of reagent needed in the former case, and better control of pH in the latter.

Possible combined schemes of different physico-chemical processes are reported in Fig. 17. Combined treatment is discussed in detail in Chapters 3.7–3.9 and 3.15.

REFERENCES

Albers, H. & Krückberg, G. (1988). Combined biological and physical/chemical treatment of sanitary landfill leachate. In *ISWA 88, 5th International Solid Wastes Conference, Proc.*, vol. 1. Academic Press, London, pp. 123–31.

Amsoneit, N. (1987). Erste Erfahrungen aus dem Betrieb der Sickerwasser—Eindampfungsanlage in Schwabach Berichte auf Wassergüterwirtschaft und Gesundheitsingenieurwesen. TU München, Germany, 74, p. 213.

Andretta, F. *et al.* (1986). Trattamento del percolato di scarico controllato di Rifiuti Solidi urbani. Sep. Pollution, Padova, Italy, 6–10 Aprile 1986.

Baldi, M., Berbenn, P., Bissolotti, G., Collivignarelli, C. & Fortina, L. (1985). L'ossidazione ad umido come trattamento dei reflui idrici. Applicazioni e prospettive di sviluppo. *Inquinamento*, **7/8**, pp. 33–46.

Bjorkman, V. B. & Mavinic, D. S. (1977). Physical-chemical treatment of a high-strength leachate. Proc. 32nd Annual Industrial Waste Conference, Purdue University, Lafayette, IN, pp. 189–95.

Braun, G. & Norrenberg, T. (1985). Sickerwasserverdampfungaktuelle Vermschergebuisse 5. Abfalritschaftliches Fachkolloquium, 25–6 April.

Chian, E. S. K. & De Walle, F. B. (1976). Sanitary landfill leachates and their treatment. *Journal of the Environmental Engineering Division ASCE*, **102**, 411–31.

Chian, E. S. K. & de Walle, F. B. (1977). Evaluation of leachate treatment—vol. 1: Characterization of leachate. EPA-600/22-77-186/b-1977, U.S. Environmental Protection Agency, Washington, DC.

Collivignarelli, C. & Bissolotti, G. (1988). Experiences in the treatment of hazardous wastes by wet oxidation. In *ISWA 88, 5th International Solid Wastes Conference, Proc.*, vol. 1. Academic Press, London, pp. 219–28.

Cook, E. N. *et al.* (1974). Aerobic biostabilization of sanitary landfill leachate. *Journal of the Water Pollution Control Federation*, **46**, 380–92.

Damhaug, T. & Jahren, P. E. (1981). Ammonia stripping from leachate by countercurrent aeration in shallow tanks. Proc., ISWA Symposium, Munich, Germany, 23–25 June, pp. 71–3.

Degrémont & Co. (1979). *Water Treatment Handbook*, 5th edn International Standard Book, Firmin-Didot S.A., Paris.

De Walle, F. & Chian, E. S. K. (1974). Removal of organic matter by activated carbon columns. *Journal of the Environmental Engineering Division ASCE*, **100**(EE5), 1089–1104.

Doedens, G. & Theilen, U. (1989). Comparison of leachate treatment concepts fulfilling the stringent requirements in the FRG. In *Sardinia 89—2nd International Landfill Symposium, Porto Conte, Proc.*, vol. 2, CIPA. Milan, Italy.

Ehrig, H. J. (1985*a*). Flocculation and adsorption as post-treatment steps for highly polluted organic wastewaters. (Unpublished report.)

Ehrig, H. J. (1985*b*). Laboratory and full scale experiments on physical-chemical treatment of sanitary landfill leachate. Proc. Conference 'New Directions and Research in Waste Management', University of British Columbia, Vancouver, Canada.

Ehrig, H. J. (1986). Flocculation as a post-treatment step for high-strength organic wastewaters, recycling in chemical water and wastewater treatment. Schiftenreihe ISWW, Karlsruhe, Germany, part 50, p. 91.

Ehrig, H. J. (1987*a*). Leachate treatment: physico-chemical processes. In *Sardinia 87, First International Sanitary Landfill Symposium, Cagliari, Proc.*, vol. 1. CIPA, Milan, Italy.

Ehrig, H. J. (1987*b*). Weitergehende Reinigung von Sickerwässern aus Abfalldeponien. Veröffentlichungen des Instituts für Stadtbauwesen, TU Braunschweig, Germany, vol. 41.

Ehrig, H. J. (1988). Flockungs- und Adsorptionsvorgänge zur Sickerwasserinigung, Zentrum für Abfallforschung, TU Braunschweig, Germany, vol. 3, p. 305.

Ehrig, H. J. (1989*a*). Leachate treatment overview. *Sardinia 89, 2nd International Landfill Symposium, Porto Conte*, vol. 1. CIPA, Milan, Italy.

Ehrig, H. J. (1989*b*). Kombination von Biologie, Flockung/Fällung und Aktivkohleadsorption zur Sickerwasserbehandlung. *Entsorgungs-Praxis*, **9**, 32–6.

Ehrig, H. J. (1989*c*). Physico-chemical treatment. In: *Sanitary Landfilling: Process, Technology, Environmental Impact*, ed. T. H. Christensen, R. Cossu & R. Stegmann. Academic Press, London, UK.

Fraser, J. & Tytler, N. (1983). Operational experience using H_2O_2 in landfill leachate treatment. *Effluent and Water Treatment Journal*, April, 149–56.

Gajewski, W., Brückner, B., Richter, F., Kraft, M., Tichaczek, K. & Hagenmaier, H. (1989). Cleaning of polluted wastewater by catalytic oxidation. In *Proceedings Berlin Recycling International Conference*, Vol. 3. EF-Verlag für Energie- und Umwelttechnik, Berlin, Germany, pp. 1914–39.

Gregor, H. P. & Gregor, C. D. (1978). Synthetic membrane technology, *Scientific Amer.*, July. Also in *Handbook of Water Purification*, Ellis Horwood, Chichester, UK.

Große, G. (1988). Betriebserfahrens durch Einsatz des Umkehrosmoseverfahrens zur Aufbereitung von Sickerwasser einer Hausmülldeponie. Tagungsunterlagen Technische Akademie, Esslingen, Germany.

Ho, S., Boyle, W. C. & Ham, R. K. (1974). Chemical treatment by coagulation and precipitation. *Journal of the Environmental Engineering Division ASCE*, **99**, 535–44.

Jans, J. M., van der Schroeff, J. & Jaap, A. (1987). A treatment concept for leachate from sanitary landfills. In *Sardinia 87, First International Landfill Symposium, Cagliari*, vol. I. CIPA, Milan, Italy.

Jordain, P. (1987). Ultrafiltration. In *Handbook of Water Purification*, ed. W. Lorch. Ellis Horwood, Chichester, UK, (2nd edn) pp. 316–72.

Karr, P. R., III, (1972). Treatment of leachate from sanitary landfill. Special Research Problem, School of Civil Engineering, Georgia Institute of Technology, Atlanta, GA.

Kayser, R. & Steensen, M. (1991). Unpublished data.

Keenan, J. D., Steiner, R. L. & Fungaroli, A. A. (1983). Chemical-physical leachate treatment. *Journal of Environmental Engineering*, **109**(6), 1371–84.

Kettern, J. T. (1989). Problems of landfill leachate treatment and proposals of solution. In *Sardinia 89, 2nd International Landfill Symposium, Porto Conte, Proc.*, vol. 2. CIPA, Milan, Italy.

Leonhard, K. (1983). Die Destillation flüssiger Sonderabfälle. *Berichte aus Wassergütewirtschaft und Gesundheitsingenieurwesen*, TU München, Germany, **43**, 112.

Leonhard, K. & Tiefel, H. (1988). Thermische Sickerwasserbehandlungserfahrungen aus dem Betrieb der Sickerwassereindampfanlage in Schwabach. Diskussionstagung 'Verfahrenstechnik der mechanischen, thermischen, chemischen und biologischen Abwasserreinigung', Baden–Baden, Germany, 17–19.10.1988.

Logemann, F. P. (1987). Behandlung von Sickerwasser mittels Umkehrosmose. Firmenunterlagen Storck Friesland BV.

Logemann, F. P. & Glas, H. (1989). Using the reverse osmosis process for leach water treatment. *Recycling International*, **3**, 1965–71.

Lonsdale, H. K., Cross, B. P., Graber, P. M. & Milstead, C. E. (1971). *J. Macromol. Sci.-Phys.*, **5B**, 167 (cited in Pohland, H. W., 1987).

Marquardt, K. (1989). Leachate treatment by reverse osmosis up to a complete disposal including gas utilization. In *Proceedings Berlin Recycling International Conference*, Vol. 3. EF-Verlag für Energie und Umwelttechnik, Berlin, Germany, pp. 1959–64.

Marquardt, K., Bäuerle, V. & Kollbach, J. St. (1988). Aufbereitung von Deponiesickerwasser mit Membranverfahren. Betriebserfahrungen, Reinigungsleistungen, Kosten, Wasser, Luft and Betrieb, 48.

Mattson *et al.* (1969). Surface chemistry of active carbon: specific adsorption of phenols. *J. Colloid Interface Sci.*, **31**, 116.

McDougall, W. J. (1980). Containment and treatment of the Love Canal Landfill leachate. *Journal WPCF*, **52**(12), 2914–24.

Merten, U. (1966). Transport properties of osmotic membranes. In: *Desalination by Reverse Osmosis*, ed. U. Marten. MIT Press, Cambridge, MA.

Pfeiffer, W. (1989). Catalytic oxidation for landfill leachate treatment. In *Proceedings Berlin Recycling International Conference*, vol. 3. EF-Verlag für Energie- und Umwelttechnik, Berlin, Germany, pp. 1940–51.

Pohland, F. G. (1976). Landfill management with leachate recycle and treatment: an overview. EPA-600/9-76-004, US Environmental Protection Agency, Washington, DC, pp. 159–67.

Pohland, H. W. (1987). Reverse osmosis. In: *Handbook of Water Purification*, ed. W. Lorch. Ellis Horwood Publishers, Chichester, UK, (2nd edn) Chapter 9, pp. 316–72.

Rauntenbach, R., Kollbach, J. & Kopp, W. (1986). Aufbereitung von Deponiesickerwasser. *Wasser, Luft und Betrieb*, **6**, 61.

Robinson, H. & Maris, P. J. (1979). Leachate from domestic waste; generation, composition and treatment: a review. Technical report TR 108, Water Research Centre, Medmenham Lab., Marlow, UK.

Ryser, W. & Ritz, W. K. (1985). Behandlung von Sickerwasser durch Membrantrennverfahren auf der Mülldeponie Uttingen, Fachtagung: Sickerwasser aus Mülldeponie. Einflüsse und Behandlung, Braunschweig, Germany.

Schwarzer, H. (1981). Abwasserreinigung mittels Wasserstoffperoxid. *Umwelt*, **6**, 482–7.

Seyfried, C. F. & Theilen, U. (1991). Leachate treatment by reverse osmosis and evaporation, effects of biological pre-treatment. In *Sardinia 91, 3rd International Landfill Symposium, Cagliari*, vol. I. CISA, Cagliari, Italy, pp. 919–28.

Spencer, G. S. & Farquhar, G. J. (1975). Biological and physical-chemical treatment of leachate from sanitary landfills. Technical Paper no. 75-2, Water Resources Group, Department of Civil Engineering, University of Waterloo, Ontario, Canada.

Steensen, M. (1989). Versuche zur Sickerwasserreinigung mit Umkehrosmose-anlagen. *Entsorgungs-Praxis*, **9**, 45–7.

Steiner, R. L. *et al.* (1977). Demonstration of a leachate treatment plant. US National Technical Information Service, Springfield, VA, PB/269/502.

Theilen, U. (1989*a*). Versuche zur Sickerwasserbehandlung mit einer Halbtechnischen Umkehrosmoseanlage. *Entsorgungs-Praxis*, **9**, 48–51.

Theilen, U. (1989*b*). Aktuelle Möglichkeiten der Sickerwasserreinigung. *Entsorgungs-Praxis*, **9**, 14–18.

Thomanetz, E. (1989). Elimination Grundwasser- und Bodenfährdender Chlororganika aus hochkontaminierten Abwässern. Aufgezeigt am Beispiel des Sickerwassers der Sonderabfalldeponie Malsch. Vertiefungsseminar-Zeitgemäße Deponietechnik III. Stuttgart, Germany, 1989.

Thorton, R. J. & Blanc, F. C. (1973). Leachate treatment by coagulation and precipitation. *Journal of the Environmental Engineering Division ASCE*, **99**, 535–44.

Tiefel, H. (1989). The disposal of concentrates and crystal from leachate evaporation. In *Proceedings Berlin Recycling International Conference*, Vol. 3, EF Verlag für Energie- und Umwelttechnik, Berlin, Germany, pp. 1872–7.

Weber, B. (1988). Control and multi-stage treatment of leachate. In *Proc.*

ENVIRO 88, Specialized Seminar on Sanitary Landfilling, Amsterdam, Vol. A3. RAI, Amsterdam, The Netherlands, pp. 1–13.

Weber, B. & Holz, F. (1989). Significance of the biological pretreatment of sanitary landfill leachate on the efficiency of the reverse osmosis process. In *Sardinia 89, 2nd International Landfill Symposium*, Porto Conte, Proc., vol. 1. CIPA, Milan, Italy, pp. 1–10.

Weber, B. & Holz, F. (1991). Disposal of leachate treatment residues. In *Sardinia 91, 3rd International Landfill Symposium, Cagliari*, vol. I. CISA, Cagliari, Italy, pp. 951–68.

Weber, W. J. (1972). *Physicochemical Processes for Water Quality Control*. Wiley Interscience, Toronto, Canada.

3.7 Combination of Aerobic Pre-treatment, Carbon Adsorption and Coagulation

HENNING ALBERS

Ingenieurbüro Dr Born – Dr Ermel, Finienweg 7, D-2807 Achim-Baden, Germany

&

GERD KRÜCKEBERG

Abfallentsorgungsbetrieb des Kreises Minden–Lübbecke, Pohlsche Heide, D-4955 Hille 1, Germany

INTRODUCTION

New landfill leachate treatment plants in Germany must comply with strict effluent requirements with regard to COD (<200 mg . litre^{-1}), BOD$_5$ (<20 mg . litre^{-1}), suspended solids (<20 mg . litre^{-1}) and AOX (<0.5 mg . litre^{-1}). Also metal concentration and fish toxicity criteria must be met. In general, biological treatment alone is insufficient for meeting the criteria, while a combination of biological and physico-chemical treatment may ensure sufficiently low effluent concentrations.

The first full-scale leachate treatment plant in Germany based on combined biological and physico-chemical processes is located near Minden in Nordrhein-Westfalen. This chapter describes the plant and its treatment performance.

TREATMENT PLANT

A process scheme of the plant at Minden-Heisterholz, located west of Hannover, is given in Fig. 1. The chemical/physical part of the plant started operation in 1977 as a result of a research project (published in Ehrig, 1983). In 1982 a biological section was added as a pre-treatment

Biological plant   Chemical / physical plant

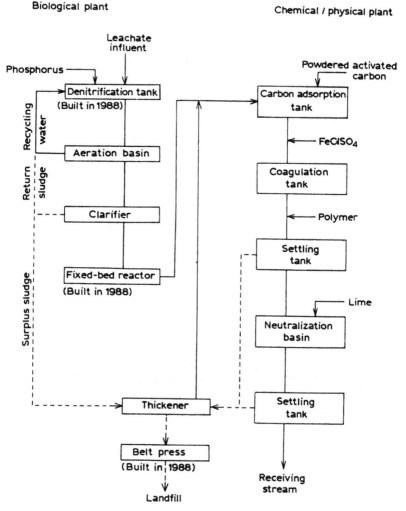

Figure 1. Process scheme of the sanitary landfill leachate treatment plant at Minden-Heisterholz, Germany.

step. In 1988 the activated sludge plant was extended by adding a denitrification tank and a fixed-bed reactor for nitrification in order to overcome problems associated with hydraulic overloading and insufficient nitrogen removal. In addition a belt filter press was installed

TABLE 1. Tank Volumes and Retention Times at Mean Flow Rates (values from 1988 on in brackets)

	Tank volume (m^3)		*Mean retention time (h)*	
Denitrification tank	—	(417)	—	(21)
Aeration basin	725	(1670)	72·5	(84)
Clarifier	19	(146)	1·9	(7·3)
Fixed bed reactor	—	(460)	—	(23)
Carbon adsorption tank	36	(80)	3·6	(4·0)
Coagulation tank	36	(30)	3·6	(1·5)
Settling tank for adsorption + coagulation	60	(277)	6·0	(14)
Neutralization basin	131	(48)	13·1	(2·4)
Settling tank neutralization	60	(filter)	6·0	—
Total	1 067	(3 128)	106·7	(157)

to reduce the water content of the sludge and to improve the leaching characteristics of the sludge before disposal on the landfill.

The plant was designed for a leachate flow of $20 \, m^3 . h^{-1}$. The mean leachate production rate from August 1985 to January 1987—the time interval for which results are presented here—was $10 \, m^3 . h^{-1}$. Tank volumes and mean retention times are presented in Table 1.

TREATMENT RESULTS

The landfill is progressing to a stable methane phase and therefore produces leachate with low amounts of biodegradable organics but high

TABLE 2. Mean Leachate Concentrations for Minden-Heisterholz and other German Landfills

	Minden-Heisterholz Aug. 1985 to Jan. 1987	*Leachates (Ehrig, 1986)*
pH	7·1	8·0
COD (mg . litre^{-1})	1 400	3 000
BOD$_5$ (mg . litre^{-1})	343	180
NH$_4$-N (mg . litre^{-1})	493	750
Cl (mg . litre^{-1})	1 130	2 100
AOX[a] (μg . litre^{-1})	742	2 000

[a] Only a few data points.

TABLE 3. BOD$_5$, COD and AOX Levels and Removal Efficiencies (Mean Values)

	BOD$_5$		COD		AOX	
	Values (mg . litre^{-1})	Removal (%)	Values (mg . litre^{-1})	Removal (%)	Values (µg . litre^{-1})	Removal (%)
Influent	343	—	1 400	—	742	—
Effluents						
Activated sludge	51	85	742	47	605	18
Powdered act. carbon	20	9	420	23	343	36
Coagulation	7	4	112	22	122	30
Lime neutralization	7	0	110	0	122	0
Total removal	—	98	—	92	—	84

levels of ammonia–nitrogen. Table 2 presents mean values in comparison to mean concentrations of leachates from other landfills in Germany.

Due to the location and the shape of the landfill and its former use as a clay pit this leachate is more diluted than is usually found in Germany. This fact has to be taken into account when looking at removal rates and effluent values in Table 3.

About 50% of total COD was biodegradable and to a high degree removed in the activated sludge plant. But still a certain amount of organics (analysed as BOD$_5$) entered the chemical/physical plant. Almost during the entire operation period the oxygen level in the aeration basin was near zero, indicating that the aeration capacity was not sufficient for the increasing leachate pollution. Due to the same reason a nitrification process could not be established.

The non-biodegradable COD was reduced by 45% due to adsorption and coagulation. No further reduction was observed in the lime neutralization step. Only a slight AOX removal was found in the activated sludge plant, mainly due to volatilization and probably to adsorption onto biomass. Main portions of AOX in sanitary landfill leachate are of non-volatile nature and can effectively be removed by both adsorption and coagulation processes (see Table 3).

A statistical approach to COD-concentrations of the influent and different effluents in Fig. 2 shows large fluctuations of influent concentrations. But in nearly all cases these peak levels could already be compensated in the activated sludge plant so that effluent concentrations

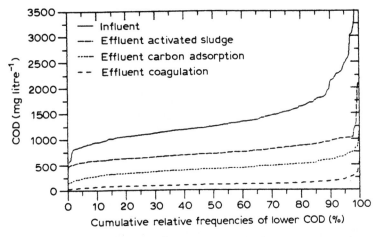

Figure 2. Cumulative relative frequencies of lower COD values for influent and effluents of the various stages.

were held at constant levels. The measured COD- and AOX-effluent concentrations are a result of carbon doses of 800–1200 g . m^{-3} leachate and chemical coagulant ($FeClSO_4$) doses of 600–1200 g Fe^{3+} . m^{-3} leachate. Costs were about US\$ 1·5–2·5 for activated carbon and US\$ 0·8–1·6 for coagulant per cubic meter of leachate, respectively.

SPECIFIC ASPECTS OF PHYSICAL/CHEMICAL TREATMENT

In the adsorption step a linear relationship was found between activated carbon addition and COD removal: with a dosage of 800 g . m^{-3} carbon about 250 mg . $litre^{-1}$ of COD were removed; the addition of 1200 g . m^{-3} resulted in a COD removal of nearly 400 mg . $litre^{-1}$. Figure 3 shows a constant capacity over the whole COD range. These results are somewhat surprising because theoretically a decrease in adsorptive capacity with lower COD equilibrium concentration had been expected.

Since this reaction could not be described by means of isotherms using the parameter COD, a super-position of competing and reversible adsorption processes might have occurred. Biochemical processes could also have caused interference because the reactor was stirred by means of air injection.

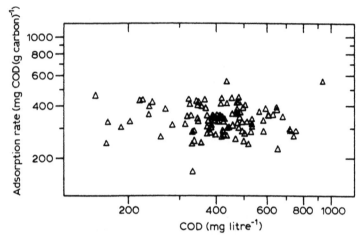

Figure 3. Adsorptive capacity versus COD equilibrium concentration in the reactor.

COD removal efficiency in the coagulation step was mainly influenced by pH. Optimum values for leachate with iron salts have been found at pH levels between 4 and 5 (Ehrig, 1983). Therefore the amount of iron required depends on the alkalinity of the leachate after biological treatment. This relation can be described by the following equation:

$$Fe^{3+} (g . m^{-3}) = 0.373 . CaCO_3 (mg . litre^{-1})$$

The data points derived from the Heisterholz plant can be described well by means of this equation. Iron salt addition was controlled by pH. The set value was between 4·8 and 5·0.

Of specific interest were interactions between the adsorption and the coagulation step. Figure 4 shows COD reductions by means of coagulation versus corresponding values for adsorption. These results indicate a higher removal rate by coagulation when, at the same time, COD concentrations after adsorption remain high. This fact is also valid for the case vice versa, so that about 500–600 mg . litre^{-1} COD were removed in total, independent of removal rates in the individual steps. It is therefore not possible to distinguish between adsorbable organic matter of the leachate and constituents which are only removable by means of coagulation.

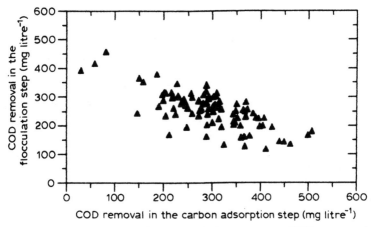

Figure 4. COD reduction by coagulation versus COD removal by adsorption.

PROCESS MODIFICATIONS AND FUTURE OPTIONS

The main improvements have been achieved by establishing a nitrification process in the activated sludge plant. To meet the effluent limits of 3 mg NH_4-N . litre^{-1} a fixed bed reactor with packed plastic material has been installed. The pre-denitrification step was introduced to improve total nitrogen removal as well as process stability. In addition a possible lack of alkalinity could be avoided. The higher alkalinity consumption due to nitrification in the biological plant will reduce the amount of iron salt needed for coagulation. It is therefore expected that chemical sludge volumes will decrease.

If a more extensive nitrogen removal using the denitrification process is required, this could be achieved in this plant in principle. But it has to be kept in mind that leachates from older landfills do not have sufficient biodegradable organic matter (BOD) for complete denitrification (Mennerich & Albers, 1988). One possibility to solve this problem is the addition of organic concentrates like methanol or acetic acid.

SUMMARY AND CONCLUSIONS

BOD_5, COD and AOX removal efficiencies and influencing factors were investigated during an 18-month test period at the biological and chemical/physical leachate treatment plant at Minden-Heisterholz. These results from the full-scale plant show that the main BOD_5-concentrations are removed in the biological plant. Further improvements in BOD_5 removal and full nitrification are expected when the new biological plant is completed.

Biological pre-treatment improves the efficiency of the following chemical/physical treatment step where carbon adsorption and coagulation with iron salts is practised. This combination removes non-biodegradable organic matter (COD) and halogenated hydrocarbons (AOX) to low levels which will meet the new German effluent standards for leachate treatment (see Chapter 3.15). The three individual treatment methods are able to react flexibly so that a high process stability as a whole is achieved.

ACKNOWLEDGEMENT

This work was carried out while Henning Albers was at the Technische Universität Braunschweig, Institut für Siedlungswasserwirtschaft, Braunschweig, Germany.

REFERENCES

Ehrig, H.-J. (1983). Großtechnische Verfahrensentwicklung zur chemisch/physikalischen Behandlung von Deponiesickerwässern (Development of a full-scale chemical/physical treatment for landfill leachates). Research Project Report, TU Braunschweig, Germany, unpublished.

Ehrig, H.-J. (1986). Was ist Deponiesickerwasser?—Mengen und Inhaltsstoffe (What is landfill leachate?—quantities and constituents). ATV-Dokumentation, Number 4, 19–36.

Mennerich, A. & Albers, H. (1988). Nitrification/denitrification of landfill leachates. *Water Supply*, **6**, 157–66.

3.8 Combination of UASB Pre-treatment and Reverse Osmosis

J. M. JANS, A. VAN DER SCHROEFF & A. JAAP

Euroconsult, PO Box 441, NL-6800 AK Arnhem, The Netherlands

INTRODUCTION

In the Netherlands at present all new landfills, or new sections of existing landfills, are equipped with an impermeable layer between waste and soil to prevent contamination of groundwater aquifers. Moreover provisions are made to collect surface run-off water. As a result of these measures all rainfall is collected and a controlled discharge of leachate as well as run-off water is obtained.

Under Dutch climatological conditions a quantitative collection of rainfall implies that the inherently available storage capacity of the waste only suffices for approximately two years. After that period discharge of leachate is required.

In general this implies that initially only highly polluted leachate is available for discharge as the self-purifying mechanism of the waste-body has not yet developed in that period. An operational two-stage treatment plant is described in this chapter that enables purification of highly polluted leachate up to the standards required for discharge to surface water (Table 1).

The treatment systems consist of:

—pre-treatment of highly polluted leachate by a high-rate anaerobic treatment system;

313

TABLE 1. Effluent Quality Standards in The
Netherlands for Discharge into Surface Water

Parameter	Concentration (mg/litre)
COD	75–150
BOD	10–20
TKN	10–20
NH_4-N	5–15
Heavy metals	0·01

—post-treatment of the effluent from the anaerobic reactor, in combination with stabilized leachate originating from 'older' landfill sections, by reverse osmosis.

The flow diagram of the treatment plant is given in Fig. 1. At present the concentrate from the reverse osmosis unit is infiltrated into the landfill, as alternative disposal techniques for this stream are still in development. The operational capacity of this plant is $2 \, m^3/h$.

PRE-TREATMENT BY UASB REACTOR

At the beginning of the 1980s the Dutch Ministry of Housing, Physical Planning and Environment promoted the investigation of anaerobic treatment of leachate on a pilot scale.

Based on two years of successful research a full-scale treatment plant has been designed and constructed at the Bavel Landfill Site. The operational data of the $60 \, m^3$ upflow anaerobic sludge bed (UASB) reactor are summarized in Table 2.

After a two-month start-up period a COD loading of 800 kg/day has already been reached and results are shown in Table 3.

The anaerobic sludge which develops in the reactor has excellent characteristics with respect to metabolic activity as well as settling properties. It must be stated, however, that according to calcium, manganese and iron concentrations, special provisions for stable operation are required. Otherwise scaling in piping and heat exchangers, and precipitation in the reactor will hamper stable operation.

A restart of the reactor, after a shut-down period required for technical modifications, provided performance data as shown in Figs 2 and 3.

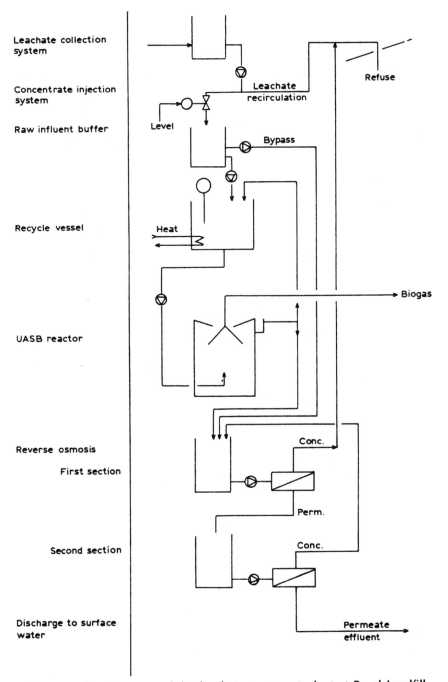

Leachate collection system

Concentrate injection system

Raw influent buffer

Recycle vessel

UASB reactor

Reverse osmosis

First section

Second section

Discharge to surface water

Refuse

Leachate recirculation

Level

Bypass

Heat

Biogas

Conc.

Perm.

Conc.

Permeate effluent

Figure 1. Flow diagrams of the leachate treatment plant at Bavel Landfill Site, The Netherlands.

TABLE 2. Operational Data from the UASB Reactor at the Bavel Landfill Site

Parameter	Values
COD loading (kg/day)	900–1 200
VFA loading (eq/day)	10 000–12 000
COD removal	
total (%)	80–85
filtered (%)	80–90
VFA removal (%)	>90
Biogas production (m³/day)	360–480
CO_2 content of biogas (%)	25–30
Temperature (°C)	33--35
Hydraulic retention time (h^{-1})	8–12

It appeared that loadings of 25 kg COD/m^3 . day (1500 kg COD/day) could be well accommodated and resulted in only a slight decrease of the COD removal efficiencies.

Under these extreme loadings the reactor temperature appeared to be a critical process parameter. In order to maintain reduction rates of approximately 80% at 25 kg $COD/(m^3$. day), a temperature of at least 30°C is required. The biogas yield of the anaerobic reactor covers all energy requirements of the treatment plant. The remaining 50% is added to the landfill gas extracted from the Bavel landfill (1500 m^3/h).

The actual operational conditions of the UASB reactor guarantee an adequate effluent quality for further treatment by the reverse osmosis unit.

TABLE 3. Results from the UASB Treatment of Leachate at the Bavel Landfill Site

Parameter	Influent	Effluent
COD total (mg/litre)	25 000–35 000	3 000–5 000
filtered (mg/litre)	24 500–34 000	2 500–5 000
HCO_3 (meq/litre)	30–70	100–150
pH	6·6–6·8	7·4–7·8
Total N (mg/litre)	1600	1550
TSS (mg/litre)	0–50	150–200

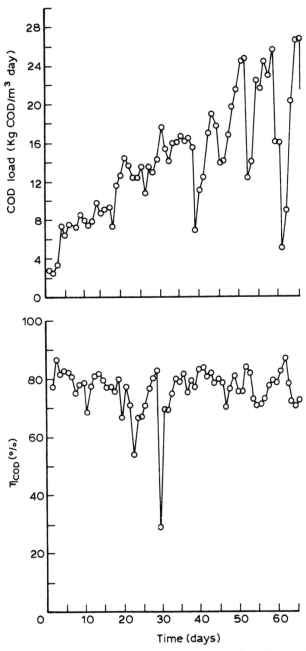

Figure 2. COD loads (a) and removal efficiencies (b) observed during 65 days of operation of the UASB reactor at the Bavel Landfill Site.

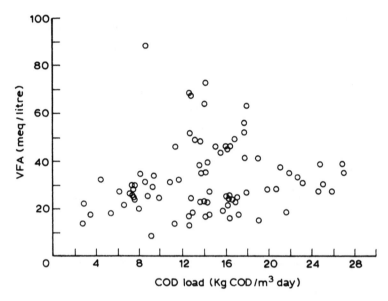

Figure 3. Volatile fatty acids (VFA) in the effuent of the UASB reactor according to COD Load.

POST-TREATMENT BY REVERSE OSMOSIS

Figure 4 shows a simple flow diagram of the reverse osmosis stage. The unit is built up of two sections, each section containing several steps. The first section is specially designed to cope with solid particles. The membranes used are tubular with an internal diameter of approximately 1 cm. By using a high linear velocity of the leachate in the system, fouling can be minimized. Each step can also be cleaned mechanically by circulating sponge balls through the tubes. The second section contains spiral wound composite membranes that have a high retention capability for dissolved molecules.

Effluent quality can be controlled by changing the ratios of concentrate flow and permeate flow, the choice of types of membrane, the process conditions such as temperature and pressure, and configuration of the steps in each section.

The configuration of the two-step treatment plant described enables treatment of stabilized leachate (directly) or acidified leachate (via UASB

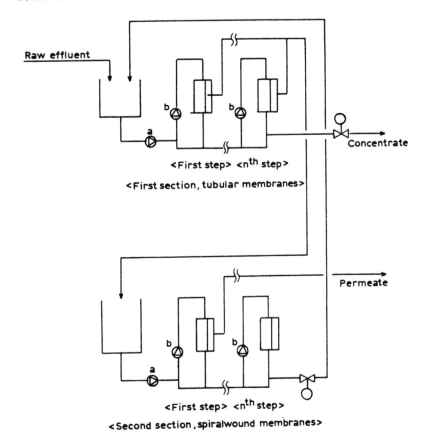

Figure 4. Flow sheet of the reverse osmosis unit at the leachate treatment plant in Bavel. a, High pressure pump; b, circulation pump.

pre-treatment). Some results of the treatment of acidified leachate are given in Table 4.

In Table 5 performance data are given on leachate originating from older landfill sections. In these landfill sections anaerobic stabilization of the leachate has already taken place.

Table 5 illustrates that reverse osmosis also forms an effective barrier for heavy metals present in leachate. The capacity of the reverse osmosis section is influenced by the temperature and Ca-concentration of the influent. The flux through the membranes increases with the temperature of the influent. In this respect the temperature of the effluent of the anaerobic reactor (30–33°C) forms a positive factor.

TABLE 4. Quality Data from the Operation of Two-Stage Leachate Treatment Plant at Bavel Landfill Site

Parameters	Raw leachate	UASB reactor effluent	Reverse osmosis effluent
COD (mg/litre)	25 000–35 000	3 000–5 000	5–8
TKN (mg/litre)	1 600	1 550	10–16
VFA (meq/litre)	180–240	5–25	n.d.
pH	6·6–6·8	7·4–7·8	

n.d., not detectable

On average 75% of the influent flow (2 m³/h) is composed as shown in Table 3, and is discharged to surface water. The pollutants 'removed' by reverse osmosis and present in the concentrate (0·5 m³/h), are infiltrated into the landfill.

CONCLUSIONS

Leachate from freshly tipped waste is heavily polluted, especially with volatile fatty acids (acid leachate); the pH is low. It is advantageous to postpone treatment and discharge of this leachate until an anaerobic methanogenic transition has taken place in the refuse. This will lower dramatically the amount of dissolved organic compounds. The pH will

TABLE 5. Performance Data of Reverse Osmosis Treatment of Old Stabilized Leachate Produced at Bavel Landfill Site

Parameters	Influent (mg/litre)	Reduction (%)
COD	1 300	99·4
BOD	110	>99·9
TKN	350	99·0
Cl⁻	1 500	99·4
P_{tot}	10	>99·9
Fe	8 800	99·9
Zn	0·65	98·5
Cu	0·15	98·7
Pb	0·12	99·9
Ni	0·11	99·9

increase causing a precipitation process of (heavy) metals in the waste. If it is not feasible to wait several years, the acid leachate can be pre-treated effectively with a UASB reactor. In all cases the effluent of the 'anaerobic treatment', either from a UASB reactor or originating from the tip site itself, cannot be discharged to surface water without post-treatment.

Treatment of anerobically stabilized leachate using reverse osmosis enables discharge of the effluent to surface water.

For the anaerobic pre-treatment step, high loading rates up to 25 kg COD/(m^3 . day) are applicable. Attention must be paid to operational temperature and Ca-content of the influent.

These two parameters are also of importance for the reverse osmosis unit. The temperature determines the fluxes applicable for a given plant while the Ca-level influences the required frequency for chemical cleansing (and thus operational costs).

3.9 Combination of Activated Sludge Pre-treatment and Reverse Osmosis

BURKHARD WEBER & FELIX HOLZ

Haase Energietechnik GmbH, Gadelander Strasse 172, D-2350 Neumünster, Germany

INTRODUCTION

Reverse osmosis is a process which gives good results in removal of trace compounds, recalcitrant organics and inorganics from leachate. Normally the process is applied to stabilized or diluted leachates.

Stabilized leachates are produced after a biological degradation which may occur in the landfill or in on-site treatment units. Diluted leachate is often produced in old landfill sites when surface water comes into contact with the waste.

As the performance of the modules and membranes controls the permeate quality in the reverse osmosis (RO) process, it is important to evaluate the necessity of an advanced biological treatment to remove biodegradable compounds still present in leachate in order to enhance the efficiency of the reverse osmosis process.

In this chapter results of several tests with reverse osmosis applied to stabilized and diluted leachates, with and without aerobic pre-treatment, are considered.

DESCRIPTION OF THE LEACHATE TREATMENT PLANT

Test runs were carried out using a semitechnical RO-plant for different kinds of leachate and for the effluent of different treatment stages of a

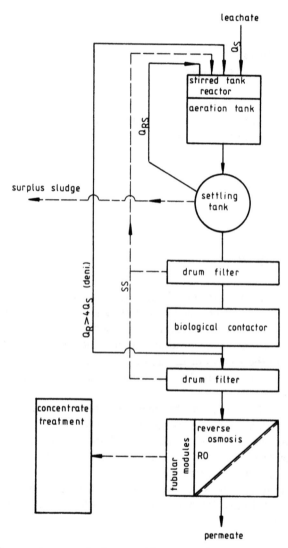

Figure 1. Flow scheme of the multi-stage pilot plant for leachate treatment.

semitechnical and a full-scale biological purification plant. Figure 1 shows the flow scheme of the multi-stage pilot plant.

The RO-plant was equipped with tubular modules. Cellulose acetate (CA) membranes with a salt rejection of 95 per cent and a thin film composite (TFC) membrane with a salt rejection of 99 per cent were tested.

Equations (6)–(8) reported in Chapter 3.6 permit interpretation of the test run results with the RO-pilot plant. Only by performing flux measurements at various concentrations and fitting a linear regression according to eqn (8) in Chapter 3.6 is it possible to obtain the constant b of the leachate to be treated. Once this has been carried out, it is possible to optimize the full-scale plant by doing computer simulations.

To simplify the performance of the influence of biological pre-treatment on leachate quality the parameter C in van't Hoff's law was replaced by the global parameter 'conductivity' C_{LF}

$$\frac{\Phi_P}{\Phi_o} = 1 - \frac{b}{\Delta P} \cdot C_{LF} \qquad (1)$$

TREATMENT OF RAW LEACHATE BY REVERSE OSMOSIS

The treatment of raw diluted leachate and of raw leachate from the methanogenic phase yielded no satisfying results (see Table 1). Minimum requirements, especially for ammonia, could not be met using either a cellulose acetate (CA) membrane or a thin film composite (TFC) membrane. The pH-value of raw leachate is 7·5–8. The ammonia rejection of the membranes used depends on the pH-value. To reach a suitable pH of 6·5, it was necessary to dose with sulphuric acid. Also precipitation inhibitors had to be added to prevent scaling effects caused by gypsum.

Raw Diluted Leachate

The treatment of permeate from the first RO-stage (tubular membranes) by a second RO-stage (spiral-wound membranes) was necessary. It is possible to use a CA-membrane in the first stage with the advantage of a

TABLE 1. Results of Leachate Treatment by Reverse Osmosis (see text for symbol definition)

Quality	Diluted leachate				Methanogenic leachate					
Quantity	$7 \cdot 5\ \text{m}^3/(\text{ha . day}) \times 20\ \text{ha}$ = 150 m³/day				$5\ \text{m}^3/(\text{ha . day}) \times 20\ \text{ha}$ = 100 m³/day					
Pre-treatment	Raw		Nit/den $B_{TS}=0 \cdot 2$ $B_{A,N}=5$		Raw		ASP $B_{TS}>0 \cdot 2$		Nit/den $B_{TS}=0 \cdot 2$ $B_{A,N}=5$	
Effluent from pre-treatment										
COD (mg/litre)	1 500		1 000		5 000		2 000		1 500	
TKN (mg/litre)	600		50		2 000		1 700		100	
NH$_4$–N (mg/litre)	500		<10		1 800		1 600		<10	
NO$_x$–N (mg/litre)	0		250		0		0		400	
AOX (μg/litre)	1 200		1 000		4 000		2 500		2 000	
RO 1st stage[a]	CA	TFC	CA	TFC	CA	TFC	CA	TFC	CA	TFC
Effluent										
COD (mg/litre)	90	50	50	30	300	175	100	60	75	40
TKN (mg/litre)	140	80	10	5	400	260	340	235	20	10
NH$_4$–N (mg/litre)	125	70	<2	<1	375	250	325	225	<2	<1
NO$_x$–N (mg/litre)	0	0	25	15	0	0	0	0	40	20
AOX (μg/litre)	160	120	150	100	600	400	400	250	300	200
Flux for $C_F = 5$ (1/m² . h)	19	17	32	29	15	13	28	25	31	27
Membrane area (m²)	300	330	175	190	250	280	135	150	120	135
Maximum C_{FV}	5	5	7·5	7·5	4	4	6	6	7	7
RO 2nd stage[b]	TFC	TFC	TFC	TFC	TFC	TFC	TFC	TFC	TFC	TFC
Effluent										
COD (mg/litre)	<15	<15	<15	<15	<15	<15	<15	<15	<15	<15
TKN (mg/litre)	15	8	<2	<1	45	30	35	25	<2	<1
NH$_4$–N (mg/litre)	10	5	<1	<1	40	25	30	20	<1	<1
NO$_x$–N (mg/litre)	0	0	2	1	0	0	0	0	3	2
AOX (μg/litre)	<50	<50	<50	<50	100	75	75	<50	<50	<50
Recommended treatment processes	ASP + nitri/deni + 1st stage TFC-RO or 1st CA–RO + 2nd TFC-RO				ASP + nitri/deni + 1st stage TFC-RO					
Minimum requirements										
COD					<200 mg/litre					
TKN/NO$_x$–N					no values to date					
NH$_4$–N					<50 mg/litre					
AOX					<500 μg/litre					

[a] 1st stage tubular membranes.
[b] Spiral-wound membranes.

higher flux rate compared to the TFC-membrane. Disadvantages are:

—no clearance to meet $C_{e,NH4-N} = 10$ mg/litre
—hydrolysis of the membrane material

Using CA-membranes the operator has to concentrate particularly on possible hydrolysis of the material. Neglecting a significant drop in permeate quality leads to irreversible damage of the second stage spiral-wound membranes.

Raw Leachate from the Methanogenic Phase

It might prove possible to meet the minimum requirements by treating raw leachate from the methanogenic phase by reverse osmosis. In this case, it would be necessary to use TFC-membranes in the first and second stages. The maximum concentration factor in the first stage is $C_F = 3·5$ and in the second stage $C_F = 5$. Here the concentration factor is not limited by the RO-process but by the requirements for the quality of the effluent. Briefly, the treatment of raw leachate from the methanogenic phase cannot be recommended.

Cleansing Procedures

Chemical cleansing *in situ* was carried out every 75 h of operation using the detergent Ultrasil and every 200 h of operation using citric acid.

TREATMENT BY REVERSE OSMOSIS OF EFFLUENT FROM AN ACTIVATED SLUDGE PROCESS

The biological treatment of raw leachate by the activated sludge process (ASP) and a sludge loading of $B_{TS,BOD} > 0·25$ kg/(kg . day) diminishes only the TOC (total organic carbon) and the AOX (adsorbable organic halogens). The reduction of the ammonia by incorporation in the sludge reaches a total of up to 10 per cent. Ammonia is the bottle-neck for rejection by reverse osmosis but does not interfere with TOC, COD, BOD or AOX; no particular benefits are obtained by applying a relatively highly loaded activated sludge process. Subordinate advantages are:

—buffer for COD/BOD-peaks
—removal of compounds that cause scaling and fouling
—enhancement of flux rate and concentration factor

ADVANCED BIOLOGICAL PRE-TREATMENT OF LEACHATE AND FINAL PURIFICATION BY RO

Advanced biological treatment of raw leachate including nitrification and denitrification can be carried out by the following processes and treatment stages:

1. Activated sludge process and biological contactor (ASP/BC-system)

(a)	stirred tank reactor	denitrification, TOC-removal
(b)	aeration tank sludge loading of (a) plus (b): $B_{TS,BOD} = 0{\cdot}2$ kg/(kg . day)	TOC-removal
(c)	settling tank surface overflow rate $q_A = 0{\cdot}6$ m/h	
(d)	drum filter surface overflow rate $q_A = 5.0$ m/h	separation of solids
(e)	rotating biological contactor surface loading rate $B_{A,N} =$ 5 gN/(m^2 . day)	nitrification
(f)	recirculation pump to (a) with $Q_R = 4 . Q$	
(g)	drum filter surface overflow rate $q_A = 5{\cdot}0$ m/h	separation of solids

2. Activated sludge process (ASP-system)

(a)	stirred tank reactor connected in plug flow with	denitrification, TOC-removal
(b)	aeration tank nitrogen sludge loading of (b):	TOC-removal, nitrification

$B_{TS,N} = 30$ gN/(kg . day)

volume of (a) is 25 per cent of (b)

(c)	recirculation pump to (a) $Q_R = 5 . Q$
(d)	settling tank
(e)	drum filter separation of solids

The task of the fabric-drum filters is to enhance the nitrification process and protect the reverse osmosis.

Advantages of system 1 (ASP/BC-system), which is part of the semitechnical treatment plant used, are:

—fixed film of nitrifying bacteria
—enhanced N-loading rate due to optimum BOD/N-ratios of less than 0·1

An additional disadvantage of system 2 (ASP-system) is the possible stripping of organic halogen compounds caused by long-term aeration. A full-scale system 2 treatment plant is operating at the Venneberg landfill site near Lingen, Germany.

Removal efficiencies and effluent values of advanced biological pre-treatment are listed in Table 1.

The effluent of the two advanced biological purification plants was treated by reverse osmosis. From one point of view the minimum requirements were met by using a one-stage RO-plant. From another point of view the operation of the RO-plant was more economical compared to the treatment of raw leachate, as the flux rate or the possible concentration factor was enhanced (see Table 1). The nitrification process diminishes buffer capacity and pH-value of the pre-treated leachate. Therefore addition of sulphuric acid to obtain pH of 6·5 is not necessary or can be achieved by using lesser quantities of acid.

Cleansing procedures using Ultrasil were carried out every 150 h of operation and every 500 h using citric acid. The greater cleaning interval is a hint of lower fouling and scaling potential of the aerobically-treated leachate compared to raw leachate.

RESULTS AND SPECIAL ASPECTS OF MULTI-STAGE TREATMENT

In addition to removal of biodegradable compounds and improvement of the permeate quality the advantages of primary biological purification can be specified as follows:

—only recalcitrant pollutants remain in the leachate;
—enhancement of the flux rate and/or enhancement of the possible concentration factor;
—diminished treatment costs;
—prevention of biofouling caused by high concentrations of organics;
—removal of ferrous and calcic compounds by aeration and precipitation to prevent scaling of the membrane surface;
—nitrogen removal to prevent problems caused by ammonia during the evaporation of the concentrate.

To summarize, biological pre-treatment has a considerable influence on the reverse osmosis feed. Therefore a change also of factor b in van't Hoff's equation was expected for the respective leachates of a landfill site. Contrary to this expectation a significant change of factor b was not observed throughout the investigations although it differed from landfill site to landfill site.

Figure 2 shows the flux decline Φ/Φ_0 in dependence on the conductivity C_{LF} for raw and biologically pre-treated leachate of a

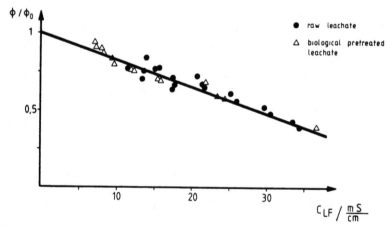

Figure 2. Flux decline as a function of feed conductivity.

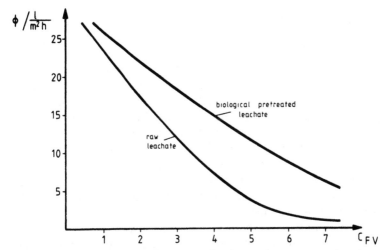

Figure 3. Permeate flux as a function of volumetric concentration factor.

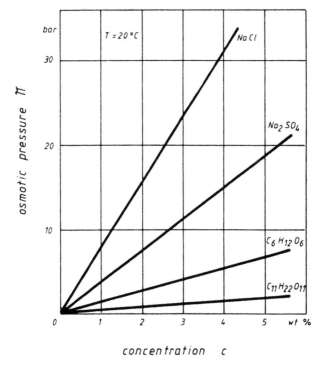

concentration c

Figure 4. Osmotic pressure of some organic and inorganic solutions.

certain landfill site. This graph was achieved by taking flux measurements at various concentrations and applying a linear regression according to eqn (1).

In spite of modification of parameter b, biological pre-treatment diminished the conductivity of each leachate tested. This offers the possibility of evaluating the effect of primary purification on the RO-process. Figure 3 shows the permeate flux versus the volumetric concentration factor C_{FV}. The more biological pre-treatment reduces conductivity, the higher the possible concentration factor.

The above mentioned phenomena could be explained by the fact that the osmotic pressure π and likewise the proportional factor b of inorganic salts are higher than π of organics or ammonia as can be seen from Fig. 4. Volatile fatty acids, which contribute to conductivity, are removed in the biological process. So the coincidence of adsorption, absorption and biodegradation diminishes the conductivity C_{LF} and the initial osmotic pressure π_o; but it does not change the proportional factor b. This effect decreases with decreasing leachate concentrations.

Due to the fact that biological pre-treatment of raw leachate enhances the effectiveness of reverse osmosis, the membrane process should only be operated as the final step of a multi-stage treatment plant.

3.10 Combination of Wet Oxidation and Activated Sludge Treatment

FRANCO AVEZZÙ

Department of Environmental Sciences, University of Venice, Calle Larga Santa Marta, Venice, Italy

GIORGIO BISSOLOTTI

SIAD SpA, Via San Bernardino 92, I-24100 Bergamo, Italy

&

CARLO COLLIVIGNARELLI

Department of Civil Engineering, University of Brescia, I-25060 Brescia, Italy

INTRODUCTION

Following the biodegradation processes which may occur in the landfill or in a biological treatment plant the leachate still contains organic substances which are resistant to biodegradation. These substances are mainly compounds such as humic and fulvic acids with relatively high molecular weight (see Chapter 2.4).

Experiences on industrial concentrated wastewaters (Collivignarelli & Bissolotti, 1988) showed that wet oxidation can considerably increase the biodegradation of recalcitrant organics by a cracking effect due to the drastic operational conditions adopted in this process.

In this chapter the results of wet oxidation tests applied to young and old leachates followed by a pure oxygen activated sludge treatment are reported. These results prove that wet oxidation increases the biodegradability of landfill leachate.

THE WET OXIDATION PROCESS

Wet oxidation is a chemical engineering 'unit process' that has found application in the sanitary engineering area. Historically, there are records of the application of this process as far back as the beginning of the century in Sweden. Only in the 1960s, however, and mainly in the USA, was wet oxidation introduced as a conditioning process for sludges on a wider scale. Even more recent is its application as a treatment process for very high strength industrial wastewaters (Cannel & Schaeffer, 1983*a*; Berbenni *et al.*, 1986; Agamennone & Pieroni, 1987). The more general process scheme is shown in Fig. 1 (Baldi *et al.*, 1985). The waste to be treated is collected in a storage basin, from which it is pumped at high pressure (1), mixed with air (2) and sent to a heat exchanger (pre-heater) (3). The heated mixture is then conveyed to the reactor (4) where oxidation occurs. The temperature in the reactor rises due to the exothermic reactions taking place, and therefore, the oxidized effluent is warmer than the influent.

A second heat exchanger (5) reduces the temperature of the waste stream, with secondary vapour production, which is utilized in a closed circuit (8). The vapour generated is used in a turbine (6), which activates the air compressors (7). The residual heat from the waste stream is utilized in the first heat exchanger (3) before discharge.

According to the type of facility and the quantity of heat produced during the oxidation phase, the typical layout shown in Fig. 1 may undergo several possible variations. In practical applications, circuit (8) is often omitted as the heat produced in the reactor is invariably barely sufficient to provide the energy required by the air compressors.

As can be observed, the process is based on the fact that the reactivity of oxygen with the wastewater (and therefore its oxidizing capacity) increases with temperature.

On the other hand, to improve oxygen diffusion and contact of the latter with the matter to be oxidized, it is essential that the process be carried out during the liquid phase. It follows that, to achieve simultaneously both high temperature and liquid phase conditions, the process must be performed at a high pressure. The temperature range in which the process is usually maintained is 150–300°C, the pressure range varies from 30 to 250 bar. The lower temperature limit is due to kinetic reasons (oxidation should be fast enough to keep reactor volumes from becoming excessively large), the higher temperature is related to the fact that it is

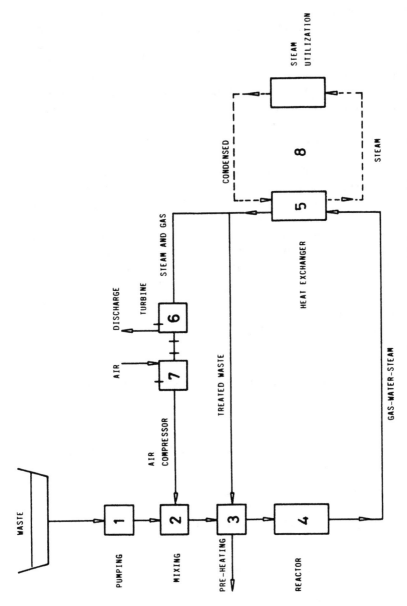

Figure 1. General flow-sheet of wet oxidation process with heat recovery and energy production.

not possible to have water in liquid phase at a temperature greater than 374°C (critical temperature). The kinetics of wet oxidation reactions may be conveniently accelerated through the use of a catalyst, the most common of which is represented by ions of heavy or transition metals associated with peroxides as radical generators (Baldi *et al.*, 1985). As far as the oxidizing medium is concerned, the reactor is in most cases fed with air, although the presence of nitrogen, which is inert for reaction purposes, implies an unavoidable oversizing of both compressors and the reactor itself. To overcome this problem the use of liquid oxygen has been proposed. The process scheme thus adopted eliminates the air compressors, and utilizes instead high pressure cryogenic pumps and an evaporator, which gasifies the oxygen to be fed into the reactor. The industrial plant used in our experiment is based on this principle.

Pure oxygen was used both for plug-flow (as in this case) and completely-mixed reactors, while air-based systems are traditionally used only in the latter configuration. Even in the absence of a wide experimental basis on plug-flow reactors, there are valid reasons for assuming that the latter should be able to achieve high removal efficiency, even at low retention times, while maintaining a higher degree of safety against the risk of explosive reactions caused by a sudden increase of temperature within the reactor during the oxidation process (Collivignarelli & Bissolotti, 1988).

WET OXIDATION TREATMENT OF LANDFILL LEACHATE

Experimental Plant

The study was carried out in a semi-industrial-scale pilot plant. The plant was made up of:

—a volumetric pump, with capacity between 150 and 600 litres/h for the introduction of the liquid waste into the reactor;
—an oxygen supply group, comprising a cryogenic pump for liquid oxygen compression and a series of tanks for oxygen storage prior to injection into the reactor;
—a liquid salt circuit with the function of reactor warm-up (to the desired temperature);
—a heater/burner group for heat transfer to the liquid salts;
—the reactor itself;

—a control display on which information regarding behaviour of the plant (pressure and temperature within the reactor, temperature in the heating and cooling circuits, oxygen pressure and flow) is shown.

The reactor is of tubular shape and built in Hastelloy C 276. It is 350-m long, with an 80-litre volume and an average diameter of 25 mm. In the first and last portion of the reactor tube, two heat exchange groups were installed. In the first the liquid salt, at a temperature of 300–400°C, heats the waste to the desired reaction temperature; in the latter first vapour and then water cool the waste to a temperature of about 30°C. Thermocouples monitor the waste temperatures in both cases.

The system works as follows: the leachate is taken from the storage tank and pumped through a pipe where it is mixed with the gaseous oxygen, injected through a diffuser. The waste/oxygen mixture is then relayed to the reactor, in the first portion of which it is heated by the liquid salt which circulates in the intermediate chamber. The liquid salt heater can be automatically disconnected if the oxidation heat generated in this first phase of the reaction is sufficient for maintenance of the desired temperature within the reactor.

The leachate then reaches an unheated portion of the reactor, which is, however, thermally insulated. The major part of the reaction occurs here.

In the latter portion of the reactor the waste is cooled both to avoid vaporization of the effluent stream and to recover the heat previously transferred by the salt or generated by the oxidation reaction. There are two cooling circuits: one for vapour production and one for hot water production. The cooled waste is then driven through a gas/liquid separator on which two pneumatic pin-type valves are inserted. The latter regulate the discharge from both liquid and gaseous phases and allow pressure control inside the reactor. The waste, which is now at atmospheric pressure, is then sent to an additional gas/liquid separator prior to discharge. A condensing separator is placed at the gaseous phase discharge point.

The plant flow-sheet is illustrated in Fig. 2.

Prior to the beginning of the experiments some hydraulic tests were conducted on the reactor. The R.T.D. curve has been reproduced through injection into the system of a saline tracer, whose concentration at discharge was continuously measured by a conductivity meter. Given the physical configuration of the reactor (long tube, very small diameter), it was reasonable to assume its hydraulic behaviour was of the plug-flow

338

Avezzù, Bissolotti, Collivignarelli

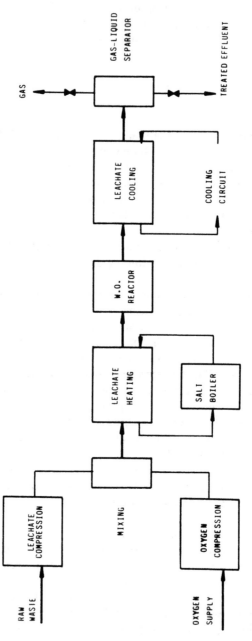

Figure 2. Flow-sheet of the wet-oxidation experimental plant.

type, with axial dispersion. This hypothesis was, in fact, validated by best-fitting of the experimental data with a disperse plug-flow reactor curve with Péclet number = 1000, where the Péclet number indicates the dispersion within the reactor. This indicates almost negligible axial dispersion, and thus a reactor behaviour which is very close to that of a perfect plug-flow reactor (Collivignarelli & Bissolotti, 1988).

Characteristics of leachates

Four different leachates from four different municipal sanitary landfills were treated using the wet oxidation process in the plant described. The leachate characteristics are described herein and analytical values are reported in Table 1.

Leachate A. This originates from a large landfill (Brescia, Lombardia) and may be considered as a 'young' leachate.

Its qualitative characterization includes a high organic matter content (COD = 19 400 mg/litre) which is largely biodegradable (BOD_{20} = 13 500 mg/litre), and with a BOD_{20}/COD ratio which is approximately 0·69, thus indicating an early degradation stage of the waste.

Characteristics of this waste are also, as usual, high concentrations of NH_4^+, Cl^-, and, among metals, Fe.

Leachate B. This originates from an urban waste landfill in Colle Umberto (TV) in the Veneto Region.

Its characteristics are not very different from those of leachate A (see Table 1). Also in this case, in fact, the originating landfill is 'young' (1–2 years old) with high BOD and COD contents and a BOD_{20}/COD ratio of 0·75. This leachate too shows high concentrations of NH_4^+ and Cl^-, while metal concentrations are definitely lower than for leachate A.

Leachate C. This comes from the sanitary landfill (urban wastes) in Fontanafredda (Pordenone) in the Friuli Region.

The waste was buried 5–6 years prior to collection of this leachate. Its organic matter content is in fact considerably lower than that of the previous cases (see Table 1), as is the BOD_{20}/COD ratio (0·35). This can also be confirmed by the high total concentration of volatile fatty acids, which is approximately one order of magnitude lower than that observed in leachate B. The concentrations of NH_4^+ and Cl^- are also definitely lower and the metal content (including Fe) becomes almost negligible.

TABLE 1. Analytical Composition of the Four Leachates Used for the Wet Oxidation Treatment Test (all values in mg/litre except pH)

Parameters	Leachates			
	A	B	C	D
pH	7·1	7·3	7·3	7·8
Alkalinity (as $CaCO_3$)	n.a.	12 100	4 500	2 600
Cl^-	1 830	2 236	710	2 201
SO_4^{2-}	450	18	4	8
Total P	28	4·9	3·9	8·6
TKN	n.a.	2 635	1 250	1 750
NH_4^+-N	2 045	1 692	641	998
NO_2^--N	n.a.	0·00	0·00	0·06
NO_3^--N	n.a.	0·32	0·25	0·3
Na	n.a.	1 200	425	1 170
K	n.a.	458	433	480
Ca	n.a.	488	96	64
Mg	n.a.	797	199·3	315·9
Fe	210	82	19·5	10
Cu	n.a.	0·42	0·40	0·75
Ni	4·5	0·66	0·18	0·35
Cr	n.a.	0·72	0·32	0·70
Cd	n.a.	0·04	0·2	0·09
Pb	5·0	0·60	0·07	0·08
Mn	n.a.	4·20	0·25	0·20
Zn	10	1·80	n.a.	n.a.
SS	15 000[a]	583	120	280
VSS	6 000[a]	146	22	26
COD	19 400	17 350	3 100	4 140
BOD_5	7 236	8 200	780	1 900
BOD_{20}	13 497	12 980	1 080	2 880
VFA	n.a.	4 260	537	1 319
Acetic acid	n.a.	853	85	231
Propionic	n.a.	1 158	135	546
Iso-butyric	n.a.	535	106	30
n-Butyric	n.a.	549	60	41
Iso-valeric	n.a.	479	87	176
n-Valeric	n.a.	686	64	295
Phenols	4	n.a.	n.a.	n.a.
Oil	140	n.a.	n.a.	n.a.

[a] Total solids (dissolved + suspended).

Leachate D. This originates from landfill in Fossalta di Portogruaro (Venice). Although the landfill itself is about 10–11 years old, it has been 'contaminated' by recent waste deposits, as can be seen by the relatively high concentrations of organic matter in the leachate. Furthermore, the BOD_{20}/COD ratio is not in the typical range that can be expected from an old landfill (0·70), and the consequences of contamination are also apparent on examination of the concentration of volatile fatty acids, which are present in much higher concentrations than in leachate C. The content of NH_4^+ and Cl^- too is fairly high while the metal content is definitely low, as shown in Table 1. A common characte 3tic of leachates B, C and D is the total phosphorus content, which is extremely low in all three cases, with BOD_5/P ratios of 1630, 200 and 220, respectively.

Experimental Procedure and Results

The experimental results of the wet oxidation tests are reported in Table 2, and described briefly below.

Leachate A. Three wet oxidation tests were performed at temperatures of 275°C, 295°C and 305°C, respectively. All three tests were carried out at a constant pressure of 10 000 kPa, a steady waste flow of 150 litres/h and an oxygen supply set at 20 kg/h.

A fourth experiment was carried out at 275°C (and flow of 150 litres/h), to verify the effect of thermal treatment alone.

The removal efficiencies for organic substances are in the range of 50%. Thermal treatment alone, however, provided negligible treatment (12–14%) for all parameters controlled.

Leachate B. Two tests were performed at temperatures of 250°C and 230°C, pressure of 9500 and 9000 kPa and waste flow of 150 and 200 litres/h, respectively. Oxygen supply was constant, and equal to 20 kg/h for both cases.

Efficiencies were generally low, with a consistent reduction in the case of the lower temperature effluent. The NH_4^+ content was also monitored during the first test: concentration increases from 1692 to 2010 mg/litre, showing a partial transformation of organic nitrogen into ammonia.

Leachate C. This leachate underwent a wet oxidation test at a temperature of 250°C, pressure of about 9000 kPa, flow of 200 litres/h and oxygen supply of 20 kg/h.

TABLE 2. Wet Oxidation Treatment of the Four Leachates: Results

	T (°C)	COD (mg/litre)	Removal (%)	BOD$_5$ (mg/litre)	Removal (%)
Leachate A					
Raw	—	19 400	—	7 236	—
No. 1 effluent	275	9 702	50	4 536	37
No. 2 effluent	295	6 664	66	3 240	55
No. 3 effluent	305	7 310	62	3 660	49
Effluent from thermal treatment (without O$_2$)	275	17 052	12	6 372	12
Leachate B					
Raw	—	17 350	—	8 200	—
No. 1 effluent	250	12 900	26	7 665	7
No. 2 effluent	230	14 850	14	—	—
Leachate C					
Raw	—	3 100	—	780	—
Effluent	250	2 269	27	1 665	—
Leachate D					
Raw	—	4 140	—	1 900	—
Effluent	270	685	83	257	86

It was observed that, while a moderate reduction in COD content was achieved, there was also an increase in BOD$_5$ concentration. This is an indication of the transformation of biorefractory organic matter into biodegradable organics. Also in this case, an increase in the ammonia content after wet oxidation treatment was noted: ammonia concentration rose from 641 to 1004 mg/litre.

Leachate D. Wet oxidation was performed on this leachate at a temperature of 270°C, a pressure of 9500 kPa, flow equal to 200 litres/h and oxygen supply at 20 kg/h.

Efficiencies were 83% for COD and 86% for BOD$_5$ removal. The effluent NH_4^+ concentration was 1610 mg/litre.

Biological Treatability of Leachates after the Wet Oxidation Process: Laboratory Investigation

The main aim of this experimental study was to investigate the changes in the biological treatability characteristics of a leachate due to preventive

TABLE 3. Long-Term BOD and Biodegradability Indices for the Four Leachates

	T (°C)	BOD_{20} (mg/litre)	$BOD_{20}/$ COD (%)	$COD -$ BOD_{20} (mg/litre)	$BOD_5/$ BOD_{20} (%)
Leachate A					
Raw	—	13 497	69	5 903	54
No. 1 effluent	275	8 580	88	1 122	53
No. 2 effluent	295	6 070	91	594	53
No. 3 effluent	305	6 910	95	400	53
Effluent from thermal treatment (without O_2)	275	10 750	63	6 302	59
Leachate B					
Raw	—	12 980	75	4 370	63
No. 1 effluent	250	11 900	92	1 000	64
Leachate C					
Raw	—	1 080	35	2 020	72
Effluent	250	2 005	88	264	83
Leachate D					
Raw	—	2 880	70	1 260	66
Effluent	270	656	96	29	39

wet oxidation treatment. For this purpose, customary laboratory analyses were carried out to define the behaviour of the following indices: BOD_{20}/COD, $(COD - BOD_{20})$ and BOD_5/BOD_{20}. The first two quantify the presence of biodegradable organic matter, compared with the total organic content; the last is an indication of the biodegradability rate.

Leachate A. As reported in Table 3, long-term BOD tests showed that the ratio BOD_{20}/COD increases in the treated leachate as compared to the raw influent (from 69% to 88–95%), and that the bioresistant fraction $(COD - BOD_{20})$ is reduced by about 90% by means of the wet oxidation process.

In the case of thermal treatment alone, however, the BOD_{20}/COD ratio decreases as compared to the values measured in raw leachate.

Leachate B. Long-term BOD tests were carried out solely on effluent No. 1. These tests show (see Table 3) that the BOD_{20}/COD ratio

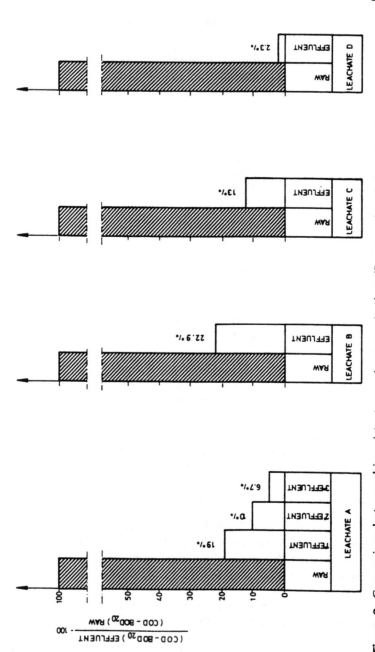

Figure 3. Comparison between bioresistant organic matter in the effluent and in the raw leachate (leachates A, B, C and D).

increases in the treated leachate, compared to the raw one (from 75 to 92%).

The bioresistant fraction ($COD - BOD_{20}$) is reduced by about 77% after wet oxidation.

Leachate C.   Also in this case, long term BOD tests were performed and results are shown in Table 3. The ratio BOD_{20}/COD increased in the treated leachate, as compared to the raw one, from 35% to 88%. The bioresistant fraction ($COD - BOD_{20}$) is reduced after treatment by about 87%.

Leachate D.   Once again, as reported in Table 3, long term BOD tests indicated that the BOD_{20}/COD ratio increases in treated leachate, as compared to raw, from 70% to 96%, and the bioresistant fraction ($COD - BOD_{20}$) is almost completely removed (by 98%) after the process.

An interesting application of the wet oxidation process is therefore suggested by the results described herein, as this process converted leachates originating even from very old urban waste landfills into almost completely biodegradable susbstrates (see Fig. 3).

BIOLOGICAL TREATMENT OF WET OXIDATION PRETREATED LEACHATES

Problems in Biological Treatment of Leachate

The high total ammonia concentration levels in leachate are inhibitors to biological oxidation as well as to biological nitrification. With total ammonia nitrogen concentration more than 200 mg/litre, at temperature = 15°C and pH 8 the percentage distribution of ammonia is more than 6% with corresponding values of 120 mg/litre of NH_3, impairing biological oxidation (Metcalf & Eddy Inc., 1979; Liberti & Petruzzelli, 1981).

Moreover a nitrification–denitrification process is required after biological oxidation in order to remove the excessive NH_4-N nitrogen concentration.

Before the activated sludge treatment, the reduction of excessive levels of ammonia in leachate pretreated with wet oxidation can be performed by addition of NaOH or Ca(OH)$_2$ to achieve pH 10·5–11, followed by air stripping. This treatment is favoured by the high NH$_3$ concentration and so is proper to a full-scale application. The sequence took advantage of the high temperature of leachate after wet oxidation treatment.

The previously described low level of phosphorus in the leachate (Table 1) contributed to inhibition of activated sludge microorganisms. The BOD$_5$:P ratio for the leachate (1673) is much higher than the value of 100:1 widely recommended (Winkler, 1981); therefore, addition of phosphorus is necessary.

The presence of heavy metals in wastewater can affect the performance of the activated sludge process. For most of the metals present in leachate B the concentrations are below the threshold inhibitory concentrations for carbonaceous BOD removal as reported in literature (McDermott *et al.*, 1962; Ganczarczyk, 1983; Lester, 1983), but the lack of uniformity of measurement techniques and the dependence on operational variables and environmental conditions make it impossible to use the process limiting concentrations in each particular case.

Description of the Experimental Plant

A pure oxygen activated sludge process was adopted for the biological treatment phase. Pure oxygen aeration was chosen because of the advantages of generating a very small amount of aerosols and opportunity of operating at high concentrations of dissolved oxygen while maintaining high dissolution efficiencies. Low aerosol production implies both economic and environmental advantages. Pure oxygen biological treatment is usually carried out in covered basins, with an additional beneficial effect on the reactor appearance.

The high concentrations achieved in pure oxygen systems (the concentration is usually 6 mg/litre, but levels of up to 40 mg/litre are possible) allow the penetration of oxygen into the whole sludge flake, since the concentration gradient between liquid phase and the inside of the flake is increased (McWhirter, 1978; Beccari *et al.*, 1987). It has been shown, in fact, that at oxygen concentrations commonly measured in air-fed reactors (usually 1–2 mg/litre), the activated sludge flakes are not able to maintain aerobic metabolic conditions throughout their entire mass. It follows that, in the case of pure oxygen systems, a higher level of activity in the sludge flakes allows higher volumetric loads to be used in the reactor, as compared to air-fed systems, or higher removal yields at the same volumetric loads.

This latter aspect was the one that suggested the use of a pure oxygen system for the following experiment. The poor biodegradability of leachate, due to its content of toxic or inhibiting compounds (e.g. metals) is well known. The possibility of achievement of better removal yields, which has already been shown on other types of wastewaters, together with better microorganism resistance to toxic or inhibiting compounds due to a more favourable living environment, were both factors which fully justified the adoption of pure oxygen for our tests.

The pilot plant used in the study had a biological oxidation volume of about 80 litres, divided into four equal and interconnected compartments. Both the upper and lower portions of each compartment were in communication with each other, through a slot opening in the middle of their separating wall.

The oxygen was supplied from the bottom of each compartment by means of porous stone and distributed into the whole volume of the liquid by means of submerged mixers. A peristaltic pump recycled the sludge from the clarifier (with a surface of $15.9\,\mathrm{dm^2}$) into the first compartment, where the leachate is also introduced. Several taps allowed sampling of the liquor from each compartment.

Experimentation

The experimental phase here described is concerned only with the biological treatment of leachates B, C and D, whose characteristics after wet oxidation treatment are indicated in Tables 2 and 3.

Pretreatment before activated sludge.

(a) Air stripping of ammonia. To remove the high level of ammonia-nitrogen, the leachate was treated with sodium hydroxide to raise the pH to the range 10–10·5 prior to air stripping. The mixture of NaOH solution and leachate was stirred in a 1000-litre holding tank with an air diffuser in the bottom for some days. The ammonia concentration level was reduced to 150 mg/litre.

Afterwards the leachate, without aeration, was allowed to settle for two hours and the sludge was drawn off the bottom of tank and placed with the waste activated sludge.

(b) Neutralization and nutrient supplementation. Hydrochloric and phosphoric acid were added to reduce the pH values of the leachate prior to biological treatment. Phosphoric acid was added to achieve a P concentration in the feed to activated sludge and maintain an effluent concentration of about 1–2 mg/litre.

Procedure used to start the pilot plant. The activated sludge seed culture used at start-up was obtained from a return sludge pipe of a domestic wastewater treatment plant. The culture contained 6000 mg/litre of SS and 4000 mg/litre of VSS. The balance of the reactor volume was made up with settled domestic wastewater.

During the first period of start-up the stirred reactor was fed with settled domestic wastewater and the working level of dissolved oxygen concentration was maintained in the range 7–9 mg/litre. The return sludge flow was maintained at about 700 litres/day in order to homogenize the biomass in the four stages and dilute the potential inhibitor effects of heavy metals and organic compounds. After five days, the acclimatization period commenced, with the continuous feed of wet oxidation products and stripping of treated leachate.

Table 5 shows the physical-chemic properties of leachate used for feeding the pilot plant.

The organic load was increased in three steps from 1 litre/day to 4 litres/day over a 20-day period and the corresponding MLSS concentrations increased to the 5000 mg/litre range (the MLSS value is the same in each stage ±3%). The organic load based on BOD_5 varied between 0·12 and 0·15 g BOD_5/g MLVSS. day. The pilot plant operated at a temperature of $16 \pm 1°C$.

The average concentration of MLVSS was maintained between 4900 and 5200 mg/litre (representing 2900–3100 mg/litre of MLVSS) by wasting sludge once a day from the reactor (Matsch & Drnevich, 1978). The corresponding mean bacterial cell residence time is about 80 days. After approximately 10 days of operation, HCl was added to reduce the increasing pH in the reactor.

PILOT PLANT PERFORMANCE

The main operational parameters and the results obtained are reported in Tables 4 and 5.

TABLE 4. Operational Parameters of Pure Oxygen Activated Sludge Plant

	Leachate B	Leachate C	Leachate D
Organic loading rate (kg COD/ kg SSV . day)	0·1–0·2	0·06–0·2	0·1
Organic loading rate (kg BOD₅/ kg SSV . day)	0·06–0·15	0·04–0·15	—
Hydraulic retention time (days)	6–20	3–13	2
SVI (cm³/g)	70–80	100–150	70–90
MLSS (g/litre)	2–5	1·6–2·0	1–2·5
MLVSS (g/litre)	1·4–3·5	1–1·5	0·7–1·4
Temperature (°C)	15–20	7–13	18–26

TABLE 5. Efficiency of the Combined 'Wet Oxidation–Biological' Process for the Three Experimental Leachates

	WO influent (mg/litre)	Biological effluent (mg/litre)	Removal Efficiency (%)		
			WO	Biological	Total
Leachate B					
COD	17 350	1 000	26	68	94
BOD₅	8 200	0	7	93	100
BOD₂₀	12 980	110	8	91	99
Leachate C					
COD	3 100	200	27	67	94
BOD₅	780	7	—	100	100
BOD₂₀	1 080	122	—	89	89
Leachate D					
COD	4 410	100	83	15	98
BOD₅	1 900	0	86	14	100
BOD₂₀	2 880	80	77	20	97

The organic loading rate was maintained between 0·06 and 0·2 kg COD/kg SSV. day; hydraulic retention time was between 2 and 20 days. The sludge volume index was always very good (150–350 cm^3/g at 7°C). The BOD_5 and BOD_{20} concentrations in the effluent were 0–7 mg/litre and 80–120 mg/litre, respectively.

It can be seen that for the leachates B and C the most important removal rate is due to the biological process, while for the leachate D about 80% of removal was due to wet oxidation.

The total removal efficiency of the combined 'wet oxidation–biological' process was, in conclusion, very high (89–100%).

CONCLUSIONS

From results obtained the following conclusions can be drawn. The wet oxidation pre-treatment applied to four leachates (different in landfill age and source) confirmed that the process is able to increase substantially the biodegradability characteristics of the waste; in particular, for leachates C and D (obtained from the older landfills) it has been observed that the BOD_{20}/COD rate increased respectively from 35 to 88% and from 70 to 96%, with an 87–98% removal of the bioresistant organic matter. These results lead to the conclusion that wet oxidation is a suitable pre-treatment for all types of leachate, independent of landfill age.

The pure oxygen activated sludge treatment of the leachates, pre-treated by wet oxidation, provided the following important results:

—high ammonia concentrations require, before the biological phase, the use of an ammonia removal treatment by means of pH correction and air stripping;

—confirmation of the necessity, in the biological treatment of leachates, of controlling the P/BOD ratio;

—the biological process showed a considerable efficiency in the removal of organic matter, in spite of the presence of potential inhibitors such as heavy metals.

Based on these observations a general flowsheet for leachate treatment can be suggested. As shown in Fig. 4 it involves the sequence of the tested processes: wet oxidation, pH correction with lime, ammonia stripping (possibly followed by an air-washing treatment with fertilizer recovery) and, finally, activated sludge.

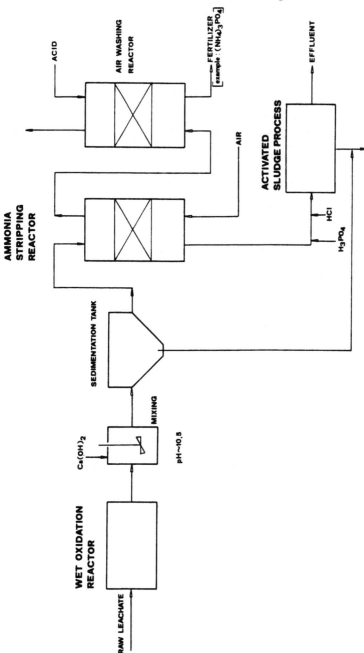

Figure 4. Proposed leachate treatment flow-sheet.

352
Avezzù, Bissolotti, Collivignarelli

REFERENCES

Agamennone, M. & Pieroni, M. (1987). Ossidazione ad umido di reflui non biodegradabili. *Ingegneria Ambientale*, **16**(5), 251–61.

Baldi, M., Berbenni, P., Bissolotti, G., Collivignarelli, C. & Fortina, L. (1985). L'ossidazione a umido come trattamento dei reflui idrici: applicazioni e prospettive di sviluppo. *Inquinamento*, (7/8), 33–46.

Beccari, M., Franzini, M. & Ramadori, R. (1987). Influenza delle resistenze diffusionali sulla cinetica di ossidazione biologica dell'ammoniaca, in reattori a biomassa sospesa. *IRSA–CNR Quaderno*, (77), 29–46.

Berbenni, P., Bissolotti, G., Collivignarelli, C. & Fortina, L. (1986). L'ossidazione a umido come trattamento dei reflui idrici: rassegna di alcune soluzioni brevettate. *Inquinamento*, (1, 2) 48–56.

Cannel, P. J. & Schaeffer, P. T. (1983). Detoxification of hazardous industrial wastewater by wet air oxidation. Presented at Spring National AIChE Meeting, Houston, TX.

Collivignarelli, C. & Bissolotti, G. (1988). Esperienze nel trattamento di reflui mediante ossidazione ad umido: fattibilità tecnico-economica. *Ingegneria Ambientale*, **17**(10), 542–53.

Ganczarczyk, J. J. (1983). *Activated Sludge Process. Theory and Practice.* Marcel Dekker Inc., New York, p. 208.

Lester, J. N. (1983). Significance and behaviour of heavy metals in waste water treatment processes. I. Sewage treatment of effluent discharge. *Sci. Total Environ.*, **30**, 1–44.

Liberti, L. & Petruzzelli, D. (1981). Rimozione e recupero di nutrienti da acque reflue mediante scambio ionico. *Inquinamento*, (2), 23.

Matsch, L. C. & Drnevich, R. F. (1978). Sludge production in oxygen systems. In: *The Use of High Purity Oxygen in the Activated Sludge Process.* CRC Press, Boca Raton, FL, Vol. 2, Chapter 2.

McDermott, G. N., Barth, E. F., Salotto, V. & Ettinger, M. B. (1962). Zinc in relation to activated sludge and anaerobic digestion processes. *Proc. 17th Ind. Waste Conf.*, Purdue Univ. Eng. Ext. Ser. No. 112, pp. 461–75.

McWhirter, J. R. (1978). Oxygen and the activated sludge process. In: *The Use of High Purity Oxygen in the Activated Sludge Process.* CRC Press, Boca Raton, FL, Vol 1, Chapter 3.

Metcalf and Eddy Inc. (1979). *Wastewater Engineering Treatment Disposal.* Tata McGraw Hill Co, New Delhi, (2nd edn), pp. 734–6.

Winkler, M. (1981). *Biological Treatment of Waste-Water.* J. Wiley, New York, p. 81.

3.11 Combination of Long-Term Aeration, Sand Filtration and Soil Infiltration

KAJ NILSSON & VLADIMIR VANEK

VBB VIAK, Division for Water & Environment, Geijersgaten 8, S-216 18 Malmö, Sweden

INTRODUCTION

In Sweden, leachate from large municipal landfills is usually treated in conventional wastewater treatment plants, together with municipal wastewater. This practice is often costly, especially if leachate is to be transported by special piping or trucks for longer distances. Slowly degradable organic compounds often pass through conventional treatment facilities without any decomposition. In addition, large quantities of cold leachate, which are produced during snowmelt periods or rainy Autumns, frequently lead to serious disturbances in the wastewater treatment process.

In order to avoid these problems, a system of on-site leachate treatment based on long-term aeration, sand filtration and infiltration ditches has been developed and tested at Måsalycke, southern Sweden, and is described in this chapter.

SITE DESCRIPTION

Måsalycke is a municipal landfill localized in south-eastern Scania (Fig. 1) and has served Simrishamn and other nearby located communities since 1975. Annual disposal rate is about 30 000 tonnes of mixed wastes and the landfilling area covers *c*. 8 ha. On an annual basis, precipitation

353

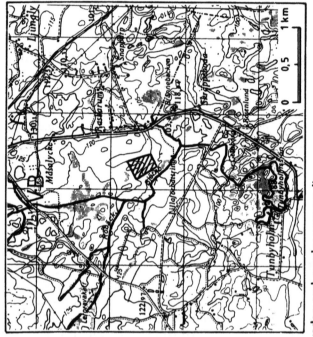

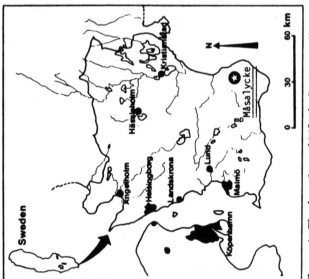

Figure 1. The location of Måsalycke in southern Sweden and nearby surroundings.

exceeds evaporation by 300–400 mm, which leads to annual production of 20 000–30 000 m³ of leachate. In late 1970s and early 1980s, several accidental discharges of leachate were observed, risking pollution of downstream areas including Björnbäcken brook and Lake Tunbyholmssjön (Fig. 1). Therefore, it was decided to undertake suitable remedial measures. An additional task was to replace transport by the trucking of leachate to the nearest wastewater

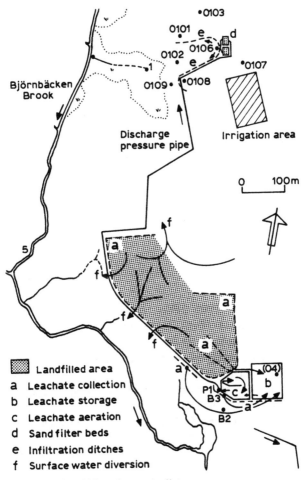

Figure 2. Måsalycke landfill (schematically).

treatment plant by another, more economical and effective treatment system.

This system consists today of six steps (Fig. 2).

(a) Leachate collection via subsurface drains, interception ditches and pumping wells.

(b) Leachate storage and flow equalization basin, volume 12 000 m^3.

(c) Long-term aeration basin, volume 3000 m^3, where leachate is oxygenated and mixed for at least 20 days. In order to optimize biological activity, phosphate is dosed manually into the basin twice a week.

(d) Two sand filter beds of 350 m^2 each, designed to remove suspended solids which might clog the infiltration ditches.

(e) Two 120-m long and 1-m deep open ditches which since 1986–88 were used intermittently to infiltrate about 60 m^3 of pre-treated leachate a day. In addition, some leachate was used for irrigation experiments at several plots vegetated with various grass and tree species and localized just south-east of the sand filter beds.

(f) In order to diminish the production of new leachate, the landfill surface is successively capped with low-permeable till. When landfilling is completed, vegetation cover is established. In addition, the unpolluted surface water originating from direct precipitation or snowmelt is diverted away from the landfill by means of a superficial drainage system.

RESULTS

Yearly mean values (1986–1988) for some chemical variables and treatment steps a to e (Fig. 2) are presented in Table 1. Specific conductance was measured in the field at all points twice a week. All other parameters are analysed every second month except samples of raw leachate (step a), where only two analyses per year are available. Step a represents a mean of two points (a pumping well and an interception ditch), whereas step e includes values from seven observation wells located at varying distances (5–20 m) from the infiltration ditches.

Raw leachate (step a) can be characterized as a low-strength leachate, with COD values ranging between 300 and 600 mg/litre, and total content of nitrogen about 50 mg/litre. The leachate was somewhat more diluted in 1987 than in 1986 and 1988, presumably due to climatic reasons.

TABLE 1. Mean Yearly Values of Analytical Parameters in the Leachate After Sequential Treatment Steps, a to e, as Described in the Text (all values in mg/litre except specific conductance in μS/cm)

Parameter	Year	Treatment step				
		a	b	c	d	e
NH₄-N	1986	48·5	—	26·4	18·5	5·7
	1987	34·4	21·7	2·5	1·8	1·3
	1988	42·8	25·0	8·1	5·7	0·5
NO₃-N	1986	0·09	—	2·8	2·6	2·8
	1987	0·05	0·4	9·4	5·4	3·4
	1988	0·04	0·8	6·6	10·0	6·8
N-tot	1986	53·7	—	34·7	22·2	9·6
	1987	36·7	24·4	16·8	9·4	6·1
	1988	53·9	42·6	27·1	17·6	11·4
Chloride	1986	79·5	—	130·7	130·0	55·1
	1987	78·0	60·5	68·0	86·7	29·9
	1988	137·5	94·1	98·9	95·2	48·0
Specific conductance	1986	1 237	—	1 252	1 244	479
	1987	998	945	879	723	317
	1988	747	1 036	592	855	237
COD	1986	592	—	371	110	75
	1987	272	173	136	95	41
	1988	434	236	147	110	58
BOD	1986	—	—	38		—
	1987	—	32	21	8	—
	1988	—	88	18	11	—
Iron	1986	157·5	—	31·7	1·4	4·7
	1987	90·0	16·7	1·3	0·2	1·1
	1988	108·7	14·3	2·3	0·5	1·7
Manganese	1986	19·2	—	9·8	5·35	3·4
	1987	16·9	6·3	2·0	0·26	2·2
	1988	18·9	5·5	3·0	1·56	1·3

In the leachate storage and equalization basin (step b, in operation since early 1987) leachate is slightly diluted by direct precipitation, and a part of the nitrogen probably evaporates as ammonia. The drop in iron content and the frequently found traces of nitrate indicate that water, at least locally, contains some oxygen. Also, the presence of algae and zooplankton as well as water chemical analyses and toxicological tests indicate that the concentrations of heavy metals or other toxic compounds are usually very low.

Water in the long-term aeration basin (step c) contains as a rule a dense population consisting of small green algae, rotifiers and copepods. The introduction of step b in early 1987 resulted in improved efficiency of iron, manganese, COD and BOD removal, and in almost complete conversion of ammonia to nitrate and organically bound nitrogen. Low nitrification rate and the disappearance of algae observed in the Spring and early Summer 1988 were probably due to insufficient dosing of phosphorus during high nitrogen loads, and/or due to the temporarily elevated concentrations of copper (up to 0·3 mg/litre) which is known to be highly toxic to algae and nitrification bacteria.

The sand filter beds (step d) removed most of the suspended solids such as metal precipitates, algae and bacterial flocs, and had also some nitrification activity.

Water quality during infiltration (step e) also improved significantly after the introduction of step b. Today water quality in observation wells differs only slightly from the background values (Table 2). Further improvement takes place during flow through the peaty, water-logged areas surrounding the Björnbäcken brook. Water from three observation wells located in this area has almost no nitrogen and only slightly elevated concentrations of chloride and COD (Table 2).

TABLE 2. Groundwater Quality Near the Infiltration Area. X, 1985 Background Values; Y, 1988 Mean Values from Seven Observation Wells Located within the Infiltration Area; Z, February 1988 Values from Three Observation Wells Located in the Water-Logged Area Between the Infiltration Area and the Draining Brook (all values in mg/litre except specific conductivity in μS/cm)

Parameter	Observation period		
	X	Y	Z
Specific conductivity	184	238	140–183
Chloride	12	48	19–30
NH_4-N	0·1	0·5	0·003–0·03
NO_3-N	2·4	6·8	0–0·5
N-tot	3·0	11·4	0·6–1·2
COD	30	58	40–65
Fe	17	1·7	0·2–2
Mn	0·6	1·3	0–0·09

The infiltrated pre-treated leachate is sooner or later drained together with local groundwater flow towards Björnbäcken, a clearwater brook hosting salmonid and crayfish populations, and exhibiting mean water flow of about 100 litre/s. Mass balance calculations indicate that the mean increase of concentrations in the brook due to the infiltration of leachate does not exceed 5% for TDS and is much less for all other constituents.

DISCUSSION

Leachate treatment by long-term aeration has been studied by several authors (Knoch, 1972; Persson, 1981; Stegmann & Ehrig, 1981; Marris *et al.*, 1984; Storhaug, 1984; Robinson & Grantham, 1988). A frequently encountered problem is low treatment efficiency (especially low nitrification of ammonia) during the Winter period, requiring long residence times of 30 days or more. Experiences with low-strength leachate from Måsalycke as well as high-strength leachate from Bösarp (unpublished data) suggest that high and stable treatment efficiency can be achieved throughout the year, provided that the following requirements are fulfilled:

(a) the aeration step is well protected against sudden large inputs of cold, highly concentrated leachate;
(b) the N:P ratio is kept sufficiently low (optimum 10:1) by the addition of phosphorus;
(c) the residence time is constantly longer than two or three weeks.

The infiltration of pre-treated landfill leachate has not yet been described in the literature. Somewhat similar approaches are leachate recycling (Pohland, 1980; Norrman & Lindfors, 1981), spray irrigation (Rowe, 1979; Menser *et al.*, 1983) and the concept of natural attenuation landfills (Mather, 1977; Bagchi, 1983). The system described here, however, has the advantage of good performance throughout the year, low adverse effects on the vegetation, and adequate possibilities for monitoring and controlling the infiltration process.

When local geological, hydrogeological and legal conditions are favourable, the infiltration of pre-treated leachate could be considered as an effective, safe and inexpensive method of landfill leachate treatment.

REFERENCES

Bagchi, A. (1983). Design of natural attenuation landfills. *J. Environ. Eng. Div. ASCE*, **109**, 800–11.

Knoch, J. (1972). Reinigung von Müllsickerwasser mit belüfteten Teichen. *Müll u. Abfall* **4**(4), 123–33.

Marris, P. J., Harrington, D. W. & Chismon, G. L. (1984). Leachate treatment with particular reference to aerated lagoons. *Wat. Pollut. Control*, **83**, 521–38.

Mather, J. D. (1977). Attenuation and control of landfill leachates. *Solid Wastes*, **67**, 362–78.

Menser, H. A., Winant, W. A. & Bennett, O. L. (1983). Spray irrigation with landfill leachate. *BioCycle*, **24**(3), 22–5.

Norrman, J. & Lindfors, L.-G. (1981). Styrning av nedbrytningsförloppen vid avfallsdeponering (Enhancement of landfill stabilization processes). Institut för Vatten- och Luftvårdsforskning (Swedish Environmental Research Institute, Stockholm), IVL report series B-publ. 590.

Persson, B. L. (1981). Lakvattenbehandling genom långtidsluftning (Leachate treatment by extended aeration). *Stadsbyggnad*, **47**(2), 12–14.

Pohland, F. G. (1980). Leachate recycle as landfill management option. *J. Environ. Eng. Div. ASCE*, **106**, 1057–69.

Robinson, H. D. & Grantham, G. (1988). The treatment of landfill leachate in on-site aerated lagoon plants: Experience in Britain and Ireland. *Wat. Res.*, **22**, 733–47.

Rowe, A. (1979). Tip leachate treatment by land irrigation. *Solid Wastes*, **69**, 603–23.

Stegmann, R. & Ehrig, H.-J. (1981). Operation and design of biological leachate treatment plants. *Wat. Sci. & Technol.*, **13**, 919–47.

Storhaug, R. (1984). Luftet lagune for rensing av sigevann. Delrapport 1: Driftserfaringer (Treatment of leachate by aerated lagooning. Report 1: Operational experiences). Norsk Institutt for Vannforskning NIVA, VA-rapport 2/84 (Norwegian Water Research Institute in Oslo, Water and Sewage Report 2/84), NIVA 0-83027.

3.12 Soil–Plant Treatment System

KENTH HASSELGREN*

Department of Civil Works, Eslöv Municipality, Box 1100,
S-241 26 Eslöv, Sweden

INTRODUCTION

The general approach to the landfill leachate problem in Sweden has traditionally been co-treatment with municipal wastewater. At about 100 sites from a total of approximately 400 sanitary landfill sites in operation, leachate is collected, equalized and directed to a sewage treatment plant. In spite of the large investments often made, the effects of leachate treated together with wastewater have not been thoroughly investigated. A general conception is that the leachate discharge is transferred to another waterbody without being really treated. Furthermore, assessments indicate disturbances in the biological treatment step due to toxic leachate ammonia concentrations (NSEPB, 1982).

Experiences gained from on-site leachate treatment by modified sewage unit processes have been discouraging. A main problem has been the low leachate temperature during the winter period together with freezing of treatment equipment.

The possible utilization of naturally overgrown land or specifically selected soil–plant systems as leachate treatment media has not been particularly considered in Sweden.

A soil–plant system (SP-system) may be defined as a physical-biological-chemical reactor including the following main active parts and

* Present address: Department of Civil Works, S-268 80 Svalöv, Sweden.

processes:

—*Soil particles* which filter suspended solids and fix dissolved components in the leachate by adsorption, ion exchange or precipitation.

—*Macro- and microorganisms* which transform and stabilize organic substances and transform nitrogen in applied leachate.

—*Vegetation* which utilizes macro- and micronutrients in the leachate for growth, maintains or increases the infiltration capacity in the soil and reduces applied leachate volumes by transpiration.

There are two principal applications of SP-systems. In one, collected leachate is pumped back to the completed and planted landfill area or gradually as parts of the site are completed. The other is that leachate is directed to surrounding areas, e.g. wooded ground, pasture land or specifically created treatment areas, where the fertilizer effect of the nutrients present in leachate (nitrogen, potassium) could be utilized.

A number of examples of natural leachate treatment have been reported in the literature. Overland flow treatment and spray irrigation of grass plots have been practised in the UK at some sites (Robinson & Maris, 1979; Rowe, 1979; Hawley, 1983). Leachate irrigation is evaluated in parallel with 'conventional' treatment in Germany (Collins & Spillmann, 1982). The effects on different types of vegetation after leachate application have been examined in the USA (Menser *et al.*, 1979, 1983; Pavacic, 1983). In Finland, leachate irrigation of willow plantations on the landfill area is used to treat leachate and increase the evapotranspiration from landfills (Ettala, 1987). General conclusions, from these and other reports in the field, are that applications of SP-systems are often cost effective in comparison with other leachate management alternatives. An important factor is also that fluctuations of flow and pollutant concentrations could be managed quite easily due to the stable and rather insensitive natural purification processes. Further, the technical equipment is rather uncomplicated and requires little supervision.

The results obtained from a completed small pilot field study and, briefly, some preliminary results from an ongoing full-scale project at an old sanitary landfill in the south of Sweden are discussed in this chapter. The objectives of the small-scale study included assessment of leachate application to some SP-systems considering potentials of evapotranspiration, biomass production and leachate treatment. The aim of the full-scale project is to demonstrate application of leachate to a completed part of the landfill area planted with willow trees and a grass mixture.

DESCRIPTION OF THE LANDFILL SITE

The Rönneholm Landfill Site where the experiments have been carried out is situated outside Eslöv City in the south of Sweden, at the edge of an exploited peat bog, upon a 1-m layer of humified peat overlying a dense calcareous soil layer. The site has been in operation since the 1940s for disposal of municipal, industrial, construction and demolition wastes. The landfilled quantities amount to about 30 000 tonnes per year. The precipitation, mainly as rain, averages 663 mm per year.

Since 1981 leachate from the 22-ha landfill area has been collected in surrounding ditches and directed to a pond from which, up until 1982, all leachate was discharged to the adjacent surface water recipient. Temporary leachate irrigation commenced in 1983 in existing clumps of birch trees and grass plots on the site. Completion of a first landfill phase, consisting of a 4-ha SP-system with leachate irrigation, was initiated in 1985. The facility became fully operational in 1986.

SMALL-SCALE STUDY

Experimental Procedures

Tests of leachate application to some soil–plant systems were carried out at the landfill during 1982–6. The types of soils and vegetation used in the tests are summarized in Table 1.

TABLE 1. Types of Vegetation and Soils Represented in the Tests on Soil–Plant Leachate Treatment System

Tests	Period	Vegetation types	Soil types[a]
Plots	1982–6	Salix viminalis, Q77082 Salix dasyclados, Q77056 Salix dasyclados, Q79052 Dactylis glomerata	Silty loam, 1 m
Lysi- meters	1983–6 1984–6	Dactylis glomerata Salix viminalis, Q77082	L1, L2, L3, L4 L3

[a] Soil profiles L1–L4 consist of a 0·2-m surface layer and a 0·3-m bottom layer where: L1, compost from household waste mixed with silty loam, 0·2 m + silty loam, 0·3 m; L2, low-humified peat mixed with silty loam, 0·2 m + clay, 0·3 m; L3, low-humified peat mixed with silty loam, 0·2 m + silty loam, 0·3 m; L4, silty loam, 0·5 m.

The selection of plant species was based on factors such as fast growth, high water and nutrient requirements, as well as tolerance to high moisture content and salinity in the root zone. Both orchard grass, *Dactylis glomerata*, and the willow species, *Salix viminalis* (osier willow) and *Salix dasyclados* (water willow), meet these requirements. They are also appropriate for revegetation purposes. Orchard grass and the selected *Salix* species are also included in the Swedish research programme on energy cultivation systems. The soil types included in the study are quite common and could be regarded as representative of Swedish soils. They are also frequently used as covering material and capping for landfills.

The test plot included 100 m² of orchard grass and 100 m² of willows. Willow cuttings, 0·25 m in length, were taken from a forest nursery and planted to a depth of 0·2 m with a density of 2·5 plants/m². The *Salix* plants were cut after the first growth period for a proper resprouting and then harvested after the third and the fifth growth period with two-year-old shoots, respectively. The grass was cut once the first year, three times the second year, four times the third year and three times during the fourth year. Harvested plants and grass were weighed and sampled for chemical analyses concerning the content of nutrients and heavy metals.

Leachate was taken from the leachate pond and sprayed throughout the growth periods on one half of the plots leaving the rest as reference areas. The spray head was placed 0·8 m above ground. The spray head nozzle diameter was 4 mm. Designed application rate amounted to 5 mm/h. Leachate was applied normally three times per week for 2–4 hours each time.

The lysimeters consisted of impermeable PVC-coated polyester tarpaulins welded together to a sacklike construction. The eight grass lysimeters each had a surface area of (2 × 2) m² and the two *Salix* lysimeters were (5 × 5) m². All ten lysimeters were filled with soil to a depth of 0·5 m and levelled to the surrounding ground surface. Double sets of lysimeters were used to compare leachate-irrigated and non-irrigated lysimeters. Grass was seeded in spring and cut once the first year, four times the second year, three times the third year and twice during the fourth year. Willow cuttings were planted in Spring with a density of 2·7 plants/m². The plants were cut after the first growth period and then left to resprout without any harvesting.

Leachate was applied to the grass lysimeters via two fixed spray heads in each lysimeter placed in opposite corners. A spraying angle of 90° and a spraying length of 2 m resulted in total coverage of the lysimeter surface. The soil surface in the *Salix* lysimeters sloped 1:30. Leachate was applied via a 5-m perforated PVC-tube placed at the higher end. The slope, the soil qualitites and the designed application rate resulted in an evenly distributed leachate application to the lysimeter. Application rates for all irrigated lysimeters amounted to 36 mm/h. Irrigation was carried out throughout the growth periods three times a week for 5–20 minutes each time, depending on the prevailing weather conditions.

Percolating water was collected by gravity from the lysimeter bottoms, where a thin layer of gravel and sand was placed in order to obtain draining conditions, as well as to prevent clogging of the collection tubes. The amount of percolating water from grass lysimeters was measured in buckets with lids and from *Salix* lysimeters with a 'tipping bucket' construction.

Applied leachate was sampled on about 10 irrigation occasions each growth period. Percolated water from the lysimeter tests was sampled proportionally to the outflow. All samples were kept in a freezer at the site and analysed on a monthly basis. Chemical analyses of water quality and plant material were carried out according to Swedish standards.

Results and Discussion

Leachate application. The mean chemical composition of applied leachate during the study period is presented in Table 2, together with the quality of the 'average' leachate in Sweden, based on data from 26 sanitary landfills (Meijer, 1980).

The quality of leachate from this landfill is weak compared with average leachate quality, indicating a far advanced methanogenic phase in the landfill. This is emphasized by a high pH and a low BOD/COD ratio. The nitrogen content is relatively high, which may be partly explained by N-release from peat mineralization in the bottom of the fill, but could also be due to the age of the landfill. For instance, a slight increase of ammonia with time has been reported from old landfills in Germany (Ehrig, 1983).

Leachate application to the test plots amounted to more than 3000 mm during the five-year study period. The results are summarized in Table 3. The large amounts applied during the first two years were chosen for assessment of development and growth of young plants at high hydraulic and nutrient loads. The following two years reflect appropriate applica-

TABLE 2. Mean Chemical Composition and Standard Deviation of Leachate Used in this Study and Leachate from 26 Swedish Landfills (mg/litre)

Parameter	Mean	± SD	Swedish leachate
Total N	98	42	80
NH_4-N	84	47	23
NO_3-N	9·0	4·8	0·6
NO_2-N	2·4	1·4	—
Organic N	1·9	1·4	—
Total P	0·19	0·09	1·1
COD	190	46	800
BOD_7	17	9	600
K	64	26	—
Ca	92	22	—
Fe	2·8	1·2	30
Mn	0·15	0·06	2·5
Cd	<0·001	—	0·005
Cr	0·010	0·005	0·05
Cu	0·074	0·068	0·05
Ni	0·025	0·011	0·05
Pb	<0·001	—	0·04
Zn	0·072	0·042	0·6
Cl	123	13	500
pH	8·1	0·4	7·1

TABLE 3. Application of Leachate and Hydraulic Load on Test Plots

Test periods		Days	Applied leachate		Rain-fall (mm)	Hydraulic load	
Year	Time		(mm)	(mm/day)		(mm)	(mm/day)
1982	7/6–3/10	119	829	7·0	227	1 056	8·9
1983	13/6–13/11	154	992	6·4	247	1 239	8·1
1984	21/5–30/9	133	468	3·5	407	875	6·6
1985	22/4–10/11	203	480	2·4	486	966	4·8
1986	20/5–10/11	175	285	1·6	338	623	3·6

tion rates under extreme precipitation conditions. The annual precipitation in 1985, for instance, was 25% higher than the annual mean in the area. One could characterize 1986 as a year when nutrient application by leachate was balanced by plant uptake of nutrients.

The results clearly indicate that extended leachate application to *Salix* and orchard grass up to field capacity could be managed without negative effects on growth. On the contrary, leachate irrigated plants developed extremely well from the start throughout the experiment, indicating the suitability of these plant species as components in SP-systems.

During some heavy storms, when the total hydraulic load exceeded 100 mm per week, ponding on the horizontal soil surface (indicating potential surface run-off) occurred on grass plots but not in *Salix* stands. This may be explained by the fact that willow roots are rather thick at the soil surface while finer roots are found deeper down in the soil, resulting in a higher infiltration capacity compared to a grass area with a shallower and fine-threaded root system. In fact, a higher hydraulic load, without ponding or run-off, as was found in a similar test with wastewater application to *Salix* stands on a clayey soil (Hasselgren, 1984), is possible. The latter reported an average application rate of 18 mm/day during the period June–September.

An abundant water supply in the root zone is of great importance to plant growth in many respects. If the plant has to economize with water, the stomata openings will close to counteract transpiration. This will reduce the necessary conditions for an optimum photosynthesis. Further, a good circulation of water in the root zone creates a uniform distribution of dissolved nutrients. This will reduce the energy needed for the plant to develop roots to search for water and nutrients in deeper soil layers in favour of above-ground biomass production. A horizontal, shallow root system is also advantageous if the soil–plant system is placed upon the landfill area, since this will reduce the risk of plant damage due to the presence, if any, of methane in the covering soil.

Evapotranspiration. Evapotranspiration rate from lysimeters was calculated as the rest term from the water balance equation (mm),

$$E = L + P - D \pm S \tag{1}$$

where

E = evapotranspiration
L = leachate application

P = precipitation

D = percolation at 0·5-m soil depth (lysimeter bottoms)

S = change in soil water storage during the time interval selected.

The total water application $(L + P)$ and the amount of percolated water (D) were measured continuously during the test periods. The change in soil water storage (S) was neglected in the calculations, considering naturally small variations in these soil profiles compared to the total fluxes. A time interval of about one month covering 10–20 irrigation occasions was used. Results, exemplified with the L3 soil profile (see Table 1), are presented in Table 4 concerning grass lysimeters and Table 5 concerning willow lysimeters.

Evapotranspiration rates from grass lysimeters showed small differences on a year-to-year basis. The variations could be explained mainly by a combination of application rates (availability of water), distribution of precipitation, number of harvests (interception capacity), grass yield, and the length of the test period. The results indicate that evapotranspiration from an orchard grass area could reach 500–700 mm per year, which is in the range of the annual precipitation.

Salix lysimeters showed high percolation and low water losses to air during the first year due to small water requirements for plants during the establishing phase. The increased growth during the second year

TABLE 4. Calculated Mean Evapotranspiration Rates from Leachate Irrigated and Non-irrigated Orchard Grass Lysimeters (notations according to eqn (1) in the text)

	Test periods			
	1983 20/6–31/10	1984 21/5–21/12	1985 19/4–13/11	1986 10/4–21/11
Days	134	215	209	227
P (mm)	247	620	509	362
L (mm)	257	347	510	345
Leachate				
D (% of $L + P$)	38	56	59	27
E (mm/day)	2·3	2·0	2·0	2·3
Control				
D (% of P)	27	44	36	27
E (mm/day)	1·4	1·6	1·6	1·2

TABLE 5. Calculated Mean Evapotranspiration Rates from Leachate Irrigated and Non-irrigated Willow Lysimeters (notations according to eqn (1) in the text)

	Test periods		
	1984 *18/6–21/12*	*1985* *19/4–13/11*	*1986* *10/4–21/11*
Days	187	209	227
P (mm)	537	509	362
L (mm)	260	750	696
Leachate			
$\quad D$ (% of $L + P$)	37	21	11
$\quad E$ (mm/day)	2·7	4·8	4·2
Control			
$\quad D$ (% of P)	33	15	4
$\quad E$ (mm/day)	1·8	2·1	1·5

resulted in increased evapotranspiration. The water loss rate and also the total evapotranspiration during the third year were lower than during the second year, in spite of an increased biomass production. This is explained by the longer test period and the drier conditions, as well as the lower leachate application amounts during 1986.

The results clearly show an extreme water consumption in a willow stand. Leachate application was carried out on a time schedule basis for practical reasons, which limited the possibility of better balancing leachate application with precipitation. Especially, the 1986 growing season was dry for long periods. A system which is operated carefully, considering a constantly high water content in the soil close to field capacity, would probably result in larger water losses than was obtained in this study. The interception capacity in a *Salix* stand is in the range of 4–5 mm on a single rain occasion (Lindroth, 1988). Thus, spray-irrigation would also result in higher evaporation rates in comparison with the results obtained here. Furthermore, the *Salix* crop needs more than three growth periods to attain a fully developed root system, before optimum above-ground biomass production could be reached. After consideration of all these factors, it could roughly be estimated that the annual evapotranspiration capacity of a *Salix* plantation in this region of the country amounts to between 1200 and 1500 mm, which is in the range of twice the annual precipitation.

Biomass production. Measurements of plant development and growth resulted in clear differences between leachate-irrigated and non-irrigated plots or lysimeters.

Leachate supplied to orchard grass lysimeters yielded between 0·6 and 1·3 kg dry matter (DM)/m², as shown in Fig. 1. This was at the same level as, or higher than, other investigated energy grass species. Fogelfors *et al.* (1981) reported, for instance, optimum production levels of 0·7 kg DM/m² for common timothy, *Phleum pratense*, while tall fescue, *Phalaris arundinacea*, which thrives in wet environments, yielded 0·8–0·9 kg DM/m². The results obtained clearly demonstrate the advantage of continuous irrigation and fertilization in energy crop production systems throughout the growth period.

The grass in control lysimeters showed obvious signs of nutrient shortages, and the grass died back during the third and fourth years. The non-irrigated lysimeter with compost in the surface layer (L1 soil profile according to Table 1) competed well during the first two years, but declined rapidly during the third growth period in spite of large rain amounts. This may indicate that the nutrient content in the compost was used or washed out during the initial years.

The results indicate that the soil type is of minor importance for the biomass yield after leachate application. For a two-year-old to four-year-old orchard grass crop, it has been assessed that a potential dry

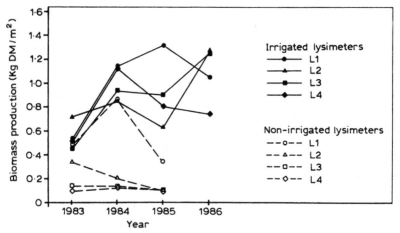

Figure 1. Annual biomass production of orchard grass in leachate-irrigated and non-irrigated lysimeters. Soil profiles L1–L4 according to Table 1.

matter production level of about 10 tonnes/ha can be reached. This corresponds to an energy potential of 50 MWh/ha at 20% moisture content. The high water content of fresh grass, generally around 80%, requires drying for conversion via conventional incineration. Thus, anaerobic digestion of harvested fresh grass with biogas production could be an interesting alternative. Assuming that 30–40% of the dry matter content is digestible, providing 3–4 tonnes of digestible matter per hectare, the energy production in terms of methane gas should correspond to 15–20 MWh/ha year.

As shown in Figs 2 and 3, the willow clones responded substantially to irrigation and fertilization with leachate. Production figures are related to the yield per square metre, since the plots were small. However, since an edge zone of at least 1 metre or three plant rows was disregarded for the production measurements, the results most likely correspond to a full-scale *Salix* plantation in comparable conditions in other respects. Sirén (1983) reported, for instance, that the edge effect was already eliminated in the second row, 0·5 m from the edge row.

The water willow clone, Q77056, was best established and showed the

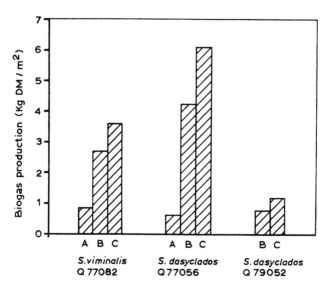

Figure 2. Biomass production of stem wood from *Salix* clones on test plots. Two-year-old shoots on three-year-old roots on control plots (A) and leachate applied plots (B), and two-year-old shoots on five-year-old roots on leachate applied plots (C).

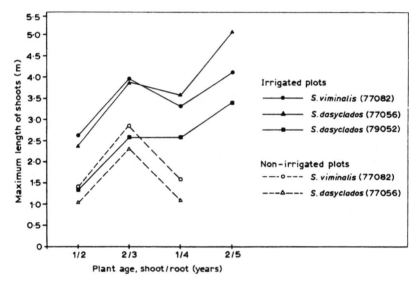

Figure 3. Maximum shoot length of *Salix* clones measured on test plots.

highest production figures. The growth during the third growing season with two-year-old shoots amounted to 36 tonnes DM/ha, which is one of the highest production figures obtained in energy forestry tests in Sweden (Olsson, 1986). Similar results were obtained in another water willow plantation, fertilized conventionally and irrigated with river water (Christersson, 1987). During the following two-year rotation period, the total production reached 60 tonnes DM/ha with similar levels during the fifth growth period with two-year-old shoots as during the third growth period. The other water willow clone produced less but could develop better with longer rotation periods between harvests. The osier willow clone yielded around 20 tonnes of dry matter per hectare during the growth periods with two-year-old shoots. Generally, a longer rotation period would possibly have resulted in higher production figures, which have been obtained in other field tests with *Salix* (e.g. Sirén, 1983; Perttu, 1987). Many plants on control plots died during the second rotation period, indicating the importance of sufficient nutrient and water supplies for relatively young *Salix* stands.

Depending on the *Salix* species or clones chosen, it can be estimated that a fully established willow plantation in this region with sufficient access to water and nutrients will produce 20–25 tonnes of dry matter per hectare annually. This corresponds to an energy production of about 100 MWh or 10 tonnes of oil equivalent per ha per year, assuming a stem water content of 50%. The stem water content in this study was in the range of 45–55%. At present, the *Salix* crop is destined to be used in wood chip incineration plants, but could be converted to fluid or gaseous fuels in the future when the technologies for fluidization and gasification have reached the commercial stage.

Leachate treatment. Nitrogen was considered as the most important component in this leachate from a resource, as well as a pollution, point of view.

Calculated N budgets for the grass lysimeters are exemplified in Table 6. The uptake of nitrogen in the grass was lower than the applied N amounts, but higher than the N reduction over the 0·5-m soil column measured as N input minus N percolation. This may indicate that the grass utilized some of the initial nitrogen content in the soil and probably also immobilized ammonia in the soil profile as a result of leachate application during the first year. The low N uptake in lysimeter L2 in comparison with the other lysimeters indicates that the availability of nitrogen for the grass was low in this lysimeter, in spite of low N percolation rates. One possible explanation is that the extent of ammonia volatilization could have been higher in this lysimeter in relation to the others, due to the greater evapotranspiration rates obtained. Many investigations have shown that evaporation of ammonia is proportional to the water losses (Fenn & Escarzaga, 1977). More likely, however, the net adsorption of ammonia to clay particles in the bottom layer could have been more pronounced in this lysimeter.

Nitrogen removal as a function of the nitrogen load for the grass systems is plotted in Fig. 4. Nitrification in the soil profile of applied

TABLE 6. Nitrogen Budgets for Orchard Grass Lysimeters During the Second Growth Period (g N/m²). Soil Profiles L1–L4 According to Table 1

		L1	L2	L3	L4
Input	(+)	34·2	34·2	34·2	34·2
Output;					
plant uptake	(−)	32·7	23·8	26·5	31·9
percolation	(−)	15·2	13·0	20·9	15·8
Difference	(±)	−13·7	−2·6	−13·2	−13·5
Uptake/input	(%)	96	70	77	93

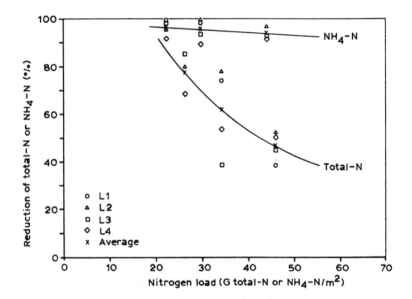

Figure 4. Removal of nitrogen in orchard grass lysimeters, measured as applied amounts minus percolated amounts, as a function of the seasonal nitrogen load. Soil profiles L1–L4 according to Table 1.

ammonia to nitrate explains the high ammonia reduction rates obtained throughout the experiment. Similar results were recorded for *Salix* with 96·8, 99·1 and 99·9% ammonia removal during the application periods of 1984, 1985 and 1986, respectively. The reduction of total nitrogen over the *Salix* lysimeter increased with biomass production and amounted to 43·4, 78·7 and 93·3%, or 8·8, 51·6 and 51·6 g N/m² during the three seasons respectively, indicating the importance of plant uptake for nitrogen removal in a soil–plant system.

The N content in percolated water at 0·5-m soil depth was highly dominated by nitrate, which may be an explanation of the relatively high leakage of nitrogen, especially from grass lysimeters. As already mentioned, the experimental design limited the possibility of properly balancing leachate application with precipitation, thus resulting in percolation of the readily mobile nitrate ions during storm events.

Uptake of ammonia or nitrate by plants was the major nitrogen removal mechanism, considering the irrigation rates used. Other removal processes such as ammonia volatilization, net adsorption of ammonia to soil particles as well as denitrification, may have occurred especially in consideration of the prevailing system conditions involved in these processes. For instance, many investigations have shown that denitrification in soil is promoted by vegetation, mainly depending on organic exudates from plant roots and oxygen consumption due to respiration processes.

Results obtained for the other components of leachate were less interesting, since the concentrations of these in applied leachate were low. It has been assessed that the removal capacity in the soil–plant systems studied is far greater than was achieved, since the concentrations of most components in percolated water from control lysimeters were often higher than from leachate applied lysimeters. Typical removal data are presented in Table 7. Pollutants in the compost material (lysimeter L1) resulted in low reduction of COD, P and the majority of metals. The reduction of BOD, K and Fe was more or less independent of soil type. For most of the parameters, the removal rates were somewhat higher in lysimeter L2, probably due to the more clayey soil material in this lysimeter. The results recorded for the *Salix* lysimeter showed the highest degrees of removal.

The nitrogen and potassium content in this leachate was sufficient from

TABLE 7. Removal of Leachate Constituents in Lysimeters During the Second Growth Period Measured as Applied Amounts Minus Percolated Amounts at 0·5-m Soil Depth (%). Soil Profiles L1–L4 According to Table 1

Parameter	Orchard grass				Salix
	L1	L2	L3	L4	L3
COD	48	62	61	58	86
BOD$_7$	89	91	88	85	94
Total P	2	83	66	63	95
K	80	94	93	91	91
Fe	83	85	84	85	90
Mn	76	97	^{a}	^{a}	86
Cd	25	66	58	60	93
Cr	9	86	34	57	90
Cu	34	26	^{a}	^{a}	84
Ni	12	57	60	27	72
Pb	8	17	0	23	84
Zn	61	83	50	68	99

^{a} Percolation from non-irrigated lysimeter greater than percolation from irrigated lysimeter.

a plant nutrition point of view, and at moderate irrigation levels. This may also be the case for trace nutrients such as copper and zinc. However, the phosphorus content is too low to balance plant uptake in a long-term perspective. In the tests, however, the initial P content in the soils was sufficient for plant development and growth as indicated in the text above. Plant analyses of two-year-old *Salix* stems resulted in NPK-relationships of 100–16–39 for the osier willow clone and 100–19–43 for the water willow clone Q77056. With the nutrient content in the leaves included, the relationship amounted to 100–12–26 and 100–17–39, respectively. The NPK-relationship for orchard grass averaged 100–12–20.

The applied leachate had a mean relationship of NPK of 100–0·19–65, indicating a substantial phosphorus deficit. Thus, if the leachate irrigation is designed to balance the crop uptake of nitrogen, which would be most likely, only 1–2% of the P requirement could be met. In consequence, phosphorus fertilizers would have to be added to neutralize the uptake by the plants. This could easily be accomplished by P dosage to the leachate in the irrigation system.

Conclusions

The leachate from this landfill site seems to be an excellent plant fertilizer. The content of nitrogen and potassium and probably also the majority of the micronutrients seem to be sufficient at normal irrigation rates. The content of phosphorus in leachate however was low, therefore addition of P fertilizers would be required in a long-term perspective to compensate for uptake in produced biomass.

The vegetation was immensely stimulated by leachate application. The biomass production was as high as in other tests with energy crop production. The strong continuous plant development indicated that the leachate content of heavy metals or trace organics was not phytotoxic during the five-year test period. However, in the long run toxic substances could accumulate in the soil.

Both orchard grass and willow seem tolerant of high soil moisture contents, which makes them appropriate as components in soil–plant systems. A higher hydraulic load and also application of larger leachate volumes could be used in SP-systems with *Salix* compared with orchard grass.

The dominating nitrogen removal process in an SP-system at normal irrigation rates is most likely oxidation in the soil of applied ammonia to nitrate, which together with ammonia is assimilated by the vegetation at rates corresponding to the biomass production. The results indicate that leachate application to an SP-system located outside the landfill area (for instance wooded ground, pasture land or energy plantation) should balance with the nitrogen uptake by the plants in the system to minimize the risk of nitrate leakage to the groundwater. On the other hand, if the SP-system is placed upon completed parts of the landfill surface, it is possible that higher nitrogen amounts could be applied, since the environment in the waste beneath the SP-system should promote denitrification of percolated nitrate from the root zone.

FULL-SCALE PROJECT

A full-scale soil–plant system was established on the site during 1985–86. The facility consists of a 4-ha completed area on the 22 ha landfill, planted with two *Salix* species (1·2 ha) and a grass mixture (2·8 ha). The area was prepared with three terraces along one side of the landfill on which 30 000 willow plants were planted in Spring 1986. Between the terraces and the collecting ditch, a grass area with a width of 20–30 m and a slope of 1:10 was created. The terraces had a width of 5 m and a slope of 1:30. The height between two terraces was 2·5 m. The soil consisted of a 0·3-m bottom layer of clay and a 0·3-m layer of topsoil mixed with excavation residues.

The leachate is taken from the pond and lifted to the SP-system via a 110-mm PVC pipeline, lateral pipelines, and some 225 solid-set sprinklers. The irrigation system is divided into 10 sections, which can be operated independently. The maximum hydraulic capacity is 42 m³/h, with a maximum application rate of 25 mm/day. The spray head nozzle diameter is 4 mm. The irrigation system operates at low pressure, 0·3 MPa. Leachate irrigation started in June 1986.

Evaluation of the system will be carried out after a five-year operational phase, mainly in terms of treatment effects, biomass production, operation and maintenance aspects, and operational costs. Some preliminary results after three years of operation are reported here.

Spray nozzles have occasionally become clogged due to transportation of coarser material from sediments on the pond bottom. A one-hour

TABLE 8. Leachate Discharges from the Rönneholm Landfill Site During 1986–88 and Reduction versus the Period 1981–85

	Annual mean 1986–88 (kg/ha)	Reduction versus annual mean during 1981–85 (%)
Total nitrogen	307	71
Total phosphorus	2·1	61
BOD$_7$	126	49

weekly supervision seems to be sufficient for control of the function of the irrigation system.

Applied leachate volumes to the SP-system have amounted to 3290 mm in total, or 600-mm equivalent for the whole landfill area, during the first three years. The evapotranspiration, measured as precipitation minus discharged leachate volume, amounted to 340 mm as an annual average, or 51% of precipitation. Corresponding values from 1981–85 were 140 mm and 19%, respectively. Thus, leachate application to the SP-system, covering 20% of the landfill area (the rest has not been completed), has increased the evapotranspiration from the landfill by 2–3 times.

Plant growth and development are similar to the results obtained in the small-scale study. It is possible that phosphorus addition will be required after three growth periods as discussed in previous sections.

The leachate discharges from the landfill have decreased since the SP-system became operational (Table 8). This is not only a result of decreased leachate volumes but also lower concentrations. For instance, the mean nitrogen concentration has declined by almost 60% during the period 1981–88, from 194 mg N/litre to 83 mg N/litre. It has been assessed that the treatment results will improve further in a year or two when the *Salix* plants are established and the root systems are fully developed and active.

REFERENCES

Christersson, L. (1987). Biomass production by irrigated and fertilized *Salix* clones. *Biomass*, **12**, 83–95.

Collins, H. J. & Spillmann, P. (1982). Lysimeters for simulating sanitary landfills. *J. Environ. Eng. Div. ASCE-EE5*, **108**, 852–63.

Ehrig, H.-J. (1983). Quality and quantity of sanitary landfill leachate. *Waste Management & Research*, **1**, 53–68.

Ettala, M. (1987). Influence of irrigation with leachate on biomass production and evapotranspiration on a sanitary landfill. *Aqua Fennica*, **17**(1), 69–86.

Fenn, L. B. & Escarzaga, R. (1977). Ammonia volatilization from surface applications of ammonium compounds on calcareous soils: VI. Effects of initial soil-water content and quantity of applied water. *Soil Sci. Soc. Am. J.*, **41**, 358–63.

Fogelfors, H., Akerberg, C., Falk, B. & Theander, O. (1981). Ecological, biological and chemical analysis of energy crops from agriculture (Ekologisk, biologisk och kemisk analys av energigrödor från jordbruket). General Report No. 35, Consultant Department, The Swedish University of Agricultural Sciences, Uppsala, Sweden. (In Swedish.)

Hasselgren, K. (1984). Municipal wastewater reuse and treatment in energy cultivation. *Proceedings, Water Reuse Symposium 3*, 26–31 August 1984, San Diego, CA. American Waterworks Association Research Foundation, Denver, CO, Vol. 1, pp. 414–27.

Hawley, D. C. (1983). Experience of leachate treatment in a national park. *Wastes Management*, **73**, 79–85.

Karlqvist, L. & Olsson, T. (1983). Hydrogeological conditions for leachate reduction at landfill sites (Hydrologiska förutsättningar för reduktion av lakvattenbildning vid avfallsupplag). SNV PM 1647, The National Swedish Environmental Protection Board, Stockholm, Sweden. (English summary.)

Lindroth, A. (1988). Private communication. The Energy Forestry Section, The Swedish University of Agricultural Sciences, Uppsala, Sweden.

Meijer, J.-E. (1980). Characteristics of leachate before and after infiltration (Lakvattenkarakteristik före och efter infiltration). Dept of Land Improvement and Drainage, Royal Institute of Technology, Stockholm, Sweden. (In Swedish.)

Menser, H. A., Winant, W. M., Bennett, O. L. & Lundberg, P. E. (1979). The utilization of forage grasses for decontamination of spray-irrigated leachate from a municipal sanitary landfill. *Environ. Pollut.*, **19**, 249–60.

Menser, H. A., Winant, W. M. & Bennett, O. L. (1983). Spray irrigation with landfill leachate. *BioCycle*, **24**(3), 22–5.

NSEPB (1982). Landfill leachate—character, consideration and control (Lakvatten—karaktär, åtgärd och kontroll). M2/1982, The National Swedish Environmental Protection Board, Stockholm, Sweden. (In Swedish.)

Olsson, T. (1986). Production results in Swedish energy forestries (Produktionsresultat i svenska odlingar av energiskog). The Energy Forestry Section, The Swedish University of Agricultural Sciences, Uppsala, Sweden. (In Swedish.)

Pavacic, J. W. (1983). A leachate recirculation project. *Public Works*, **114**, 68–70.

Perttu, K. (ed.) (1987). The energy forestry project—final report for the period April 1984–March 1987 (Projekt energiskog—Slutrapport för perioden 1984-04-01–1987-03-31). The Energy Forestry Section, The Swedish University of Agricultural Sciences, Uppsala, Sweden. (In Swedish.)

Robinson, H. D. & Maris, P. J. (1979). Leachate from domestic waste—
generation, composition and treatment. Technical Report 108, Water Research
Centre, Stevenage, UK.

Rowe, A. (1979). Tip leachate treatment by land irrigation. *Solid Wastes,* **49,**
603–23.

Sirén, G. (1983). Energy forestry (Energiskogsodling). Project Results NE
1983/11. The National Energy Administration, Stockholm, Sweden. (English
summary.)

3.13 Leachate Recirculation: Full-Scale Experience

CHRIS BARBER* & PETER J. MARIS‡

Water Research Centre plc, Medmenham, PO Box 16, Marlow, Bucks, UK, SL7 2HD

INTRODUCTION

Pilot-scale experimental studies in the USA (Pohland, 1976; Robinson *et al.*, 1982; Tittlebaum, 1982) have shown that a major benefit of leachate recirculation or recycle through the landfill is the production of a leachate with low organic strength in a relatively short period of time (*c.* 18 months). These leachates were similar in their organic composition to liquids produced by 'aged' wastes, for example domestic wastes that have been emplaced for five years or more. The solid wastes were also reported (Pohland, 1976) to have degraded and stabilized more rapidly due to the increased moisture content in the landfill obtained by leachate recycle.

Similar experiences have been made in Germany (Doedens & Cord-Landwehr, 1989).

Research in the UK (Robinson *et al.*, 1982) also found that, in addition to reducing the organic strength of leachate, particularly the concentration of volatile acids, its volume could also be reduced by evaporation if leachates were recycled by spraying onto the surface of the experimental landfills. However, it was concluded in the latter study that recycle by itself would not be a complete answer to leachate problems for

* Present address: CSIRO, Director of Groundwater Research, Wembley, WA 6014, Australia.
‡ Present address: Monitor Environmental Services, 3 Dewpond Close, Stevenage, Herts, UK, SG1 3BL.

the following reasons:

1. Under UK climatic conditions, rainfall exceeds potential evaporation/evapotranspiration. The volume of leachate available for recycle would consequently increase with time and leachate would need to be discharged off-site at some stage for further treatment (e.g. to sewer) or treated on-site before discharge.
2. Although the organic fraction of leachate can be greatly reduced by recycle, other constituents of leachate are not significantly removed (ammonia, chloride and metals, in particular).

It was concluded that, despite these drawbacks, reducing the volume and organic strength of leachate by recycle could benefit landfill site operations by reducing costs of further treatment, for example at a sewage treatment works.

In order to determine the practicalities of obtaining these benefits on a large-scale, investigation was initiated at the Seamer Carr landfill, North Yorkshire, UK.

THE SEAMER CARR LANDFILL EXPERIMENT

The Seamer Carr site is 2 ha in area, lined with a 3-mm thick, high density polyethylene liner and is drained by four porous tile drains, which traverse the lined area in a north–south direction (Fig. 1). The site contains a 4-m depth of pulverized domestic wastes at a density of between 800 and 1000 kg/m³.

Experimental Work

Measured leachate recirculation was carried out by spray irrigation through perforated 'layflat' tubing over a 1-ha area of the site, an adjoining area being used as a 'control' where no leachate is sprayed (Fig. 1). The flow of leachate from the recirculation and control areas has been measured using calibrated V notch weirs installed in the site drainage system (manholes 2 and 5, Fig. 1). Water levels in the weir chambers have been measured every hour using ultrasonic sensors. The sensors convert water height to voltage output, which is transmitted to a data logger for storage on magnetic tape and for eventual processing using a VAX computer. More recently, because of the low flows

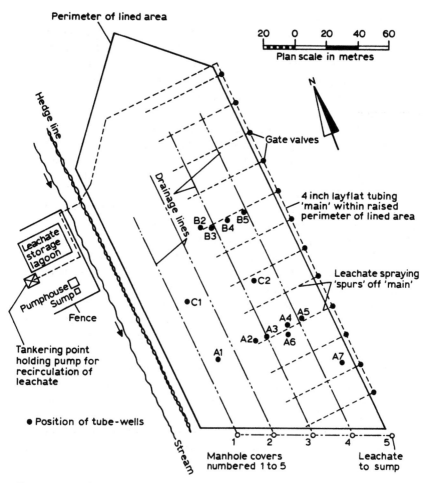

Figure 1. Scale plan of Seamer Carr Landfill Site, showing control and recirculation areas, leachate recirculation system and position of bore-holes. Reproduced by permission of the Geological Society from *Quarterly Journal of Engineering Geology*, **17** (1984) 19–29.

encountered from the control area, a tipping-trough flow meter has been installed to measure leachate flow from this area more accurately. Total flow per month is recorded mechanically using this equipment. Rainfall within the area of the site is estimated from data recorded at four Meteorological Office stations within 6 km of the landfill.

The volume of leachate sprayed on the site was calculated from the recorded time of operation of a submersible pump in the holding lagoon feeding the spray system, and the average volume of liquid pumped per unit time, which was determined by flow gauging at monthly intervals. 3025 m^3 of leachate were recycled over the landfill in the first (weeks 31–52), 3756 m^3 in the second, and 11 572 m^3 in the third year of investigation.

The development of saturated conditions within the landfill was monitored initially using three 50-mm diameter boreholes extending 4 m through the wastes to the liner. In the second year of investigation a further eight 150-mm diameter boreholes and two further 50-mm diameter monitoring points were installed; locations of all these are shown in Fig. 1. Samples of waste were recovered during installation of these boreholes for moisture determination (Table 1).

Water level recorders were placed on two 150-mm diameter boreholes, A1 and B3 (Fig. 1) also in the second year of investigation, to record variations in water level in the area receiving sprayed leachate (borehole B3) and in the control area receiving only rainfall (A1).

Samples of liquid have been recovered for analysis at two-week intervals since the beginning of the experiment. These have been taken from discharges in manholes 2 and 5, and from areas of (perched) saturation within the landfill from the boreholes A1–A7, B2–B5 and C1–C2 at less regular intervals. Determinations of pH value, COD, BOD, total organic carbon (TOC), volatile acids, ammonia, organic N, oxidized N, orthophosphate, chloride and metals (Na, Mg, K, Ca, Cr, Mn, Fe, Ni, Cu, Zn, Cd, Pb) have been carried out on these samples. Analyses of samples from manhole 2 give a direct indication of the composition of leachate draining from the 'control' area. The composi-

TABLE 1. Results of Moisture Determination on Samples Recovered from Borehole B5, Seamer Carr. (Perched water table at *c.* 2·4 m depth, no sample taken from this region)

Approximate sample depth (m)	Sample dry weight (g)	Moisture loss on drying (g)	% Moisture (wet wt.)	% Moisture (dry wt.)
1	904·6	768·4	45·9	84·9
2	866·2	697·2	44·6	80·5
3	965·5	854·1	46·9	88·5
4	869·1	1 020·6	54·0	117·4

tion of leachate draining from the 'recirculation' area is calculated (eqn (1)) from the analyses of leachate from the whole site (samples from manhole 5) and from the control area (manhole 2), and from the respective volumes of liquid (total flow and flow from the control area).

$$C_r = \frac{C_T V_T - C_C V_V}{V_T - V_C}$$

(1)

where C_r = concentration of C in leachate from recirculation area, C_T = determined concentration of C in total flow (manhole 5), C_C = determined concentration of C in leachate from control area, V_T = total flow from whole site at time of sampling (monitored continuously) and V_C = flow from control area at time of sampling (monitored continuously).

Calculation of the composition of leachate from the 'recirculation' area, rather than direct determination, has been carried out because it is not possible to separate the recirculation area flow from the total flow within the drainage system (control flow mixes with the recirculation flows in manholes 3 and 4). The calculated composition, however, is considered to be representative of recirculation area leachate, as the control flow is a relatively small proportion (10–20%) of the total flow.

Results

Determination of moisture content of domestic solid waste recovered during installation of boreholes showed that these were at field capacity and would consequently allow free drainage of leachate (Table 1).

A perched water table has developed within the site at the depth of 2–2·5 m, coinciding with a layer of intermediate cover (clay soil) within the wastes. Continuous records of water level changes in borehole B3 (in the spray area) indicated on a number of occasions that rapid recharge of the perched water table took place during spraying. An example of these fluctuations is shown in Fig. 2, where daily spraying of 2 mm of leachate at midday over 20 min produced sharp increases in water levels followed by a decay over a period of a few hours. Direct measurements of water levels before and during spraying on one occasion showed responses within 20 min of commencement of spraying, confirming the continuous records.

The thickness of saturation has increased with time, particularly over the winter periods of the second and third years of investigation (Fig. 3).

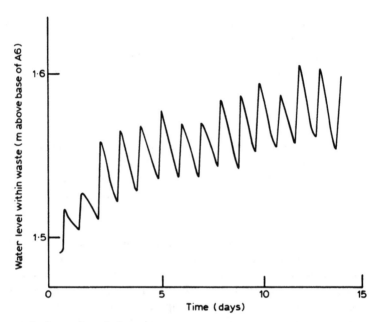

Figure 2. Example of fluctuation in water level in perched areas of saturation within the landfill, caused by spraying leachate, February–March (third year). (Data from continuous water level recorder installed in borehole B3). Reproduced by permission of the Geological Society from *Quarterly Journal of Engineering Geology,* **17** (1984) 19–29.

It is difficult to estimate accurately the thickness of saturated wastes within the landfill because little is known of the precise position of the intermediate clay soil cover which is acting as a leaky aquiclude. However, it is clear from the array of boreholes in the site that saturated conditions extend over the whole lined area. A general indication of the thickness of this zone can be obtained from two boreholes emplaced only to a depth of 2 m. These show that this has increased from approximately 0·6–1 m in the middle of the second year to 1·6–2 m in the third year. The most obvious increases in the saturated zone occurred in the Winters of the second and third years (Fig. 3).

The rate of flow of leachate from the base of the site from both the control and recirculation areas has increased with time. There is a relatively rapid response of increased flow rate to increased precipitation and leachate spraying (within a week) in the recirculation area, although

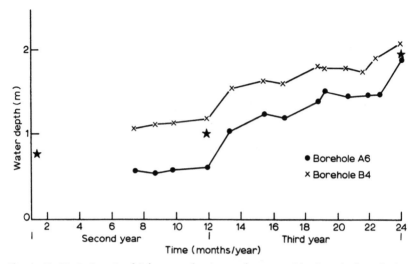

Figure 3. Variation in thickness of saturated zone with time in boreholes A6 and B4 referred to common datum of base of borehole A6 which roughly coincides with position of intermediate soil cover. Stars mark expected thickness of Saturated Zone, based on water balance (see text and Table 3). Reproduced by permission of the Geological Society from *Quarterly Journal of Engineering Geology,* **17** (1984) 19–29.

response in the control area to increased precipitation is not apparent (Fig. 4).

Throughout most of the study period low rates of flow of leachate were found from the control area relative to those from the recycle area. This was probably due in part to some lateral movement of leachate along the base of the site from the northern and eastern edges of the control area into the recirculation area (the base of the site has a shallow fall in a NW–SE direction—see Fig. 1). This has inevitably affected the water balance for the site, decreasing the apparent infiltration from the control area and increasing that from the recirculation area. The estimated differences in composition of leachates from the recirculation area relative to liquids draining from the control area will also be affected. As these differences were expected to be large once recirculation of leachate had been carried out for a long period of time, it was considered that a relatively small contribution of 'control' leachate to liquid draining from the recirculation area would not unduly affect conclusions of the experimental work. In fact, the organic strength of leachate from the

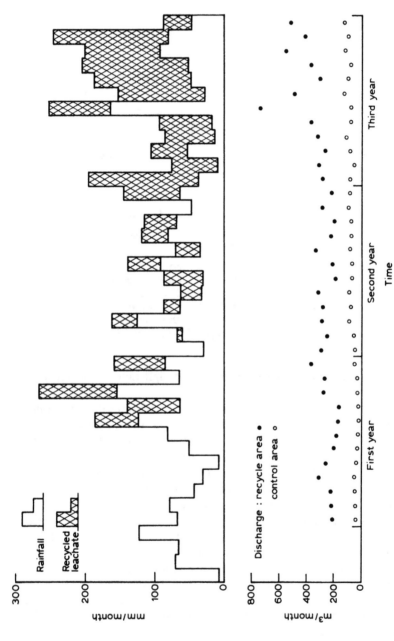

Figure 4. Rate of liquid input (rainfall + recycled leachate) and leachate discharge. Reproduced by permission of the Geological Society from *Quarterly Journal of Engineering Geology,* **17** (1984).

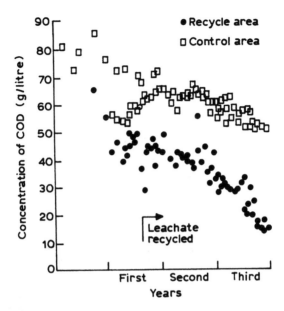

Figure 5. Variation in COD of leachates draining from recirculation (recycle) and control areas. Reproduced by permission of the Geological Society from *Quarterly Journal of Engineering Geology*, **17** (1984) 19–29.

control area (as COD) has varied little compared with the relatively large variations in the COD of leachate collected from the base of the recirculation area (Fig. 5). The COD of leachate from the recirculation area has decreased markedly relative to that from the control area since the middle of the first year when recirculation began, and particularly during the latter part of the experimental period (Fig. 5). Chloride and ammoniacal N showed little tendency to decrease with time in leachate from either area, which indicates that the organic fraction of leachate is being removed relative to other constituents (Figs 6 and 7).

The generally low strength of leachate in saturated areas within the site (i.e. collected from boreholes) contrasts with that collected from the base of the site (in the drainage system), shown in Table 2. Significant differences in strength of leachate within saturated parts of the site are also apparent, this generally increasing in an easterly direction within the mid-section of the site. However, samples taken from the same borehole over the period of monitoring show quite large variations and no obvious

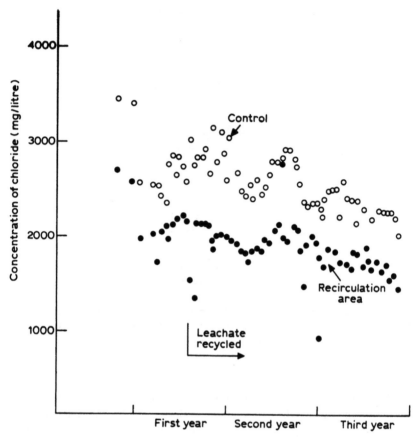

Figure 6. Variation in concentration of chloride in leachates draining from recirculation (recycle) and control areas. Reproduced by permission of the Geological Society from *Quarterly Journal of Engineering Geology*, **17** (1984) 19–29.

correlation with variations in leachate composition in other boreholes across the site.

APPRAISAL OF RESULTS AND DISCUSSION

Practical Considerations

Spraying of 3025 m³ of leachate in the latter half of the first year, 3756 m³ in the second year, and 11 572 m³ in the third year, although giving rise

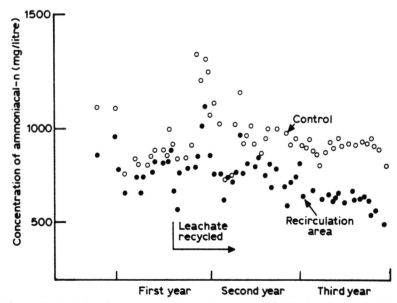

Figure 7. Variation in concentration of ammoniacal N in leachates draining from recirculation (recycle) and control areas. Reproduced by permission of the Geological Society from *Quarterly Journal of Engineering Geology*, **17** (1984) 19–29.

to increased drainage from the site, has also given rise to problems from reduced infiltration through the waste surface by formation of a hard-pan of solids. Breaking up of the surface by furrowing using a site vehicle greatly increased infiltration in the second year and removed surface ponded liquid from the site, although problems again developed in the third year following heavy rainfall in March/April. It is concluded that, in order to reduce run-off of liquid and ponding in low-lying areas of the site, the surface should be broken up or furrowed at regular intervals. Careful consideration should also be given to the use of cover materials (soils) on areas where spraying is to take place. Although most soil top cover was removed before recirculation, small residual amounts of soil greatly reduced infiltration in parts of the site at Seamer Carr in the recirculation area, but apparently not on the control area which receives rainfall only.

The use of intermediate cover within the deposited wastes at Seamer Carr landfill has also produced 'perched' areas of saturation within the wastes. The development of this zone of saturation, within both the

TABLE 2. Comparison of Compositions of Leachate Draining from the Landfill at Seamer Carr and in Boreholes Within the Site (sampled September of 2nd year of trial)

	Leachate from control area	Leach-ate from recycle area	Bore-hole A5 liquid	Bore-hole B5 liquid	Bore-hole A6 liquid	Bore-hole A1 liquid
pH	5·97	5·77	5·95	7·15	7·13	7·32
COD	62 400	43 981	26 400	5 550	1 300	1 300
BOD	38 000	21 780	3 950	1 710	120	210
TOC	19 800	11 003	8 250	1 950	450	410
TVA[a]	14 609	10 056	8 445	714	41	38
Acetic	4 481	3 161	1 680	240	14	10
Propionic	3 794	2 190	1 654	418	19	22
i-Butyric	436	286	289	16	nd	nd
n-Butyric	4 690	3 040	1 750	5	8	3
i-Valeric	118	279	312	17	nd	3
n-Valeric	470	309	1 887	6	nd	nd
i-Caproic	nd	nd	12	6	nd	nd
n-Caproic	620	790	751	6	nd	nd
Ammoniacal-N	990	729	670	735	250	375
Org-N	770	234	180	12	25	20
Ox-N	3·5	3	nd	1	nd	nd
O-Phosphate	1·4	0·78	1·6	0·47	0·15	1·7
Chloride	2 760	2 059	1 750	1 935	405	1 300
Na	2 400	1 850	1 320	760	330	840
Mg	420	324	230	265	46	171
K	2 050	1 198	780	1 090	120	620
Ca	4 100	2 725	1 820	260	125	260
Cr	1	0·66	0·43	0·12	0·03	0·07
Mn	250	133	70	1·55	0·7	0·8
Fe	2 050	1 225	750	20	8	29
Ni	1·65	0·83	0·29	0·11	0·07	0·1
Cu	0·05	0·05	0·1	0·03	0·05	0·03
Zn	130	99	27·3	2·75	0·3	0·75
Cd	0·003	0·003	0·003	0·004	0·004	0·006
Pb	0·61	0·04	0·32	0·19	0·07	0·12

[a] TVA = total volatile acids as C.
Results in mg/litre except pH.
nd, not detectable.

control area and the area receiving recirculated leachate, was greatest in the latter area but may be giving rise to lateral movement of liquid across the site from areas of high infiltration (recycle area) to areas of low infiltration (control area). However, water levels in boreholes showed little or no hydraulic gradient between these two areas.

Lateral seepages of leachate over the edge of the liner have also developed and, although of negligible extent in the second year, these increased, particularly in the middle to the end of the third year, as the areas of perched saturation have developed (Fig. 3). This necessitated remedial work (construction of clay bunds on the edge of the site to help to reduce and eliminate these troublesome discharges). It is concluded that, at sites where recirculation is being contemplated, a thorough appraisal of the hydrology of the landfill is required to anticipate problems from the development of saturated areas and lateral seepages associated with intermediate cover. Experience at other landfills suggests that this phenomenon is not unique, and may be quite widespread.

Water Balance

Development of perched areas of saturation within the wastes has made it difficult to estimate a water balance for the site, because it is only possible to estimate approximately the thickness of saturated waste (because of uncertainty as to the position of the intermediate cover) and amounts of liquid in storage (becase of the unknown void space within the wastes).

Despite these difficulties, it was considered that it would be useful to estimate a water balance, even if this was only approximate, to allow comparison of the volumes of liquid within the various constituent parts of the landfill water cycle. This has been carried out on an annual basis for the three years of investigation.

For the purposes of this balance, it has been assumed that:

1. Unsaturated wastes are at field capacity.
2. Saturated wastes have a void space of 50% of their volume filled with water.
3. Intermediate cover is at a depth of 2 m below surface, and saturated conditions have developed across the whole site above this depth.
4. Discharge of liquid from the site only takes place via the drainage system, and additional surface discharges are small enough to be ignored (i.e. are within the flow measurement errors). Lateral discharges from the control area into the recirculation area are also ignored (see above).

394   Barber, Maris

TABLE 3. Water Balance for the 3 Years of Investigation Given by Rainfall
(R) + Recycled Leachate (L) = Run-off (RO) + Discharge (D) + Evaporation
(E) + Saturated Storage (SS)

	First year		Second year		Third year	
	R	C	R	C	R	C
Input						
Rainfall (m³)	7 933	5 790	6 736	4 938	5 930	4 328
Recycled leachate (m³)	3 025	0	3 756	0	11 572	0
Total input (m³)	10 958	5 790	10 522	4 938	17 502	4 328
Discharge						
Run-off (m³)	238	—	914	—	219	—
Drainage (m³)	2 999	460	3 099	961	4 840	1 107
Total discharge (m³)	3 237	460	4 013	961	5 059	1 107
% Input discharged	30	8	38	20	29	26
Evaporation (m³)	3 730	2 704	4 649	2 588	4 865	2 307
% Input loss by evaporation	34	47	44	52	28	53
Saturated storage (m³)	3 991	2 627	1 860	1 388	7 578	914
% Input in storage	36	45	17	28	43	21
Expected thickness of saturated zone[a]	0·7		1·0		1·85	
Measured thickness of saturated zone	0·6–1·1		1·01·6		1·5–2·0	

R was measured at local Meteorological Office stations; RO and D were continuously
measured on-site; E was estimated from Meteorological Office data; SS within 'perched'
areas of saturation was estimated from the 'water balance' by difference, giving expected
thickness of saturated zone, assuming 50% void space. Expected and measured thickness of
saturated zone compare well.
R, Recirculation area (9 340 m²); C, control area (6 817 m²).
[a] Additive years 1–3 inclusive.

5. Evaporative loss is similar to that from a bare soil surface and can
be estimated using average monthly data.

A summary of the water balance for the site is shown in Table 3,
where it can be seen that 82%, 75% and 200% of liquid discharged was
recirculated by spraying in the latter part of the first, in the second and
in the third year, respectively. A large proportion of discharged leachate
from the landfill was consequently recirculated back through the wastes
over a three-year period, although some of this liquid was derived from
run-off from other parts of the site, which accounts for the excess
amount of leachate recirculated to that discharged in the third year.

The amount of leachate discharged from the site has tended to increase with time, particularly in the last year when large volumes of leachate were recirculated (Table 3). Estimated evaporative losses of rainfall and recirculated leachate are based on average data for potential evaporation from a Meteorological Office station some 75 km distance from the site (the nearest station in a similar geographical position to Seamer Carr). These amount to *c.* 50% of influent rainfall on the control area, and 28–44% of influent liquids on the recirculation area (Table 3).

The difference between expected infiltration (total input − evaporation) and actual measured discharge of leachate has been used to estimate the volume of liquid in saturated storage within the landfill (Table 3). These storage terms are additive, and would give rise to water level rises of *c.* 0·7 m at the end of the first year rising to 1·85 m at the end of the third year, assuming that 50% of the volume of wastes is occupied by liquid in the saturated parts of the site. This figure has been arbitrarily chosen as attempts to determine this in the field were unsuccessful. Common soils often contain 40% by volume of moisture at saturation, whilst pulverized refuse might be expected to contain greater quantities, hence the use of the 50% figure.

This is justified to some extent by measured water levels in those boreholes terminated at 2–2·5 m depth in the first year, which are similar to the predicted level (Fig. 3). These indicate that the estimates of saturated storage, although approximate, are in reasonable agreement with observed values.

In general, it is concluded that although the volume of leachate requiring discharge off-site has been considerably reduced by evaporation of a proportion of recycled leachate, and by increase in storage within the landfill, the discharge of leachate from the site will continue to increase with time, and increasing volumes of leachate will need to be discharged off-site. This study, therefore, confirms the conclusions reached from smaller-scale studies that recirculation of leachate cannot be a total answer to the elimination of surface discharges of leachate, although the benefits in reducing the volume of leachate are obvious.

Quality of Leachate

The quality of leachate draining from the base of the site beneath the control area has remained reasonably constant over the three years despite large variations in infiltration.

In contrast, leachate draining from beneath the area receiving recirculated leachate has decreased in 'strength' since recirculation began,
with only minor reversals of this trend during the unusually cold winter
of the second/third year (Fig. 5). COD, total organic carbon (TOC) and
particularly BOD and volatile acids have shown a marked decrease in
concentration in leachates from the recycle area over the period of the
last two years, although chloride and ammonia do not show this trend.
These data indicate that processes within the wastes receiving recycled
leachate are developing for the removal from leachate of organic carbon,
principally volatile acids. Further confirmation of this is shown by the
decrease in COD relative to that of chloride in leachate with time (Fig.
8).

The relative concentrations of volatile acids are also different in
leachate from the control area (butyrate > acetate > propionate) compared with those from the recycle area (acetate > butyrate > propionate).

Samples of leachate from boreholes in the site derived from perched
areas of saturation within the wastes decrease in strength from the east
(Boreholes A5, B5) towards the central areas (Boreholes A2, B2) (Fig. 1).
The COD of leachates from the latter two boreholes is approximately
50–60 times lower than that in leachate draining from the base of the
site. Also, in general, acetate and propionate are much lower and greatly
predominate over butyrate in these low strength leachates (Table 2, Fig.
9).

A comparison of the composition of a range of leachate types, from
beneath and within the site, is shown in detail in Table 2 and the volatile
acid concentrations (acetate, propionate and butyrate) are compared in
Fig. 9. These illustrate the differences between high strength and low
strength leachates and gradations between the two extremes within the
site. In addition to the large differences in the organic fraction of these
leachates, pH value generally increases with decreasing strength, this also
being associated with a decrease in the concentration of total volatile
acids, Fe, Mn, Zn, Ni, Pb and Cr. The decreases in concentration of
these parameters are greater than the decrease in concentration of
chloride and are, therefore, not due simply to dilution.

It is concluded that increasing the moisture content of wastes by
recirculation of leachate has modified the leaching process and biochemical reactions within the wastes relative to those in the control area
receiving infiltration only from rainfall. It is apparent that conditions
within the unsaturated parts of the control area essentially remain within
the 'acid production' stage of breakdown of solid wastes, where their

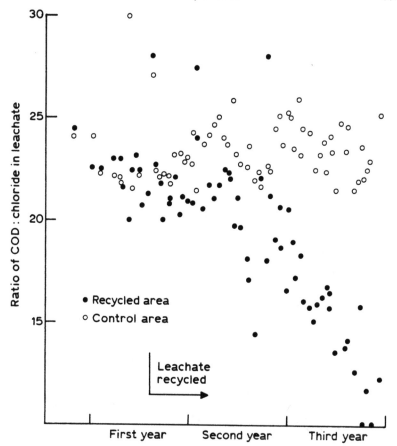

Figure 8. Variation with time in the ratio of COD:chloride in leachates draining from recirculation (recycle) and control areas. Reproduced by permission of the Geological Society from *Quarterly Journal of Engineering Geology*, **17** (1984) 19–29.

liquefaction gives rise to high concentrations of volatile acids in leachate. In the unsaturated parts of the recycle area, methanogenic conditions are developing. Biochemical reactions here involve the conversion of butyrate and propionate to acetate. Acetate, initially present in leachate and also formed from other volatile acids, is converted to methane and carbon dioxide by methanogenic bacteria. In saturated parts of the site, particularly the central recycle areas, methanogenesis is well established; this is shown by the extensive loss of volatile acids and production of

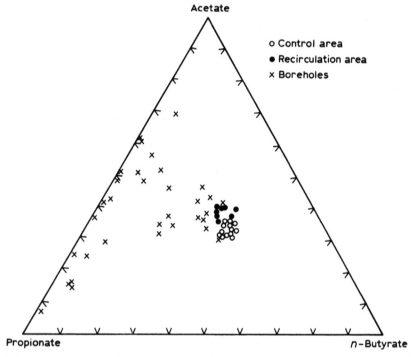

Figure 9. Variation in the relative amounts of acetate, propionate and *n*-butyrate in leachates expressed as percentages. Reproduced by permission of the Geological Society from *Quarterly Journal of Engineering Geology*, **17** (1984) 19–29.

methane. The higher pH value, which may in part reflect a loss of volatile acid, would also aid development of methanogenesis, which is inhibited at low pH values. The rapid development of methanogenesis in saturated parts of the site is considered to be largely brought about by the very high moisture content.

These variations in composition of leachates and conditions within the site are summarized in Fig. 10, where COD of leachate (normalized to concentration of chloride to remove 'dilution' effects) is plotted against pH value, for data collected between September of the second year and February of the third year of investigation. Further development of methanogenesis within the recycle area would produce low strength leachates of similar composition to those found in saturated central parts

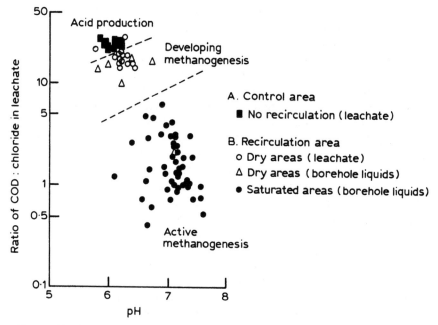

Figure 10. Interpretation of variation in conditions within landfill, related to ratio of COD:chloride and pH value of leachate. Reproduced by permission of the Geological Society from *Quarterly Journal of Engineering Geology*, **17** (1984) 19–29.

of the site. Recirculation of leachate will therefore be continued for a further period in an attempt to aid this development.

CONCLUSIONS

1. Results from the research programme show that the benefits of recirculation of leachate found in smaller studies can be obtained at full-scale, although longer periods of leachate recycle may be required (in excess of 2–3 years rather than the 18 months found in small-scale work) to produce low strength leachate.

2. Following commencement of spraying, between 80 and 100% of discharged leachate has been recirculated. Problems from surface ponding which could occur when final soil cover is used have been alleviated by furrowing at regular intervals.

3. A water balance for the site shows that, although there is a reduction in volume of leachate by evaporation, by far the greater contribution to limiting discharge of leachate is development of saturated storage in a perched water table on intermediate cover at a depth of 2 m within the landfill. This has given rise to lateral seepages of leachate over the liner, particularly in the third year as water levels increased, and this was minimized by construction of clay bunds around the edge of the site. It is clear that any study of recirculation at other sites should consider carefully the hydrology of the landfill and, in particular, the development of perched water tables within the wastes. Perched areas of saturation could develop above intermediate low-permeability cover.

4. Reduction in the strength of leachate can be achieved by increasing the moisture content of wastes, with greatest reduction occurring where wastes are fully saturated. Leachates similar to those from wastes emplaced for five years or more can be produced within a very short period of time in saturated areas of a site.

5. Although the degradable organic fraction of leachates can be reduced by leachate recirculation, the concentration of residual COD, ammonia and chloride may still be high, which would make it difficult for direct discharge of this low-strength leachate to a surface water course, unless considerable dilution was available.

6. Recirculation of leachate thus offers benefits in reducing the volume and strength of leachate, but cannot be considered to offer a complete answer to surface leachate discharge by itself. Other treatment of leachate at some stage may be required.

ACKNOWLEDGMENTS

The authors would like to acknowledge the help and encouragement of colleagues at WRC, particularly H. D. Robinson, C. P. Young and N. C. Blakey. The site study at Seamer Carr was largely funded by the EEC. The Department of the Environment funded this research in the last year; their financial assistance is gratefully acknowledged.

The site study could not have been carried out without the cooperation and help of the Scarborough District Council, North Yorkshire County Council and Yorkshire Water Authority. In particular we would like to thank Mr. J. Marshall, site Superintendent, and Mr T. Tuffs and staff at Seamer Carr. This chapter is published with the permission of the Director, WRC Environment.

REFERENCES

Doedens, H. & Cord-Landwehr, K. (1989). Leachate recirculation. In *Sanitary Landfilling: Process, Technology and Environmental Impact*, ed. T. H. Christensen, R. Cossu & R. Stegmann. Academic Press, London, UK.

Pohland, F. G. (1976). Landfill management and leachate recycle and treatment: an overview. In *Gas and Leachate from Landfills: Formation, Collection and Treatment*. ed. D. J. Genetelli & J. Cirello. Proceedings of a Research Symposium held at Rutgers University, New Brunswick, NJ, March 25 and 26, 1975. EPA-600/9-76-004. US Environmental Protection Agency, Cincinnati, OH, pp. 159–67.

Robinson, H. D., Barber, C. & Maris, P. J. (1982). Generation and treatment of leachate from domestic wastes in landfill. *Wat. Pollut. Control*, **81**, 465–78.

Tittlebaum, M. E. (1982). Organic carbon content stabilization through landfill leachate recirculation. *J. Wat. Pollut. Control Federation*, **54**, 428–33.

3.14 Co-treatment of Leachate with Sewage

LUDWIGA AHNERT & HANS-JÜRGEN EHRIG
*Universität Wuppertal, Abfall- und Siedlungswasserwirtschaft,
Pauluskirchstrasse 7, D-5600 Wuppertal, Germany*

INTRODUCTION

Biological co-treatment of sewage and sanitary landfill leachate is the
most common practice of leachate treatment in many countries. But in
contrast to this general practice, published operational descriptions and
treatment results are rare and often only presented as a qualitative
evaluation. On the other hand the potential environmental effect of
co-treatment is the subject of vehement discussions. The most important
discussion points are:

—to what degree is the leachate biologically degraded or diluted by
  sewage;
—the effects of leachate addition on nitrogen, BOD_5, and COD
  effluent concentrations;
—the behaviour of hazardous substances (heavy metals, organic
  micropollutants) during the treatment process and their potential
  toxic effects.

This chapter presents an overview of published data and summarizes the
results.

BIOLOGICAL CO-TREATMENT EXPERIMENTS

Knoch (1974) operated several laboratory-scale plants to compare
combined biological sewage and leachate treatment with biological

TABLE 1. Pollution of Leachate Used for Co-treatment
Experiments (Knoch, 1974)

Test 1	
BOD$_5$	5 800–6 500 mg/litre
COD	9 590–9 800 mg/litre
BOD$_5$/COD	0·60–0·66
TOC	3 200–3 920 mg/litre
Test 2	
BOD$_5$	15 500–17 240 mg/litre
COD	20 100–25 350 mg/litre
BOD$_5$/COD	0·68–0·77

leachate treatment. The additions of leachate during co-treatment experiments were 1, 2, and 5% by volume. Some leachate quality data are
presented in Table 1. The tests were run with two different leachate
qualities (Test 1 and Test 2). The activated sludge plants were operated
with an F/M-ratio of 0·16–0·32 kg BOD$_5$/kg MLSS . day. The lower
value represents sewage treatment and the higher value sewage with 5%
leachate addition. In every case no increase of BOD$_5$-effluent concentrations could be observed due to different percentages of leachate addition.
The increase of COD-effluents is shown in Fig. 1 for both tests. This
increase results from the higher final effluent COD-concentrations of
biologically-treated leachate (i.e. final COD-effluent concentration of
sewage ~50–100 mg COD/litre and of leachate ~800–1500
mg COD/litre). The COD-effluent concentration of a sewage treatment
plant increases due to the percentage of leachate addition. For Test 1 the
increase is similar to the calculated values using the COD-effluent
concentrations from leachate treatment experiments (activated sludge
process). The second test shows large differences between measured and
calculated values. These values originate from a biological batch plant in
laboratory scale where relatively high BOD$_5$-effluent concentrations of
40 mg/litre had been measured. The COD-effluent from continuous flow
systems operated under optimum conditions may be lower.

The difference between the measured effluent concentration of
1750 mg COD/litre from the batch test and the recalculated effluent
concentration of 600 mg COD/litre from the co-treatment experiment is
high and can partly be explained by an increase of the biodegradability of
the leachate when treated together with sewage.

Chian & DeWalle (1977) also described laboratory-scale biological

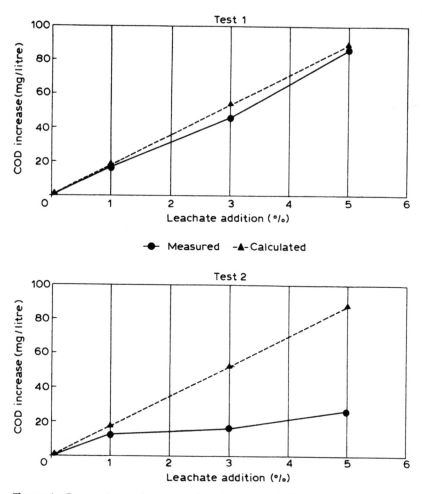

Figure 1. Comparison of measured and calculated COD-increase from two series of co-treatment experiments (calculated values from biological leachate treatment experiments) (Knoch, 1974).

co-treatment experiments. The leachate used was collected from a laboratory-scale lysimeter with BOD_5-concentration of 24 700 mg/litre and COD-concentration of ~49 300 mg/litre. The activated sludge treatment systems were operated at a constant F/M-ratio of 0·3 kg BOD_5/kg MLVSS . day. Only during the period of 2% leachate addition did the F/M-ratio increase to 0·6 kg BOD_5/kg MLVSS . day. The influence of leachate addition on BOD_5- and COD-effluent values is shown in Fig. 2, where the BOD/P-ratio in the effluent of the activated sludge plant is also presented. The highest BOD_5- and COD-effluent values during 4% leachate addition could also be the result of phosphorus deficiency, since our experiments show a decrease in biological degradation when BOD_5/P-ratios exceed 200.

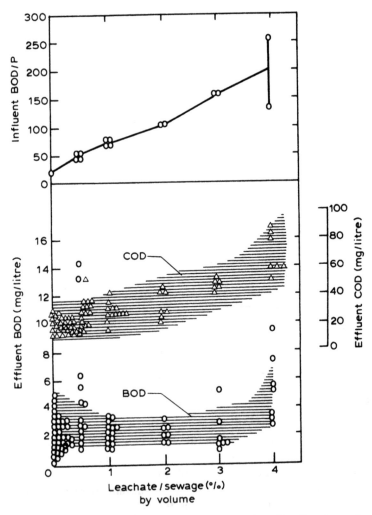

Figure 2. BOD_5- and COD-effluent values and BOD_5/P-ratios of co-treatment experiments (Chian & DeWalle, 1977).

Overall no significant increase of BOD_5-values could be observed, while the COD-effluent concentrations increased with increasing leachate addition. From this increase a theoretical COD-effluent concentration range of 600–900 mg/litre for separate leachate treatment could be calculated. This range corresponds with results from aerated lagoons with sufficient nutrient addition (Chian & Dewalle, 1977).

In all co-treatment experiments the sludge settling velocity is higher when leachate addition take place.

Co-treatment experiments using activated sludge plants were also presented by Rantala & Lehtonen (1980). Leachate quality and some relationships between parameters are shown in Table 2. Table 3 presents the operational conditions of the different units. The loading rates are similar to those used by Knoch (1974). After a period of different problems associated with Unit 2, Unit 1 and Unit 2 operated well with average BOD-reductions of 95% and COD-reductions of 85%. With Unit 3 the degradation decreased significantly after 30 days of operation. Rantala & Lehtonen (1980) concluded that the initial problems with Unit 2 and the failure of Unit 3 after 30 days were the result of the high loading rates. But in Unit 3 in the authors' opinion the high BOD_5/P-ratio of 198 may be the reason for this failure. Our experiments with biological treatment of similar low polluted leachate with sufficient nutrient addition have never shown any problems (see also Chapter 3.1).

TABLE 2. Quality of Leachate Used for Co-treatment Experiments (Rantala & Lehtonen, 1980)

	Mean	*Range*
BOD_7[a]	1 480	940–2 100
COD[a]	2 260	1 400–3 400
Total N[a]	71·6	50–101
Total P[a]	1·2	0·5–3·2
BOD_5[a]	1 268	Calculated[b]
BOD_5/COD	0·56	
BOD_5/P	1 057	
BOD_5/P	37	10% leachate addition
BOD_5/P	193	50% leachate addition

[a] All values in mg/litre.
[b] BOD_5 calculated from BOD_7 using a standard BOD-curve.

TABLE 3. Operational Conditions of Co-treatment Experiments (Rantala & Lehtonen, 1980)

Unit	1	2	3
Sewage (%)	90	50	0
Leachate (%)	10	50	100
Detention time (h)			
Mean	12·4	15·6	21·4
Range	11·8–12·9	14·4–16·7	20·4–24·6
F/M-ratio (kg BOD_7/kg MLSS . day)			
mean	0·2	0·42	0·35
range	0·11–0·27	0·29–0·66	0·26–0·64
F/M-ratio (kg BOD_5/kg MLSS . day) calculated[a]			
Mean	0·17	0·36	0·30

[a] BOD_5 calculated from BOD_7 using a standard BOD-curve.

On the other hand at lower temperatures the loading of 0·3 kg BOD_5/kg MLSS . day could be too high.

The nutrient requirement for aerobic biostabilization of landfill leachate was investigated by Temoin (1980). For those experiments Temoin also practised co-treatment in laboratory-scale aerated lagoons. The leachate quality is presented in Table 4. The detention time of these aerated lagoons was in all cases 20 days. The quantity of leachate addition, the calculated F/M-ratios and the BOD_5/P-ratios are presented in Table 5. The increase of BOD_5- and COD-effluents shown in Fig. 3 again could be the result of nutrient deficiency and in addition of biologically non-degradable organic compounds. The COD-increase calculated with effluent data from separate leachate treatment experiments

TABLE 4. Pollution of Leachate Used for Co-treatment Experiments (Temoin, 1980)

BOD_5[a]	18 025
COD[a]	28 830
total N[a]	560
BOD_5/COD	0·63
BOD_5/P	1 582

[a] All values in mg/litre.

TABLE 5. Operational Conditions and BOD$_5$/P-Ratios of Co-treatment Experiments (Temoin, 1980)

Unit	A	B	C	D	E	F
Leachate addition (%)						
	0	1	3	6	10	20
F/M-ratio (kg BOD$_5$/kg MLSS . day)						
	0·05	0·06	0·08	0·08	0·08	0·08
F/M-ratio (kg BOD$_5$/kg MLVSS . day)						
	0·06	0·09	0·09	0·12	0·12	0·13
BOD$_5$/P	23	94	222	385	345	833

corresponds to a high degree with the measured effluent COD-concentrations. On the other hand the higher BOD$_5$-effluent concentrations from separate leachate experiments give an indication that the basis value for this calculation could be too high. This tendency could be to a certain degree the result of nutrient deficiency. Analyses of the sludge produced show a significant increase in some metal concentrations, such as cadmium, chromium, lead, and zinc.

Kelly (1987) reported pilot-scale co-treatment experiments using activated sludge systems. The leachate was of relatively low concentration with COD = 1167 mg/litre, BOD$_5$ = 373 mg/litre and NH$_4$-N = 71 mg/litre. Operational conditions and COD-effluent values are shown in Table 6. During these experiments the temperatures in the plant went down to 4·6°C. The strongest increase in COD-effluent concentrations occurred during Test 2. For this series the undetermined loading rates could be estimated from the first series as nearly twice as high. This could be the reason for the high increase in the COD-effluent concentrations. There are no separate leachate treatment experiments with which to compare the increase of the effluent COD-concentrations. But for such low concentrated leachate with a BOD$_5$/COD-ratio of 0·32 theoretically the increase must be low. During these experiments the metal content in the sludge produced was also measured, but no significant increase could be observed in the co-treatment systems. This is due to very low metal concentrations in the leachate.

Doedens & Cord-Landwehr (1984) reported results from a full-scale sewage treatment plant where temporarily leachate was added. This small plant with an F/M-loading of about 0·05 kg BOD$_5$/kg MLSS . day

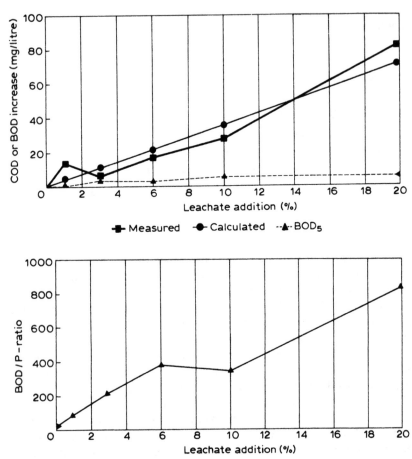

Figure 3. Measured COD- and BOD$_5$-concentrations, calculated COD-increase and BOD$_5$/P-ratios derived from co-treatment experiments (calculated values from separate biological leachate treatment experiments) (Temoin, 1980).

accepted a leachate volume of 3·2% of the sewage over a period of 10 days. The low polluted leachate (COD = 1552 mg/litre, BOD$_5$ = 80 mg/litre) had no effect on COD-effluent concentrations. The mean effluent values (respectively 26 mg COD/litre with and 31 mg COD/litre without leachate addition) were nearly the same. Based on our laboratory-scale leachate treatment experiments using the same leachate without sewage an increase of up to 30 mg COD/litre could be estimated.

TABLE 6. Leachate Addition, Operational Conditions and COD-Concentrations (Influent and Effluent) of Co-treatment Experiments (Kelly, 1987)

Test	1	2	3	4
Leachate addition (%)	4	4	4	16
Detention time (days)	8	20	15	14/12[a]
F/M-ratio (kg BOD_5/kg MLVSS . day)				
control	0·15	ND	0·06	0·18
test	0·17	ND	0·10	0·22
COD (mg/litre) (control unit)				
influent	267	297	187	200
effluent	60	72	60	74
COD (mg/litre) (test unit)				
influent	276	457	236	357
effluent	80	173	107	110

[a] Differences between control and test unit.
ND, not determined.
Control unit = pure sewage treatment.
Test unit = co-treatment experiments.

RESULTS OF CO-TREATMENT EXPERIMENTS

Most experiments of sewage and leachate co-treatment deal with the degradation of organic leachate compounds and the remaining undegradable COD-values in effluents. In exactly defined laboratory-scale experiments an increase of COD-effluents could be observed. But in some cases the increase is not so high as expected from separate leachate treatment experiments. Measuring this increasing effect at plants with variable flow rates and/or influent concentrations is much more difficult. Higher effluent concentrations could also result from increasing loading rates or from nutrient deficiencies. An increase of COD reduction due to co-treatment with sewage in a relatively small manner cannot be excluded from some experiments, but the reason for this observation might be advanced biological processes as well as additional precipitation and/or adsorption processes.

In some publications operational problems like unstable processes or sludge bulking are discussed. But it cannot be concluded whether these problems are primarily a consequence of leachate addition or the result of neglected basic biological treatment conditions. In many cases, as a result of leachate addition, the F/M-ratios increase sharply. But the plants are operated in the same manner as without leachate addition. As already mentioned, sufficient nutrients for biological treatment have to be provided. A typical result of nutrient deficiencies is a very light sludge (Temoin, 1980). A consequence of an increase in F/M-ratios could be a decrease of the nitrification rate resulting also in an unstable biological process.

Only a few investigations deal with the fate of metals during biological co-treatment. It could be expected that with higher leachate additions and high heavy-metal concentrations in the leachate (during the acetic phase of a landfill), a certain proportion of these metals remain in the sludge (Temoin, 1980). No previous investigations have been made on the behaviour of organic micropollutants (e.g. adsorbable organic halogens) during co-treatment.

CONCLUSIONS ON BIOLOGICAL CO-TREATMENT

The published data of experiments show that co-treatment is a possible method of leachate treatment. On the other hand worldwide experience on full-scale plants indicates that this is a proven way of leachate treatment that works without problems if the percentage of leachate addition to the sewage is low (e.g. $\leq 0.5\%$ by volume). If the loading is higher, the plant has to be operated and/or reconstructed in such a way that the same loading conditions are maintained as they were without leachate addition. If this in the case no increase in nitrogen- and/or BOD_5-effluent concentrations can be expected. A restriction of leachate addition to sewage may be necessary if the sewage is to be used as a carbon source for denitrification of the nitrogen of both of the wastewaters (sewage and leachate from old landfills with low organics). Table 7 shows the quantity of sewage (expressed as wastewater equivalents for a certain number of inhabitants) which is necessary as a carbon source for the full denitrification of leachate. It should be mentioned that costs may increase owing to increasing process control required by increasing ammonium influent concentrations for nitrification. Nitrification without expensive monitor-

TABLE 7. Necessary Amount of Sewage (Expressed in Inhabitants) for the Denitrification of Nitrogen from Low Concentrated Leachate (from Landfills in the Methanogenic Phase)

Mean sewage data (Germany)	
Sewage production rate	200 litres/inhabitant per day
BOD$_5$	300 mg/litre
Nitrogen	60 mg/litre
Mean leachate data (old landfills)	
BOD$_5$	300 mg/litre
Nitrogen	1 000 mg/litre

Activated sludge plant:

with pre-settling, sewage from approximately 700 inhabitants is necessary to denitrify the nitrogen of 1 m^3 leachate
without pre-settling, sewage from approximately 120 inhabitants is necessary to denitrify the nitrogen of 1 m^3 leachate

ing systems could be operated in a stable manner up to 100–150 mg ammonium-N/litre in the influent. Using the data of Table 7 this maximum is reached with approximately 4–10% by volume of leachate addition.

The nutrient requirement of the biomass also must be considered for many other wastewaters. Using sewage data from Germany in respect of the phosphorus content produced by each inhabitant per day approximately 8–9 kg leachate BOD$_5$ could be biologically degraded without nutrient deficiency (maximum BOD$_5$/P-ratio = 200). The increase of nondegradable leachate components (residual COD, halogens, metals) is a function of dilution reduced by additional precipitation and/or adsorption effects. Table 8 presents an estimation of possible leachate addition (in %) for every 10% COD- or AOX-increase during biological co-treatment. The concentrations of heavy metals are also relatively low in non-treated leachate. Only in leachates from acetic phase landfills with high organic concentrations could zinc values also be high. During biological treatment most of the zinc precipitates and accumulates in the sludge. In calculating the increase of zinc concentration in the sludge, the sludge production from leachate degradation (approximately 0·6–0·8 kg MLSS/kg BOD$_5$-degraded) also has to be considered.

In most cases it is necessary to co-treat leachate in an existing sewage treatment plant. In this case the first step has to be the comparison of the

TABLE 8. Calculation of Maximum Leachate Addition for a COD- or AOX-Effluent Increase of 10%

Mean effluent concentrations from biological sewage treatment plants (Germany)

COD	60 mg/litre
AOX	0·05 mg/litre

Mean leachate effluent concentrations from biological leachate treatment plants

COD	500–1 500 mg/litre
AOX	0·3–2·7 mg/litre

Maximum volumetric leachate addition for 10% COD- or AOX-effluent increase without respecting precipitation and/or adsorption effects

COD	0·4–1·2%
AOX	0·2–1·7%

AOX, adsorbable organic halogens.

actual and estimated future loading with the designed loading rate (nitrogen- and BOD_5-loading). The difference in loading could be used for co-treatment of leachate. If there are no differences between both values, in some cases the flexibility of biological systems could be used. That means overloading the plant by approximately 10% may have, in

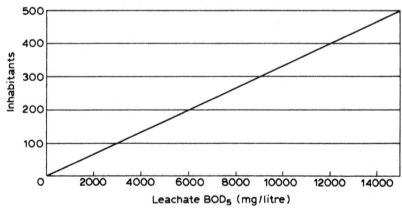

Figure 4. Size estimation of a sewage treatment plant (expressed in inhabitants) for the daily acceptance of 1 m³ leachate with the given BOD_5-concentrations in order to prevent overloading of more than 10%.

general, no negative effect on the biological treatment processes. As an example, for the treatment of 1 m³ of leachate per day with 1000 mg nitrogen/litre, an activated sludge plant sized for approximately 170 inhabitants is necessary. A similar calculation for a flow rate of 1 m³ leachate per day with BOD_5-concentrations up to 15 000 mg/litre is shown in Fig. 4. For each 1 m³ of leachate per day a sewage treatment plant sized for 500 inhabitants is necessary to prevent overloading by more than 10%.

As a conclusion it can be stated that biological co-treatment of sewage and sanitary landfill leachate is a proven technology and operates well, if the treatment plant is carefully designed and operated. But the demand today for more extensive leachate treatment in some countries (such as Germany) shows the need for more sophisticated treatment systems with much higher effluent standards (see also Chapter 3.16).

REFERENCES

Chian, E. S. K. & DeWalle, F. B. (1977). *Evaluation of Leachate Treatment*, Vol. II: Biological and Physical-Chemical Processes. University of Illinois, Urbana, IL.

Doedens, H. & Cord-Landwehr, K. (1984). Sickerwasser-Kreislaufführung auf Deponien—neue Erkenntnisse und betriebliche Varianten (Leachate recirculation—new experiences and operational methods). *Müll und Abfall*, **16**, 68–27.

Kelly, H. G. (1987). Pilot testing for combined treatment of leachate from a domestic waste landfill site. *Journal WPCF*, **59**, 254.

Knoch, J. (1974). Gemeinsame biologische Reinigung von Müllsickerwasser und kommunalem Abwasser (Biological co-treatment of sewage and leachate). Proceedings: Reinigung von Sickerwasser aus Mülldeponien Kolloquien und Seminare im Hause Edelhoff. Vol. 3.

Rantala, P. & Lehtonen, E. (1980). Leachate from landfill sites and its possible purification. *Vesitalous*, **21**, 16–19, 37.

Temoin, E. P. (1980). Nutrient requirements for aerobic biostabilization of landfill leachate. The University of British Columbia, Canada.

BIBLIOGRAPHY

Jank, B. E. (1980). Impact of leachates on municipal treatment systems. Proceedings: Leachate Management, University of Toronto, Canada, November 1980.

Kayser, R. (1986). Leistungsfähigkeit kommunalen Kläranlagen für die Sickerwasserbehandlung (Capacities of sewage treatment plants for leachate treatment). Proceedings: Deponiesickerwasser-behandlung, Aachen, Germany.

Kayser, R. & Ehrig, H.-J. (1985). Grenzen und Möglichkeiten der Einleitung
von Deponiesickerwässern in kommunale Kläranlagen (Limitations and
possibilities of the discharge of leachate into sewage treatment plants).
Proceedings: Sicherheit und Risiko wassertechnischer Anlagen, 18. Essener
Tagung 1985, Gewässerschutz–Wasser–Abwasser, Vol. 75.

Lenzen, J. & Schuck, P. (1986). Deponiesickerwasser in der kommunalen
Kläranlage—ein Erfahrungsbericht (Leachate in sewage treatment plants—a
report of experiences). Proceedings: 4. ATV-Dokumentation: Deponiesickerwasser 10.11.86, Essen, Germany.

3.15   Effluent Requirements and Related Leachate Treatment Processes

HEIKO DOEDENS & ULF THEILEN

Institut für Siedlungswasserwirtschaft und Abfalltechnik Universität Hannover, D-3000 Hannover, Germany

INTRODUCTION

Many of the bioresistent and non-degradable substances which are present in landfill leachate (see Chapter 2.4), such as organic halogenated compounds (AOX) and heavy metals, are today classified as 'hazardous components'. In Germany, 'state of technology' is required for the wastewater treatment of such hazardous substances. For BOD_5, COD and NH_4-N, however, the requirements are made according to the 'generally accepted rules of technology'. On the contrary, 'state of technology' is defined as advanced processes or operational modes by means of which it can be ensured that the emissions are limited to a certain degree which has to be fixed. This means that processes which have been proven in other fields of application also belong to the state of technology.

An overview of the effluent requirements prescribed in the legislation of different European countries is given in Table 1.

MEASURES TO CONTROL LEACHATE QUANTITY AND QUALITY

A leachate treatment plant can only work at optimum efficiency if the leachate effluent is relatively constant with regard to amount as well as concentration.

417

TABLE 1. Requirements in Germany for the Discharge of Leachate Compared with Other States, Direct (into Rivers) and Indirect (into Municipal Wastewater Treatment Plant)

	Germany		Switzerland		Italy	Netherlands		Austria	
	indirect discussed (1989)	direct discussed (1989)	indirect (1988)	direct (1988)	direct (1976)	indirect	direct	indirect (1981)	direct (1981)
COD (mg/litre)	400[b,c]	(150) 200[b]	—[d]	—[d]	160	—	—	—	75,90 max
BOD₅ (mg/litre)	—	20[b]	—[d]	20	40	300–400[d]	7–20[d]	—	20,25 max
SS (mg/litre)	20[a]	20[a]	—[d]	20	80	—	—	—	30,(50)
Cl⁻ (mg/litre)	—	—	—[d]	—[d]	1 200	—	—	—[d]	—[d]
SO₄²⁻ (mg/litre)	—	—	—[d]	300	1 000	300	500	—	—[d]
TKN (mg/litre)	—	—	—[d]	—[d]	—	300	8–15[d]	—	—[d]
NH₄-N (mg/litre)	—	(10), 50[b]	—[d]	—[d]	12	200	4–8[d]	—	—[d]
NO₃-N (mg/litre)	—	—	—[d]	—[d]	20	—	—	—	—[d]
NO₂-N (mg/litre)	—	(10)[b]	3	0·3	0·6	—	1–4[d]	9	1·5
Pb (μg/litre)	500[a]	500[a]	500	500	200	200	50	1 000	1 000
Cd (μg/litre)	100[a]	100[a]	100	100	20	10	2·5	100	100

Cr(IV) (μg/litre)	—[a]	—	500	100	200	375	75	100	100
Tot-Cr (μg/litre)	500[a]	500[a]	2 500	2 100	2 200	—	—	2 100	2 100
Cu (μg/litre)	500[a]	500[a]	1 000	500	100	250	50	1 000	1 000
Ni (μg/litre)	500[a]	500[a]	2 000	2 000	2 000	170	100	2 000	2 000
Hg (μg/litre)	50[a]	50[a]	10	10	5	2	0·5	10	10
Se (μg/litre)	—	—	—	—	30	—	—	—	—
Zn (μg/litre)	2 000[a]	2 000[a]	2 000	2 000	500	1 000	200	3 000	3 000
Sn (μg/litre)	—	—	2 000	2 000	10 000	—	—	2 000	2 000
Fe (mg/litre)	—	—	20	2	2	—	—	—[d]	2 000
AOX (μg/litre)	500[a]	500[a]	—	—	—	—	—	—	—
EOX (μg/litre)	—	—	—	—	—	5	5	—	—
Phenol (μg/litre)	—	—	5 000	50	500	—	—	20 000	100
Tot-HC (mg/litre)	—	—	20	10	—	—	—	—	—
Fishtox. class	(2)[a]	2	—	0–5[d]	2	—	—	—	—

[a] Requirement meets the 'state of technology' (modification by advanced processes possible).
[b] Requirement meets the 'generally accepted rules of technology'.
[c] Or reduction of 75% by biodegradation.
[d] Depends on quality and size of the river or the wastewater treatment plant.

The considerable variations which are observed in landfill leachate quantity (Canziani & Cossu, 1989) and quality (Chapter 2.4) cannot be easily equalized in a leachate treatment plant. Thus internal steps in the landfill are necessary to achieve a balance and a minimization effect. The following measures can be undertaken:

(A) Measures to decrease and regulate leachate quantity.

—surface sealing of closed landfill parts with the possibility of leachate recycling underneath the sealing to counteract drying up;
—spraying of stabilized leachate with very little smell ($BOD_5/COD <$ 0·1) with amounts <5 mm/day to utilize evaporation during the six months of summer;
—seasonal storage of leachate peaks (specific storage capacity combined with spraying: 500 m^3/ha recommended).

(B) Measures to decrease and regulate leachate concentrations. The aim of these measures is to accelerate the acid phase of the biodegradation process. Among the different measures, installation of a bottom layer of pre-composted domestic waste or compost, deposition of wastes in thin layers, and water concentration adjustment (50–60%) by leachate recirculation are often applied in Germany.

PROCESS COMBINATIONS FOR LEACHATE TREATMENT BASED ON STATE OF TECHNOLOGY

The process combinations under consideration in Germany, on the basis of the current 'state of technology' needed in order to respect the stringent discharge limits previously discussed, are graphically shown in Figs 1, 2 and 3, and discussed hereafter.

The process combinations include different steps as follows:

Combination A (Fig. 1).

—internal anaerobic pre-treatment;
—biological treatment (nitrification, denitrification) using external organic carbon sources for denitrification;

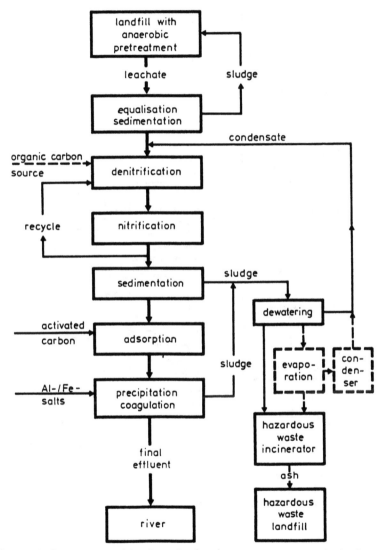

Figure 1. Process combination A: leachate treatment with biological-physical-chemical combination, pre-treatment in the landfill, nitrification, denitrification, activated carbon, precipitation, coagulation; including final sludge treatment.

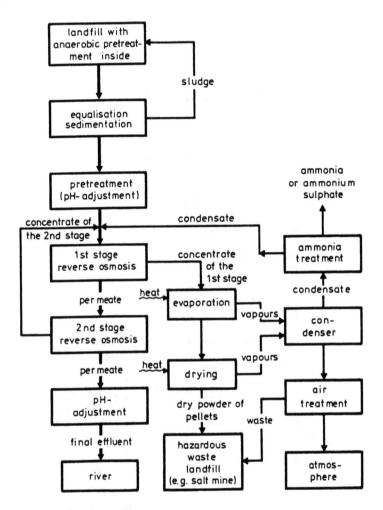

Figure 2. Process combination B: leachate treatment with biological-physical combination, pre-treatment in the landfill, two-stage reverse osmosis, evaporation and drying of the concentrate.

—adsorption by activated carbon (powdered carbon);
—precipitation/coagulation;
—sludge treatment (combustion in a hazardous waste incinerator plant).

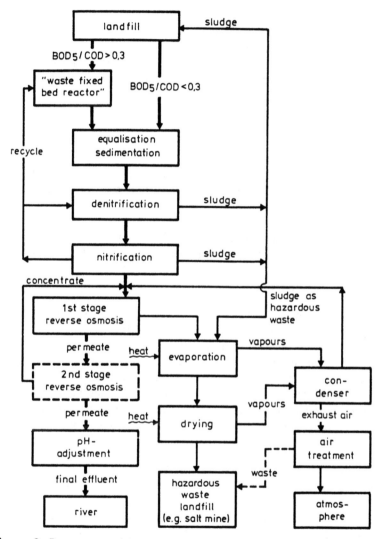

Figure 3. Process combination C: leachate treatment with biological-physical combination, anaerobic pre-treatment in a 'waste-fixed-bed reactor' if organic influent concentrations are high (BOD$_5$/COD > 0·3), biological treatment with nitrification and denitrification (partly in the waste-fixed-bed reactor), one- or two-stage reverse osmosis, evaporation and drying of the concentrate.

Combination B (Fig. 2).

—internal anaerobic pre-treatment;
—two-stage reverse osmosis plant;
—vaporization of the concentrate until it is converted into a dry solid substance (e.g. pellets) so that it can be stored in a hazardous waste landfill;
—vapour and condensate treatment.

Combination C (Fig. 3).

—biological pre-treatment (nitrification, denitrification) using the organic substances of the waste as carbon source in an external 'waste-fixed-bed reactor';
—one- or two-stage reverse osmosis plant;
—vaporization of the concentrate until it is converted into a dry solid substance (e.g. pellets) so that it can be stored in a hazardous waste landfill;
—vapour and condensate treatment.

The process combinations described above were tested at the University of Hannover and the results are shown in Table 2.

In contrast to reverse osmosis, which acts only physically, the biological treatment with nitrification and denitrification has the decisive advantage that the substances are readily converted and not concentrated, especially concerning organic compounds and ammonia.

The combination of biological treatment with activated carbon adsorption and precipitation/coagulation has the following remarkable disadvantages:

—For high denitrification rates even with leachate from the methanogenic phase, external organic carbon sources (e.g. methanol) have to be used.
—For undiluted leachate the COD and AOX limits can only be achieved if activated carbon is used to a considerable extent (up to 20 kg/m^3 of leachate).
—Using precipitants, the leachate (which is very saline itself) undergoes an even higher concentration of salt. Therefore, it is questionable whether the fish toxicity class 2 can be achieved.
—There are large amounts of sludge (especially activated carbon) which have to be disposed of as hazardous waste.

TABLE 2. Comparison of the Various Effluent Values which Can Be Achieved Using Different Process Combinations According to 'State of Technology' (Test Results of the Authors), Compared with German Requirements

Parameter	Leachate after int. pre-treatment	Nitri./deni. + pre-cipitation/coagulation (FeClSO$_4$) and a-carbon		Reverse osmosis conc. fact. 5		Nitri./deni. reverse osmosis		Requirements discussed in Germany direct
		(5 g/litre)	(20 g/litre)	1-stage (pH 6·8)	2-stage (pH 6·8)	1-stage	2-stage	
COD (mg/litre)	4 000	750[a]	500[a]	<100	<20	<50	<10	150–200
BOD$_5$ (mg/litre)	500	<10	<10	<30	<10	<10	<2	20
TKN (mg/litre)	2 000	130	130	400	50	<20	<5	—
NH$_4$-N (mg/litre)	1 800	<5	<5	360	40	<5	<1	10–50
NO$_3$-N (mg/litre)	<10	400	400	<5	<2	<35	<5	—
NO$_2$-N (mg/litre)	<1	<2	<2	<1	<1	<1	<1	—
AOX (µg/litre)	4 000	550[a]	300[a]	<800[b]	<200[b]	<500[b]	<50[b]	500
Cl$^-$ (mg/litre)	3 000	3 500	3 500	300	100	300	100	—
Fe (mg/litre)	10	>10		<2	<1	<2	<1	—
Zn (mg/litre)	1	<2		<0·5	<0·1	<0·5	<0·1	2
Cu (µg/litre)	<100	<25		<0·5	<0·1	<0·5	<0·1	500
Ni (µg/litre)	<200	<150		<5	<1	<5	<1	500
Cr (µg/litre)	<300	<200		<10	<2	<10	<2	500
Pb (µg/litre)	<50	<150		<10	<1	<10	<1	—
Hg (µg/litre)	<5	<1		<0·2	<0·1	<0·2	<0·1	50
Cd (µg/litre)	<5	<1		<0·2	<0·1	<0·2	<0·1	100
Fishtox.	?	?	?	<2	<1	<2	<1	2

[a] Effluent values determined on bench scales, probably improvable on large-scale plants.
[b] AOX effluent value with reverse osmosis dependent on molecule size.

—The specific costs including the disposal of residues as hazardous waste are higher than for a reverse osmosis treatment with evaporation of the concentrate. This is caused by the high demand for activated carbon.

Reverse osmosis treatment without biological pre-treatment raises the following problems:

—The ammonium retention is limited so that even with a two-stage RO-plant it cannot be guaranteed that the limit of 10 mg NH_4-N/litre is maintained.
—High concentrations of organic substances as well as ammonium within the concentrate lead to process engineering problems during the concentrate treatment (evaporation): foaming, material problems, exhaust air problems (hydrocarbon).
—The danger of fouling and scaling at the membranes caused by high concentrations of organic and precipitating inorganic components must be taken into account.

When using the combination 'biological pre-treatment with RO-post-treatment' the advantages of both treatment processes are used and the disadvantages of the single RO-treatment are avoided. The 'waste-fixed-bed reactor' can be used as anaerobic pre-treatment *and* denitrification reactor without external carbon sources. The effluent values which are achieved with this combination remain to all effects under the quality standards required in Germany.

The combination of biological and chemical-physical treatment is adopted in the sanitary landfills of Minden-Heisterholz and Minden-Pohlsche-Heide, Nordrhien-Westfalia (Albers & Mennerich, 1987; Albers & Krückeberg, 1988).

A two-stage reverse-osmosis plant with post-treatment for the concentrate including vapour treatment is planned for a few landfills in Germany, for example the landfills Karlsruhe-Ost, Baden-Württemberg, and Aachen (landfills Maria Theresa and Warden) in Nordrhien-Westfalia (Kollbach, J.St. (1989), Enviro-Consult, Aachen, Germany, pers. comm.).

As an example of a combination of biological pre-treatment and reverse osmosis post-treatment the leachate treatment plant at Mechernich landfill in the region of Euskirchen in Nordrhien-Westfalia, Germany, should be mentioned (Kocks & Seyfried, 1987; Doedens & Theilen, 1989).

COSTS

A question that still remains for all processes concerns the final storage of the residues. If in Germany the decision is reached that no sort of residues resulting from the leachate treatment may flow back to its landfill 'of origin'—with regard also to landfills which are sealed according to the state of technology—the amount of residues that must be disposed of as hazardous waste will become the main problem of leachate treatment. Then, the sludge amounts resulting from precipitation/coagulation and, especially, activated carbon treatment lead to considerably higher costs than the disposal of a solid from evaporation of the concentrate resulting from reverse osmosis treatment.

It is questionable whether by means of the dosage of activated carbon of 10 kg/m³ leachate, or higher as mentioned in the following cost estimation, the future limits can be maintained, especially in respect of COD (see Table 2). With higher dosages the costs for these treatment processes rise correspondingly.

Statements about reliable costs and cost comparisons can be indicatively estimated (reference year 1990). Estimates for a plant with 100 m³/day maximum capacity resulted in the following specific costs per m³ of leachate:

—Combination biology/a-carbon/precipitation–coagulation (10 kg a-carbon/m³).
    Without hazardous waste incineration 45–55 DM/m³, 25–30 $/m³
    With sludge incineration 70–80 DM/m³, 35–40 $/m³
—Two-stage reverse osmosis with concentrate evaporation.
    Without hazardous waste disposal 45–55 DM/m³, 25–30 $/m³
    With hazardous waste disposal 60–70 DM/m³, 30–35 $/m³
—Combination biology with two-stage reverse osmosis and evaporation of the concentrate.
    Without hazardous waste disposal 60–70 DM/m³, 30–35 $/m³
    With hazardous waste disposal 75–90 DM/m³, 40–50 $/m³

For a production of 2000 m³ leachate/10^4m² . a, costs varied from approximately 100 000 up to 180 000 DM/10^4 m² . a or 55 000 up to 100 000 $/$10^4$m² . a, depending on the process combination.

These considerable costs of leachate treatment according to the state of technology in Germany show that it is absolutely essential to minimize and regulate the amount of leachate.

428 *Doedens, Theilen*

REFERENCES

Albers, H. & Krückeberg, G. (1988). Combined biological and physical/chemical treatment of sanitary landfill leachate. Proceedings of the 5th International Solid Wastes Conference, ISWA, Copenhagen. Vol. 1, pp. 123–30.

Albers, H. & Mennerich, A. (1987). Chemisch/physikalische Nachreinigung von Deponiesickerwasser in Minden-Heisterholz (Physical-chemical post-treatment of landfill leachate). *Müll and Abfall*, (8), 326–37.

Canziani, R. & Cossu, R. (1989). Landfill hydrology and leachate production. In: *Sanitary Landfilling: Process Technology and Environmental Impact*, ed. T. H. Christensen, R. Cossu & R. Stegmann. Academic Press, London.

Doedens, H. & Theilen, U. (1989). Stand der Sickerwasserdiskussion (State of discussion on leachate problem). Lecture on Vertieferseminar Zeitgemäße Deponietechnik III, Universität Stuttgart, Germany.

Kocks & Seyfried, C. F. (1987). Genehmigungsentwurf für die Sicker-wasserbehandlungsanlage der Deponie Mechernich des Landkreises Euskirchen. (Design proposal for the leachate treatment plant in the landfill Mechernich Landkreises Euskirchen). Kocks Consult GmbH, unpublished.

4. ENVIRONMENTAL ASPECTS OF LEACHATE

4.1 Long-Term Leachate Emissions from Municipal Solid Waste Landfills

HASAN BELEVI & PETER BACCINI

Swiss Federal Institute of Water Resources and Water Pollution Control, CH-8600 Dübendorf, Switzerland

INTRODUCTION

The 'Guidelines to waste management in Switzerland' (EKA, 1986) were set in 1986. One essential objective of these guidelines is that all waste management procedures have to produce materials which either are recyclable or can be disposed of in a landfill without an adverse environmental impact for long-term periods. This kind of landfilling is defined as 'final storage' and the wastes in the final storage have 'final storage quality'. Emissions from a final storage are compatible with the environment without any further treatment. Another important objective of the Swiss waste management policy is that each generation handles its waste to a status of final storage quality. As a consequence of this policy, municipal solid waste landfills should reach the final storage quality within about 30 years after disposal.

The final storage concept focuses mainly on the solid waste rather than the artificial or natural barriers around the landfill body. The landfill body has to reach an 'inert' state so that the emissions from the landfill are compatible with the environment for long-term periods irrespective of the retardation and attenuation capacities of surrounding materials. However, this inert state depends on geochemical properties of the landfill site. The compounds in the inert landfill body may become 'mobile' when the physical and chemical conditions in the landfill change. Consequently, the interaction of the landfill with the environment must also be taken into account.

Even though the final storage concept mainly focuses on the landfill body and chemical, microbiological and physical processes in the landfill body, natural and artificial barriers are also a part of the final storage concept. A proper lining and a proper geological environment are basic prerequisites for final storages. They are necessary for containment, for monitoring and, last but not least, for security reasons (Baccini, 1989).

Disposed of in a landfill, municipal solid waste (MSW) will contact water, which enters the landfill continuously through precipitation. As a result of this contact, several chemical and microbiological reactions occur. Organic compounds can be transformed to other organic compounds or inorganic compounds. Inorganic compounds can undergo several chemical reactions and can be transformed to other inorganic compounds. The products of these reactions and parts of non-reacted MSW can be transported by leachate and by gas. Furthermore, many physical processes, such as adsorption, dissolution, precipitation, etc., can take place simultaneously. Consquently, a MSW landfill can be regarded as a 'partly continuous chemical and microbiological fixed bed reactor'. In the present chapter, it is also considered as a treatment facility where the objective it is to obtain a landfill body of final storage quality.

Significant research and monitoring of landfills is a recent development, controlled landfills having existed for 20 years or so. The existing controlled landfills are therefore still in the intensive reactor phase in which intensive microbiological decompositions occur. The behaviour of landfills in this period can be assessed more or less accurately (Baccini *et al.*, 1987). However, no experience exists with regard to the long-term behaviour (>20 years) of MSW landfills. Because of the highly complex nature of the processes, an exact prediction of long-term behaviour of MSW landfills is almost impossible. In the present chapter, a rough estimation is made to assess this behaviour. Leachate flux measurements, element concentrations in leachates of MSW landfills and leaching experiments in the laboratory with drilling core samples from MSW landfills are used for this purpose. The time required to reach the final storage quality is then assessed.

METHODS

The leachate and the gas leaving the landfill can contain many environmental pollutants, such as xenobiotics and heavy metals. The important gas production, however, lasts only about one or two decades (Stegmann, 1978/79; Ehrig, 1986). The gas must be treated to minimize environmental impact during this period. After gas production has reached a very low level, the gas-borne pollutants are negligible and the leachate becomes the most important flux with environmental impact. Consequently, the long-term emission potential of a landfill is determined by the further evolution of leachate fluxes and leachate pollutants.

The evolution of gas and leachate fluxes in the intensive reactor phase was studied by investigations of four comparable MSW landfills of different ages over a one-year period. These investigations are described elsewhere (Baccini *et al.*, 1987). They indicated that the leachate mass flux is stabilized after the filling stops. Assuming a more or less constant annual precipitation, the annual leachate mass flux will also remain constant after the intensive reactor period.

Non-metals (carbon, nitrogen, fluorine, phosphorus, sulphur and chlorine) and metals (iron, copper, zinc, lead and cadmium) are chosen for the characterization of the leachate. The evolution of the concentrations of these elements in leachates and a comparison with the quality standards are used to discuss the final storage quality of the MSW landfills. It is assumed that the concentrations in the leachate decrease with a first order remobilization rate. A zero order rate could also be chosen. The results would be different, but the conclusions would be the same. For a first order remobilization rate, the concentrations in the leachate can be described by eqn (1) (Belevi & Baccini, 1989a).

$$c = c_o . \exp\left(-V . c_o . t/m_o\right) \tag{1}$$

where c is the annual flux-weighted mean element concentration in the leachate (mg litre^{-1}), c_o is the annual flux-weighted mean element concentration in the leachate at the end of the intensive reactor period (mg litre^{-1}), V is the specific annual mean leachate flux (litres year^{-1} kg^{-1} MSW), t is the time (years) and m_o is the element concentration that can be mobilized (mg kg^{-1} MSW).

The final storage quality is reached when the element concentration c is equal to the quality standard chosen. By arranging the equation above, the time t_{FS} required to reach the final storage quality with respect to the leachate is given by eqn (2),

$$t_{FS} = (m_o/V . c_o) \ln\left(c_o/c_{QS}\right) + t_{IRP} \tag{2}$$

where c_{QS} is the chosen quality standard (mg litre^{-1}) and t_{IRP} is the time period of the intensive reactor phase (here about 10 years). c_{QS} is chosen on the basis of Swiss quality standards for surface waters (Swiss Ordinance on Waste Water, 1975) and appears in Table 3. The chosen values are of the same order of magnitude as the quality standards (i.e. not higher than 10 times the quality standards). c_o and V are measured by field experiments. t_{FS} can then be calculated by estimation of m_o. However, the estimation of m_o is the most speculative part of the assessment procedure.

In order to estimate m_o, drilled core samples are used. Drilling was carried out in the MSW landfills, where the element concentrations in leachates were measured. Landfill samples of 50–100 kg were taken from different layers. Five well-mixed samples of 8–12 kg from every layer were dried in the laboratory at 105°C and five fractions were sorted by hand. They were metals, stones and ceramics, glass, plastics, wood and paper. The remaining residue was ground and screened. The residual fraction had a particle size of less than 0·5 mm. It was then analysed and used for leaching experiments: 30 g of the residual fraction were mixed in an Erlenmeyer flask with 300 ml distilled water and extracted at 150 rpm for 0·5 h on a rotary-type shaker table. The suspension was filtered through a membrane filter ($0·45\mu$m) and the extract analysed. The filtration residue was leached again under the same conditions for five hours. This consecutive reaction procedure was carried out four times giving eventually four consecutive extracts of the same sample with leaching times of 0·5, 5, 50 and 288 h. The element concentrations were measured in the extracts and the fractions of the elements in samples that can be mobilized were calculated by summation of the amounts in the extracts.

The analytical techniques for concentration measurements in the residual fractions and in the extracts are described elsewhere (Baccini *et al.*, 1987; Belevi & Baccini, 1989*a*).

RESULTS AND DISCUSSION

In order to estimate the element concentrations m_o in MSW that can be mobilized, two samples of 14 years and 11 years of age were chosen. The ages of the samples were determined by the input records of the landfill company and by the dates of newspapers and magazines found in the drilled cores (Obrist, 1986). Table 1 shows the composition of the samples. The residual fraction is used for the determination of m_o, since

TABLE 1. Composition of Drilled Core Waste Samples from Two Landfills

Composition	Sample 1 (age : 14 years) (% w/w)	Sample 2 (age : 11 years) (% w/w)
Stone and ceramic	23	31
Plastic	3	9
Wood and paper	5	3
Glass	1	4
Metal	2	2
Residue	66	51

it has the highest surface/volume ratio. The contribution of the other fractions to m_o are nelgected so that the estimation of m_o is rather too low than too high.

Table 2 shows the element concentrations in the residual fractions and a comparison with the concentrations in bulk MSW. The latter concentrations were found in MSW in Switzerland, and were determined by several mass balance studies of MSW incinerators (Brunner & Ernst, 1986) and from MSW compost data (W. Obrist, pers. comm.). All element concentrations are in the same range as the concentrations in MSW except the carbon concentrations. This coincides well with the element balance investigations in the MSW landfills (Baccini *et al.*,

TABLE 2. Element Concentrations, c_{MSW}, in MSW and Element Concentrations, c_{RF}, in the Residual Fractions from Drilling Core Samples

Element	c_{MSW} (g kg^{-1})	c_{RF} (g kg^{-1})	
		Sample 1	Sample 2
C	290	120	130
N	4	3·1	3·5
F	0·2	n.d.	n.d.
P	1	0·6	0·8
S	2	1·5	1·2
Cl	7	n.d.	n.d.
Fe	40	48	44
Cu	0·3	1·1	0·3
Zn	1·4	1·7	1·2
Pb	0·4	1·3	0·5
Cd	0·01	0·006	0·006

n.d., Not determined.

1987). Element balances have shown that more than 22% of the initial carbon content of MSW was exported as gas within 10 years of disposal. More than 90% of non-metals and more than 99·9% of metals still exist in the landfill after the intensive reactor period. The remaining differences can be explained by the inaccuracy of the values due to the high inhomogeneity of the drilled core samples and by the sorting procedure. Consequently, the residual fractions can be considered as representative materials for the landfill body at the end of the intensive reactor period.

Table 3 represents the concentrations of elements, m_o, that can be mobilized by the extraction procedure described earlier, the concentrations, c_L, in the extracts of the leaching experiments with a remobilization time of 0·5 h and a comparison of the concentrations, c_o, in the leachates of MSW landfills at the end of the intensive reactor period with the quality standards, c_{QS}. Two parallel leaching experiments were carried out with samples taken from each residual fraction so that four data sets were obtained with two drilled core samples. Since a statistical evaluation with four data points is not reasonable, the observed ranges of

TABLE 3. Fractions of Elements, m_o, that Can Be Mobilized, Concentration, c_L, in the First Extract of the Leaching Experiments, Concentration, c_o, in the Leachates of an MSW Landfill at the End of the Reactor Period and Quality Standard, c_{QS}

Element	m_o		c_L (mg litre^{-1})	c_o (mg litre^{-1})	c_{QS}[c] (mg litre^{-1})
	(mg kg^{-1})[a]	(%)[a]			
TOC[b]	2 100–7 100	0·7–2·4	200–800	750	20
N	200–310	5–8	22–28	1 200	5
F	n.d.	n.d.	n.d.	0·65	1
P	5–33	0·5–3·3	0·3–3·7	6·8	0·4
S	n.d.	n.d.	n.d.	2·7	30
Cl	1 000–1 500	14–21	70–180	1 300	100
Fe	20–39	0·04–0·08	1·6–3·0	8	10
Cu	1·0–6·7	0·3–1·7	0·08–0·64	0·1	0·1
Zn	14–98	1·2–8	0·8–10	0·6	2
Pb	0·1–2·5	0·03–0·6	0·08–0·15	0·07	0·5
Cd	$(7–24) \times 10^{-3}$	0·06–0·22	$(3–24) \times 10^{-3}$	0·002	0·05

[a] Based on kg MSW.
[b] TOC = total organic carbon.
[c] Chosen on the basis of Swiss quality standards for surface waters (Swiss Ordinance, 1975).
n.d., Not determined.

m_o are given. Thus, this approach emphasizes the uncertainty of the estimation of m_o. Another method for the estimation of m_o for the elements C, N, P and Cl based on the chemical speciation in MSW (Belevi & Baccini, 1989a) leads to m_o-ranges of these elements similar to the ranges obtained by leaching experiments carried out in the present work. The concentrations in the first extract are of the same order of magnitude as the concentrations in leachates, except the nitrogen and chlorine concentrations. Since chloride and ammonium are usually highly mobile ions, their concentrations in leachates are one or two orders of magnitude higher than those in the extracts. However, the concentrations in the extracts depend strongly on the solid/liquid ratio of the leaching experiments in addition to other parameters. Consequently, the concentrations in MSW landfill leachates cannot be directly estimated by leaching experiments. It is more reasonable to determine mobile fractions of elements by leaching experiments and to use the real concentrations from field measurements for long-term assessments.

The estimation of m_o by leaching experiments presumes that the conditions in the landfill, such as pH, redox potential, ionic strength and complexing agents, remain constant over the estimated period (here about 1000 years). A detailed discussion of the conditions in landfills over long-term periods is given elsewhere (Belevi & Baccini, 1989a). However, the most important parameter determining the long-term remobilization processes is pH. A decrease of pH to less than 7 may remobilize many metals in MSW landfills.

The total organic carbon concentration in the drilled core samples at the end of the reactor period is less than $100 \, g \, kg^{-1}$ (Obrist, 1986). Provided that 50% of this organic carbon can still be oxidized to CO_2 (worst case) in the next 1000 years or so, less than 4 mol of weak acids can be produced (Brunner et al., 1987).

The precipitation (50 ml precipitation with pH 4 per kg MSW per year) would provide a proton flux of less than 10^{-5} mol per kg residue per year. In a time period of about 1000 years, this proton source would be still two orders of magnitude lower than the proton production potential of the landfill body itself.

The alkalinity of residual fraction, defined operationally by the amount of the acid required to reach pH 7, was determined in eleven parallel batch titrations. The pH of the suspensions of 1 g pulverized residual fraction in 1000 ml aqueous solution with increasing aliquots of HNO_3 was measured after 48 h of rotary-type shaking (100 rpm) in an Erlenmeyer flask. The pH of the solution without acid addition was

438

Belevi, Baccini

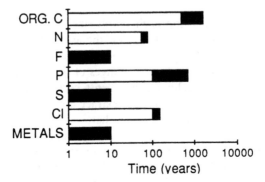

Figure 1. Time required to reach the final storage quality.

about 8·3. By adding approximately 6 mol proton per kg residual fraction, pH 7 was reached. Consequently, the alkalinity of the residual fraction (6 mol kg^{-1}) is higher than the proton production potential of the landfill body (<4 mol kg^{-1}). A higher remobilization rate of some metals due to lower pH is unlikely at least during many centuries.

Figure 1 illustrates the time ranges t_{FS} (black parts of the bars) required to reach the final storage quality as calculated by eqn (2). Fluorine, sulphur and metals can reach the final storage quality within the intensive reactor period. Higher nitrogen concentrations than the quality standards are expected over at least several decades. The longest period is needed for chlorine, phosphorus and organic carbon, i.e. at least one to several centuries. These results indicate that organic compounds are very important for long-term considerations. Our knowledge about the organic compounds leaving the MSW landfill is still incomplete. Thousands of xenobiotics are deposited in the landfill by various components of MSW. Some of them may be non-hazardous to the receiving aquifer, some of them may be decomposed before reaching the aquifer, but some of them may pollute the groundwater. Furthermore, many organic compounds may be also transformed to toxic compounds, even though they are initially not toxic. Since we do not know enough about the ecotoxicology of these compounds, the quality standards should be set at a low-level concentration for the bulk parameter DOC (dissolved organic carbon) in leachates of MSW landfills.

CONCLUSIONS

(1) According to the guidelines on waste management in Switzerland, non-recyclable wastes have to be transformed to an inert form to prevent

an adverse environmental impact for long-term periods. The controlled municipal solid waste landfills have existed for about 20 years and have not yet reached this inert state, the so-called final storage quality.

(2) The leachate is the most important mass flux after the intensive reactor period (10–20 years), i.e. after the period of a relatively high gas production rate. Higher non-metal concentrations are expected than the quality standards chosen on the basis of Swiss quality standards for surface waters (not higher than 10 times standard values) over centuries after disposal.

(3) A remobilization of metals due to a pH-decrease is not expected for at least many centuries, since the alkalinity of the landfill body after the intensive reactor period is high enough to neutralize protons produced by organic carbon degradation to CO_2.

(4) The objective of the Swiss waste management policy, that future generations should not be forced to deal with wastes of their ancestors, cannot be reached for MSW landfills. The leachates of MSW landfills have to be treated much longer than a human generation (~30 years) to prevent an adverse environmental impact, which could be caused by synthetic organic compounds existing in the MSW.

REFERENCES

Baccini, P. (ed.) (1989). The landfill, reactor and final storage. *Lecture Notes in Earth Sciences*, Vol. 20. Springer, Heidelberg, Germany.

Baccini, P., Henseler, G., Figi, R. & Belevi, H. (1987). Water and element balances of municipal solid waste landfills. *Waste Management & Research*, **5**, 483–99.

Belevi, H. & Baccini, P. (1989a). Long-term behavior of municipal solid waste landfills. *Waste Management & Research*, **7**, 43–56.

Belevi, H. & Baccini, P. (1989b). Water and element fluxes from sanitary landfills. In: *Sanitary Landfilling: Process, Technology and Environmental Impact*, ed. T. H. Christensen, R. Cossu & R. Stegmann. Academic Press, London, Chapter 6.2.

Brunner, P. H. & Ernst, W. (1986). Alternative methods for the analysis of municipal solid waste. *Waste Management & Research*, **4**, 147–60.

Brunner, P. H., Müller, M. D. & McDow, S. R. (1987). Carbon emissions from municipal incinerators. *Waste Management & Research*, **5**, 355–65.

Ehrig, H. J. (1986). Untersuchungen zur Gasproduktion aus Hausmüll (Study of gas production from municipal solid waste). *Müll und Abfall*, 5/86, 179–83.

EKA (Eidgenössiche Kommission für Abfallwirtschart) (1986). Leitbild für die Schweizerische Abfallwirtschaft (Guidelines to waste management in Switzerland). Schriftenreihe Umweltschutz Nr. 51, BUS, Bern, Switzerland.

Obrist, W. (1986). EAWAG Report 30-411, Swiss Federal Institute of Water Resources and Water Pollution Control, Dübendorf, Switzerland.

Stegmann, R. (1978/1979). Gase aus geordneten Deponien (Gases from controlled landfills). *ISWA-Journal*, **26/27**, 11–24.

Swiss Ordinance on Waste Water (1975). Schweizerischer Bundesrat, Bern, Switzerland.

4.2 Attenuation of Leachate Pollutants in Groundwater

THOMAS H. CHRISTENSEN

Department of Environmental Engineering, Technical University of Denmark, Building 115, DK-2800 Lyngby, Denmark

INTRODUCTION

Contamination of groundwater by landfill leachate posing a risk to the quality of downstream surface waters and wells is considered to constitute the major environmental concern associated with landfilling of wastes. The number of cases of groundwater pollution at old landfills with no measures to control leaking into the groundwater, and the resources spent in remediation, support the concern about leachate entering the groundwater. Groundwater pollution at properly engineered landfills with leachate collection systems have also been observed, although in far fewer cases.

Understanding the processes governing leachate pollutant attenuation in aquifers is mandatory for evaluation of environmental risks associated with leachate entering the groundwater, for coherent interpretation of groundwater data sampled for mapping of an existing leachate plume, and for determining appropriate remedial action. The purpose of this chapter is to give an introduction into this complex and rapidly developing area of contaminant hydrogeology.

Focusing on the common type of landfill receiving a mixture of municipal and commercial waste but excluding significant amounts of concentrated specific industrial waste, the landfill leachate may be

441

characterized as a water-based solution of four groups of pollutants:

1. Organic matter, expressed as chemical oxygen demand (COD) or total organic carbon (TOC), including volatile fatty acids (in particular in the acid phase of the waste stabilization) and more refractory compounds (for example, fulvic-like and humic-like compounds).
2. Specific organic compounds originating from household or industrial chemicals and present in relative low concentrations in the leachate (usually less than 1 mg litre^{-1}). These compounds include among others a variety of aromatic hydrocarbons, phenols and chlorinated aliphatics.
3. Inorganic macrocomponents: calcium (Ca^{2+}), magnesium (Mg^{2+}), sodium (Na^+), potassium (K^+), ammonium (NH_4^+), iron (Fe^{2+}), manganese (Mn^{2+}), chloride (Cl^-), sulphate (SO_4^{2-}), and hydrogen carbonate (HCO_3^-).
4. Heavy metals: cadmium (Cd), zinc (Zn), lead (Pb), copper (Cu), nickel (Ni) and chromium (Cr).

Other compounds may be found in leachate from landfills, e.g. borate, sulphide, arsenate, selenate, barium, lithium, mercury and cobalt. But in general these compounds are of only secondary importance.

Although some of the leachate pollutants of groups 1 and 3 appear in high concentrations (see Ehrig, 1983) water constitutes of the order of 97% of the leachate. Only where large amounts of chemicals are disposed of may organic compounds appear as free phases (e.g. non-aqueous-phase liquids as solvents). The following discussion of leachate pollutant attenuation is devoted to dissolved pollutants in aquifers. Introductions are given to free phase transport in aquifers by Dorgarten (1989) and to diffusive pollutant migration in clay membranes and aquitards by Rowe (1992).

DILUTION

All compounds in leachate leaking into an aquifer will be subject to dilution as the leachate mixes with unpolluted groundwater. For the non-reactive components, of which chloride is the dominant component, dilution is the only attenuation mechanism. The other attenuation mechanisms involving reactions, as later discussed, are seen usually as having only positive effects, since they reduce the concentration of the unwanted pollutants. Dilution also reduces the pollutant concentrations, but at the same time larger volumes of groundwater are being polluted, increasing the difficulties and cost of tracing the leachate plume and abating the pollution. Therefore dilution as such does not have only a positive effect. Only in cases where small quantites of leachate are released (e.g. a small local crack in the liner) can dilution be seen solely as a positive attenuation mechanism.

Leachate and Groundwater Flow

Dilution is the interaction of the leachate flow in the aquifer with the flow of unpolluted groundwater. As such, it is mandatory to understand the processes governing the flow of water in an aquifer, but a discussion of these processes is beyond the scope of this chapter (see Freeze & Cherry, 1979). However, it should be emphasized that the leachate migration should be seen as a three-dimensional plume developing in a three-dimensional geological structure, where gradients, permeabilities and physical boundaries (geological strata, infiltration, rivers, abstraction wells, etc.) determine the position and migration velocity of the plume. However, the flow of leachate may differ physically from the flow of groundwater with respect to at least three aspects.

Local water table gradients just below and around the landfill may differ from the general gradients because the landfill may have a different hydrogeology than the surrounding strata. For example, an unlined landfill with a low quality top cover may result in a larger infiltration than in the surrounding soil, and if leachate is not removed, this may result in a local water table mound affecting the local gradients. In contrast to this, lined landfills may allow very little infiltration into the aquifer, except at possible leaking points where substantial infiltration may take place.

The density of the leachate may be significantly higher than the density of the groundwater. Leachate with a total dissolved solids concentration of 20 000 mg litre^{-1} is not uncommon, corresponding to a density that is about 1% higher than the groundwater density. This may locally lead to density flow and/or instabilities resulting in deviating flow patterns. Not much is known about leachate density effects in field situations (Kimmel & Braids, 1974) but density differences may have significant effect on the vertical positioning of the plume just below the landfill. After minor dilution (a few times), density effects are not believed to be significant. In some cases (e.g. MacFarlane et al., 1983) the leachate shows higher temperature than the groundwater which results in a small negative effect on the density, but not to a level counterbalancing the density effects of the dissolved solids.

The viscosity of the leachate may theoretically deviate from the viscosity of the groundwater. Very little data exist on this aspect, but Christensen et al. (1985) reported on the measured viscosity of 10 Danish leachate samples and concluded that the effect of the leachate viscosity was marginal relative to water (1–15% increase at equal temperatures).

Dispersion

Dispersion is the mathematical term in the solute transport equation (see Freeze & Cherry, 1979) accounting for dilution or mixing according to concentration gradients. The hydrodynamic dispersion coefficient D (L^2T^{-1}) is often substituted by the product of the pore flow velocity V (LT^{-1}) and the dispersivity α (L), neglecting the contribution by the molecular diffusion coefficient.

The dispersivity α has a longitudinal component (in the flow direction), a vertical component and a horizontal, transverse component. The dispersivity is difficult to measure in field situations, and often the dispersivity has been used as a calibration parameter accounting for insufficient information on the spatial distribution of the hydraulic conductivity/permeability. Traditionally, this has suggested longitudinal dispersivities in the order of tens to hundreds of metres depending on the length scale of the problem in question. However, recent detailed field studies have shown that the dispersivity is actually very small (centimetres to hundreds of centimetres) and, for practical purposes, independent of scale (Freyberg, 1986; Garabedian et al., 1991; Jensen et al., 1992). In particular, the transversal dispersivities are much smaller than usually assumed.

The mixing of a leachate plume with the surrounding groundwater is primarily due to local heterogeneities with respect to hydraulic conductivity. This means that dilution is limited in homogenous aquifers, as illustrated in Fig. 1, which shows a chloride plume transect 35 m downgradient of an experimental leachate release (Bjerg et al., 1991). (Note also the distinct sinking of the plume). However, most aquifers contain substantial heterogeneities and will usually provide significant

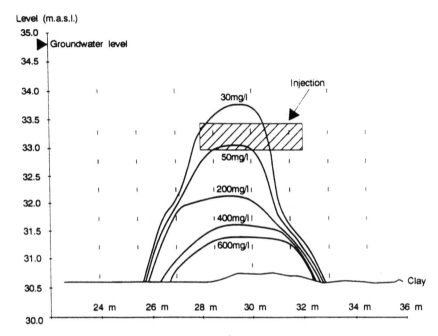

Figure 1. Chloride contours (mg litre^{-1}) of a transect 66 days after an artificial leachate release. The transect is transversal to the flow directions. The injection took place in injection filters located 35 m upstream from this transect (shaded area) (Bjerg *et al.*, 1991).

mixing, not necessarily accounted for by a large dispersivity coefficient but by the actual configuration of the hydrogeological heterogeneities. This complicates a reliable prediction of the local plume movement, which typically is in a length scale of a few hundred metres, while readily available hydrogeological data usually is of a larger length scale. This would call for a substantial effort in mapping the local hydrogeology.

Dilution may also be caused by seasonal variations in flow direction, for example, from changing infiltration and abstraction patterns. This would correspond to a larger dispersion of the pollutants, but the variations may have a systematic pattern that makes it unaccountable just by increasing the dispersivity in the solute transport equation.

Observed Pollution Plumes

The actual plume development is the result of all the abovementioned aspects and, to give an appreciation of actual plumes, Figs 2, 3 and 4

446 Christensen

contain examples of plumes reported in the literature. Three-dimensional mapping in the field is very costly and the results are often difficult to present in an illustrative way. Therefore, the individual investigations have often given priority to either horizontal plume mapping or vertical profile mapping. An exception to this is the Borden Landfill in Canada.

The Borden Landfill (Fig. 2) is located in a homogeneous, unconfined, sandy, glaciofluvial aquifer and was active from 1940 to 1976 (Fig. 2

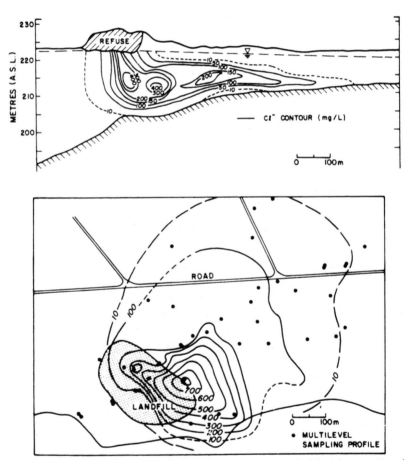

Figure 2. Horizontal and vertical transect contours of chloride (mg litre^{-1}) in the Borden Landfill plume (Canada). The porewater flow velocity is of the order of 20–50 m year^{-1} (MacFarlane et al., 1983).

dates back to 1979). The plume, as described by chloride contours, is extremely wide (~600 m) compared to the width of the landfill, supposedly owing to seasonal variations in flow directions. The length of the plume is 700 m and the maximum depth 20 m, governed downwards by less permeable strata. About 600 samples were taken for the plume delineation.

Fig. 3 shows horizontal chloride contours of the plume of an unnamed Bavarian landfill (Germany). The landfill is located on a coarse glaciofluvial aquifer and was active from 1954 to at least 1970, when the samples presented in Fig. 3 were collected. The landfill is very large (4×10^6 m^3). The plume is long and narrow, stretching about 3000 m without much widening. The plume delineation is based on approximately 20 sampling points.

The Vejen Landfill (Denmark) is also located in a homogeneous, unconfined, sandy, glaciofluvial aquifer and was active from 1962 to 1981 (Fig. 4 dates back to 1988). The plume is about 400 m long and the maximum depth is 20 m. The upstream groundwater passing under the clays lenses below the landfill is forced upwards further downgradient because of a rising clay stratum. This apparently provides substantial

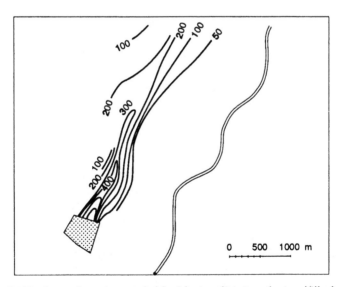

Figure 3. Horizontal contours of chloride (mg litre⁻¹) ... ⸱ ⸱ landfill plume in Bavaria (Germany). The porewater flow is of the order of 150–400 m year⁻¹ (Exler, 1972).

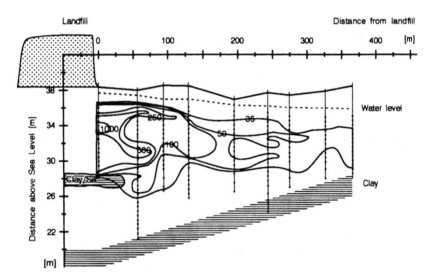

Figure 4. Vertical contours of chloride (mg litre^{-1}) of the Vejen Landfill plume (Denmark). The porewater flow is of the order of 150–200 m year^{-1} (Lyngkilde & Christensen, 1992a).

mixing. The plume delineation is based on approximately 160 groundwater samples.

Other plumes have been reported by Golwer *et al.* (1969), Kimmel & Braids (1974), DeWalle & Chain (1981), Reinhard & Goodman (1984), Lesage *et al.* (1990) and Duijvenbooden & Kooper (1981).

REDOX ZONES

The leaking of strongly reduced landfill leachate, high in organic matter, into a shallow, presumably aerobic aquifer creates a very complicated environment owing to redox processes, biodegradation, dissolution/precipitation, complexation, ion exchange and sorption processes. The sum of these processes creates a sequence of redox zones downstream from the landfill. These redox zones will strongly influence the attenuation of the pollutants in the plume and are discussed below.

The main process responsible for the development of a redox zone sequence is the degradation of leachate organic matter according to available oxidants (electron-acceptors). Assuming a volume of leachate being mixed with a volume of aerobic aquifer (like a batch), the easily degradable part of the organic matter of the leachate will degrade, utilizing in sequence free oxygen and nitrate in the groundwater, followed by oxidized manganese and iron compounds associated with the sediment. (Actually, sulphide and reduced iron may also utilize part of the oxygen in the aquifer.) Continued degradation may utilize sulphate and finally methane may be produced by fermentation and by carbon dioxide reduction. See Box 1 for representations of the chemical reactions involved in the governing redox environments. During a continuous leachate release (over months or years) these processes will lead to development of a methanogenic zone in the aquifer close to the landfill, where the most easily degradable organic matter (e.g. volatile fatty acids) is degraded. According to the thermodynamics of the redox processes,

Box 1. Redox reactions involved in degradation of organic matter (expressed as the model compound CH_2O) in different environments.

Conversion of organic matter, represented by the model compound CH_2O, in different redox environments may be represented by the following stoichiometric reactions and the corresponding Gibbs free energy (kcal mol^{-1}) at pH 7 (Champ *et al.*, 1979)

Methanogenic, fermentative:

$$CH_2O \rightarrow \tfrac{1}{2}CH_3COOH \rightarrow \tfrac{1}{2}CH_4 + \tfrac{1}{2}CO_2, \qquad \Delta G°(W) = -22$$

Note: For organic matter deviating from the model compound used, the fermentation will lead to generation of H_2, which may be oxidized by CO_2 reduction according to: $CO_2 + 4H_2 \rightarrow CH_4 + 2H_2O$.

Sulphate reduction:

$$CH_2O + \tfrac{1}{2}SO_4^{2-} + \tfrac{1}{2}H^+ \rightarrow CO_2 + \tfrac{1}{2}HS^- + H_2O \qquad \Delta G°(W) = -25$$

Iron (ferric) reduction:

$$CH_2O + 4Fe(OH)_3 + 8H^+ \rightarrow CO_2 + 4Fe^{2+} + 11H_2O \qquad \Delta G°(W) = -28$$

Manganese (manganic) reduction:

$$CH_2O + 2MnO_2 + 4H^+ \rightarrow CO_2 + 2Mn^{2+} + 3H_2O \qquad \Delta G°(W) = -81$$

Nitrate reduction:

$$CH_2O + \tfrac{4}{5}NO_3^- + \tfrac{4}{5}H^+ \rightarrow CO_2 + \tfrac{2}{5}N_2 + \tfrac{7}{5}H_2O \qquad \Delta G°(W) = -114$$

Aerobic respiration, oxygen reduction:

$$CH_2O + O_2 \rightarrow CO_2 + H_2O \qquad \Delta G°(W) = -120$$

zones of sulphate reduction, ferric reduction, manganic reduction and nitrate reduction will develop downstream from the landfill, if these electron-acceptors are present in the aquifer. Finally, where the oxygen demand of the organic matter in the leachate plume no longer exceeds the supply of free oxygen (through mixing and diffusion), the plume will become aerobic like the surrounding aquifer.

A scheme for characterization of redox environments based on concentrations of redox-sensitive compounds of groundwater samples (Table 1) has been proposed by Lyngkilde & Christensen (1992a). Direct measurement of the redox potential was not considered reliable. Instead, the governing redox environment was identified, paying attention to the redox chemistry of the components. Application of this scheme to 160 groundwater samples from the Vejen Landfill pollution plume (Denmark) identified a full sequence of redox zones as shown in Fig. 5. A small methanogenic zone (actually <20 m according to the detailed characterization given in Lyngkilde & Christensen, 1992b) was found just next to the landfill, followed by a non-continuous sulphate reducing zone. A later investigation did not recover the sulphate reducing zone. This could be due to depletion of the sulphate pool or to the fact that sulphate reduction had been only temporary owing to a temporary mixing into the leachate plume of demolition waste leachate from the neighbouring landfill cells. The largest zone was the iron and manganese reducing zone reaching more than 300 m downstream.

In the more dilute parts of the plume, nitrate reduction is the dominating redox process. The substantial nitrate reducing zone in the

TABLE 1. Groundwater Criteria for Redox Zone Identification Applied by Lyngkilde & Christensen (1992a) (All concentrations in mg litre^{-1})

Parameter	Aerobic	Nitrate reducing	Mangano-genic	Ferro-genic	Sulfido-genic	Methano-genic
Oxygen	>1·0	<1·0	<1·0	<1·0	<1·0	<1·0
Nitrate	—	—	<0·2	<0·2	<0·2	<0·2
Nitrite	<0·1	—	<0·1	<0·1	<0·1	<0·1
Ammonium	<1·0	—	—	—	—	—
Mn(II)	<0·2	<0·2	>0·2	—	—	—
Fe(II)	<1·5	<1·5	<1·5	>1·5	—	—
Sulfate	—	—	—	—	—	<40
Sulfide	<0·1	<0·1	<0·1	<0·1	>0·2	—
Methane	<1·0	<1·0	<1·0	<1·0	<1·0	>1·0

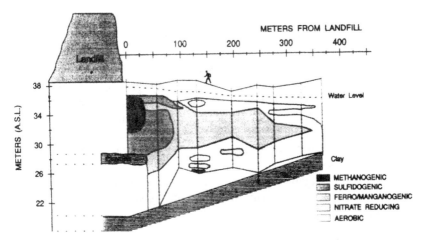

Figure 5. Redox zones identified in leachate pollution plume at Vejen Landfill (Denmark) (Lyngkilde & Christensen, 1992a).

bottom of the aquifer is natural and not necessarily created by the landfill leachate plume. Nearly 400 m away from the landfill, the plume becomes aerobic.

The redox processes creating these redox zones are supposedly microbial. Not much is known about the microbiology of leachate plumes (Beeman & Suflita, 1987), but current investigations support the view that the redox zones at Vejen Landfill are microbially active. About 15 years have elapsed since leachate migration started at Vejen Landfill, providing a long period for adaptation and build-up of microbial consortia able to mitigate the redox processes in the actual redox environments. However, the actual extent and activity of the individual redox zones is a question not only of microbial consortia but also of the degradability of the substrate (the degradability of the leachate organic matter supposedly decreases as the degradation proceeds) and of the availability of electron acceptors (e.g. iron oxides consist of many different compounds of varying solubility and reactivity).

ORGANIC MATTER

Organic matter in leachate is a bulk parameter covering a variety of organic degradation products ranging from small volatile acids to refractory fulvic-like and humic-like compounds. Discussion of basic

attenuation processes is difficult without specifying the organic compound in question. But with respect to landfill leachate, we usually have
very little information on the organic matter composition and we simply
have to accept that differences in composition are not reflected in the
employed bulk parameters. However, it may be appropriate, when
possible, to distinguish between volatile fatty acids, dominating acid
phase leachate, and the more refractory organic matter making up the
major part of well stabilized methanogenic leachate (see Chian &
DeWalle (1977) and Weis *et al.* (1989) for more comprehensive
characterizations).

Attenuation of organic matter (e.g. expressed as TOC or COD) in the
leachate plume, other than by dilution, owes to sorption and
degradation.

Sorption

Sorption of leachate organic matter on to aquifer material seems to be of
only minor significance according to column experiments reported in the
literature: relative solute velocity (see Box 2) for acid phase leachate
COD was of the order of 0·8–1·0 (Hoeks *et al.*, 1979; Kjeldsen &
Christensen, 1984) and for methane phase leachate of the order of
0·7–1·0 (Hoeks *et al.*, 1984; Kjeldsen, 1986). This means that the

Box 2. Relative solute transport velocity.

The relative solute transport velocity in the plume ($V_s \cdot V_w^{-1}$) expresses the
transport velocity of the pollutant, V_s, relative to the transport velocity of
the water, V_w:

$$V_s \cdot V_w^{-1} = (1 + (\rho \cdot \varepsilon^{-1}) \cdot K_d)^{-1} = R^{-1}$$

where

V_s = solute transport velocity (LT^{-1})
V_w = water transport velocity (LT^{-1})
ρ = bulk density of the aquifer material (ML^{-3})
ε = effective porosity (L^3L^{-3})
K_d = distribution coefficient (L^3M^{-1})
R = retardation factor

For many aquifers an approximate equation may apply:

$$V_s \cdot V_w^{-1} = (1 + 5 \cdot K_d)^{-1} = R^{-1}$$

leachate COD front practically moves with the water front or is just slightly retarded and that retardation cannot be viewed as a significant attenuation mechanism in most aquifers.

Kjeldsen (1986), however, observed for two acid subsoils, where the methanogenic leachate caused a substantial rise in the pH of the soil columns, that for a short time period the soils released COD, increasing the solute COD to levels substantially above the COD content of the leachate. The significance of this additional, dissolved organic matter is not known, but apparently this has been observed only under dramatic pH changes.

Degradation: Laboratory Experiments

Degradation of leachate organic matter has been studied in several column studies, where leachate has been loaded under anoxic conditions on to soil columns.

For acid phase leachate, Table 2 lists reported experimental results in terms of half-life values in days assuming a steady-state first order kinetic reaction. The reported results show substantial degradation of leachate COD for the acid phase leachate, dominated by volatile fatty

TABLE 2. Summary of Literature Data on Degradation of Acid Phase Leachate COD in Laboratory Column Experiments

Material	Hydraulic retention time (days)	Inlet COD (mg litre $^{-1}$)	Relative outlet concentration	Temperature (°C)	$T_{1/2}$ (days)
Sand[a]	23	29 600	0·92	20	190
Sandy soil[a]	17	6 640	0·08	20	4·6
Sandy soil[a]	40	6 640	0·04	20	8·7
Sandy soil[a]	20	7 750	0·01	20	5·3
Sandy soil[b]	9·3	5 700	0·14	?	3·3
Sandy soil[c]	7·7	7 400	0·91	10	60
Sandy soil[c]	7·4	7 400	0·95	10	100
Loamy sand[c]	7·3	7 400	0·87	10	36
Loamy sand[c]	9·6	7 400	0·78	10	27

[a] Hoeks *et al.* (1979).
[b] Soyopak (1979).
[c] Kjeldsen & Christensen (1984).

acids. Half-life values range from 5–10 days (20°C) to 30–100 days (10°C).

For methanogenic leachate the reports are few and negative. Hoeks *et al.* (1984) found no COD degradation during experimental retention times of 19–25 days and Kjeldsen (1986) found for retention times up to 240 days no degradation, except what could be attributed to volatile fatty acids making up a small part of the methanogenic leachate.

The reported column studies confirm that volatile fatty acids are relatively easily converted under anaerobic conditions. In general, the redox conditions have not been adequately described in the conducted experiments, probably because transient conditions have been dominant for loading of strongly reduced leachate on to aerobic soil columns. Since the time frame is usually very limited for such experiments (e.g. corresponding to 10 pore volumes or maybe 200 days) adequate time may not have been available for stable environments to develop. However, the experiments do indicate that methanogenic leachate COD is not readily degraded under anaerobic conditions in unadapted environments.

Degradation: Field Experiences

From plume mapping data, degradation of organic matter may be evaluated by comparing organic matter plumes with chloride plumes, the latter expressing the degree of dilution. Such comparison assumes that the organic matter does not significantly sorb on to the aquifer material and that the organic matter : chloride ratio has been constant in the infiltrating leachate during the period in question. The first assumption seems acceptable, according to the previous discussion on sorption, while the second assumption is more questionable (see Barker *et al.*, 1986). However, if focus is on methanogenic leachate and the aim is to observe degradation at order of magnitude level, comparison of organic matter concentration and chloride concentration seems acceptable.

Figure 6 shows a plot of organic matter in the Vejen Landfill leachate plume (here expressed as non-volatile organic carbon, which approximates TOC in this plume) as a function of distance from the landfill. The organic matter concentrations are corrected for dilution according to chloride concentration in samples and in unpolluted groundwater. The decreasing curve indicates a strong degradation of organic matter in the plume; the majority of the dissolved organic carbon is degraded within the first 100 m of the plume, and at a distance of

NVOC x dilution [mg Ľ]

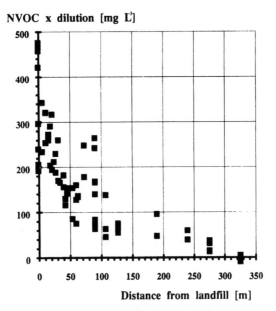

Figure 6. Organic matter concentration (non-volatile organic carbon) in the Vejen Landfill leachate plume as a function of distance from the landfill after correction for dilution (see text) (Lyngkilde & Christensen, 1992*b*).

300 m the organic carbon concentration just slightly exceeds the background value, which is of the order of 2 mg C litre^{-1}. The substantial turnover of dissolved organic carbon in the plume is also what creates the redox zones previously presented in Fig. 5. Lyngkilde & Christensen (1992*a*) compared the NVOC-fate curve in Fig. 6 with a detailed redox characterization of samples and concluded that some degradation took place in the methanogenic/sulphate-reducing zone, but the major degradation was under iron-reducing conditions. The iron concentrations in this zone were above the iron concentration in the leachate.

Only a few other reports exist on the fate of dissolved organic matter in leachate plumes. DeWalle & Chian (1981) found at the Army Creek Landfill (Delaware) that in a plume stretching about 700 m with regard to chloride, the majority of dissolved carbon (as COD) was degraded within the first few hundred metres from the landfill. Barker *et al.* (1986) found that the North Bay Landfill (Ontario, Canada) had contaminated the groundwater throughout the 700 m flow system from the site to a discharge zone at a creek. The plume was anaerobic, showing elevated concentration of dissolved organic matter (TOC) as far as 500–600 m. Exler (1972) found elevated concentrations of organic matter (determined as permanganate demand) as far as 3000 m in a narrow leachate plume from a unnamed landfill in Bavaria, Germany. Dissolved organic carbon at the Borden Landfill (Ontario, Canada) showed elevated concentrations as far as 500 m from the landfill (Nicholson *et al.*, 1983). These examples do not support any clear conclusion as to the experiences on organic matter degradation in leachate plumes. The data by Lyngkilde & Christensen (1992*b*) convincingly demonstrate significant degradation over short distances. The data by DeWalle & Chian (1981) may support this observation, while the other reports do not allow for a separation of the degradation and dilution contributions to organic matter attenuation in the plume. However, some of the reported long travel distances for organic matter show that degradation is not always very significant and we do not currently understand the governing factors.

SPECIFIC ORGANIC COMPOUNDS

Specific organic compounds constitute only a few per cent of the total dissolved organic carbon in leachate and leachate contaminated ground-water, but much concern is associated with this group of pollutants owing to their potential health risks (Brown & Donnelly, 1988) and the low drinking water quality standards enforced in many countries.

The term specific organic compounds (here excluding specific volatile fatty acids which are considered under the section on organic matter) covers a large number of organic chemicals. Of the order of 1000 different organic chemicals have so far been identified in groundwater contaminated by landfills (including chemical and industrial waste disposal sites). Experiences in the USA and in Germany, where a large number of sites have been investigated (Arneth *et al.*, 1989), however, indicates that only a few organic substances are frequent and can be considered important groundwater contaminants. Figure 7 shows a ranking of the 15 most frequent specific organic compounds found in contaminated groundwater at waste disposal sites. Chlorinated aliphatic compounds, BTEX (benzene, toluene, ethylbenzene and xylene), chloro-benzenes and vinyl chloride are the most frequent compounds. The data do not allow for a linking of the occurrence of specific organic

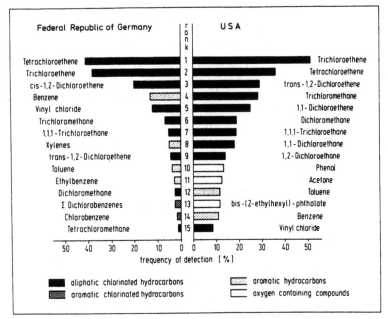

Figure 7. The 15 most frequently detected specific organic compounds in groundwater at waste disposal sites in Germany and the USA (Arneth *et al.*, 1989).

compounds to characteristics of the waste. However, it is believed that the majority of the data originate from chemical and industrial sites, and thus do not necessarily represent compounds common at landfills with a large volume of domestic waste.

Attenuation of specific organic compounds is primarily caused, besides by dilution, by sorption and degradation. For some compounds volatilization may also potentially occur, but this will only be of importance where the plume is located at the water table, which usually only occurs next to the landfill. In cases where the leachate percolates through an unsaturated zone, volatilization may be a substantial attenuation process.

Sorption

Sorption of specific organic chemicals on to soils is well described in the literature (e.g. Karickhoff, 1984; Brusseau & Rao, 1989). More information is available for non-polar components, while the sorption of polar components (e.g. organic acids and bases) is much less researched.

Sorption of non-polar components on to soil can be described as a partitioning into the solid organic carbon of the soil. The sorption process is considered to be kinetically affected and often described by a bicontinuum model assuming half of the sorption to be instantaneous and half to be governed by a slow first order kinetic reaction, representing diffusion into the solid carbon (Brusseau *et al.*, 1991*a*). Estimation of equilibrium distribution coefficients, K_d, can be made by empirical regression equations, paying attention to the organic carbon content of the soil and to the hydrophobicity of the specific organic compound expressed through its octanol–water partitioning coefficient (see Box 3). These regression equations are well established for organic carbon contents in excess of 0·1% C.

Information on sorption on to aquifer materials, usually having solid organic carbon contents less than 0·1%, is much more scarce (Boucher & Lee, 1972; Curtis *et al.*, 1986; Abdul *et al.*, 1987; Lee *et al.*, 1988; Brusseau & Reid, 1991; Larsen *et al.*, 1992*a,b*). Brusseau *et al.* (1991*b*) showed for 8 specific organic compounds that the bicontinuum model also could be applied to aquifer materials with low organic carbon. However, it was concluded that in most practical cases an equilibrium model would be a fair approximation. Larsen *et al.* (1992*a*) showed that sorption increased for increasing hydrophobicity (K_{ow}) for 12 specific organics as shown in Fig. 8. However, no simple correlation to aquifer

Box 3. Estimation of distribution coefficients for non-polar organic compounds.

The distribution coefficient for non-polar organic compounds may be estimated from regression equations like the following (Schwarzenbach & Westall, 1981):

$$K_d = 3 \cdot 1 \cdot f_{oc} \cdot K_{ow}^{0 \cdot 72}$$

where

K_d = distribution coefficient ($L^3 M^{-1}$)
f_{oc} = fraction of organic carbon in the aquifer material ($M M^{-1}$)
K_{ow} = octanol–water distribution coefficient

This equation is valid primarily for $f_{oc} > 0 \cdot 001$. For $f_{oc} < 0 \cdot 0005$ the regression equation tends to underestimate K_d.

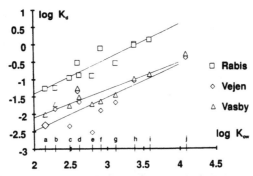

Figure 8. Sorption of 12 specific organic compounds on to 3 low organic carbon aquifer materials expressed as laboratory determined distribution coefficients (K_d) as a function of compound octanol–water distribution coefficient $(\log K_{ow})$ (Larsen *et al.*, 1992*b*): (a) benzene, (b) trichloroethylene, (c) 1,1,1-trichloroethane, (d) tetrachloroethylene and tetrachloromethane, (e) toluene, (f) indene, (g) *o*-xylene, (h) 1,4-dichlorobenzene and 1,2 dichlorobenzene, (i) naphthalene, (j) biphenyl.

material solid carbon content exists, according to Larsen *et al.* (1992*b*), who investigated 20 aquifer materials ranging from 0·06% to 0·21% C. It seems that the traditional regression equations, e.g. the one shown in Box 3, underestimate the sorption of specific organic compounds at solid organic carbon contents less than 0·05%. Apparently, other solid components are also active in the sorption process. However, for many of the volatile organic compounds (e.g. $\log K_{ow} < 3$), sorption will only retard the compound very little. For example, benzene having a K_d value of 0·01 litre kg^{-1} (according to Fig. 8, Vasby aquifer material) the transport velocity is 95% of the water transport velocity. For more hydrophobic compounds, as for example naphthalene and dichlorobenzenes, sorption may be substantial and their migration velocity may be of the order of only 10% of the water transport velocity.

Since specific organic compounds partition into solid organic carbon, as documented above, it has been hypothesized that specific organic compounds also partition into dissolved organic matter, for example, in terms of leachate organic matter. This was investigated by Larsen *et al.* (1992*c*) considering several leachates and aquifer materials in flow and batch laboratory experiments. The interaction between aquifer material, leachate and specific organic compounds was very intricate. In some cases the presence of leachate increased the sorption of specific organic compounds onto the aquifer material; in other cases the effect was a decrease. However, it was concluded that, with respect to estimating the relative transport velocity of the specific compounds, neglecting the presence of leachate would supposedly yield a bias of not more than ±20%. Larsen *et al.* (1992*c*) did not find this detrimental compared to all the other uncertainties involved in estimating transport velocities of specific organic compounds in pollution plumes.

Degradation

Twenty years ago, aquifers were considered to be microbially inactive, while we today see aquifers as complex microbial ecosystems (Ghiorse & Wilson, 1988). However, much still remains to be learned about the ability of these ecosystems to adapt to pollution and to degrade specific organic compounds under different redox conditions.

The term degradation has been used very liberally and with different significance by different scientific disciplines, ranging from compound disappearance (compared to no compound disappearance in control experiments), through positive determination of end products (e.g. collection of CO_2 generated by degradation of labelled compounds), to revealing actual degradation pathways, intermediary compounds and active bacterial species. Degradation experiments may also be designed according to specific aims. For example, determining the degradability of a compound under certain redox conditions (optimal conditions with respect to microorganisms, substrate, nutrients, electron-acceptors and absence of inhibitors), determining the ability of the actual microbial population to degrade a specific compound (optimal conditions with respect to substrate, nutrients, electron-acceptors and absence of inhibitors) or determining the actual degradation rates (optimal or natural substrate conditions).

Table 3 summarizes some of the information available about degradability (in its most liberal form) of relevant organic compounds in aquifers under different governing redox conditions. The table does not claim to be exhaustive. The purpose is to show in general terms the current level of information. For specific applications, the relevant literature must be consulted. Table 3 reveals that many compounds have been shown to be degradable under aerobic conditions, the chlorinated aliphatic compounds being an exception (only degradable under aerobic conditions co-metabolically with methane). Far fewer compounds have been investigated under anaerobic conditions, and much less degradability is

TABLE 3. Summary of the Degradability of Specific Organic Compounds under Different Redox Conditions in Aquifers

Compound	Aerobic	Anaerobic			
	O_2	NO_3	$Fe(III)$	SO_4	CH_4
BTEX					
Benzene	+	÷	?	?	(+)
Toluene	+	+	+	+	(+)
Ethyl-benzene	+	+	?	?	(+)
o-Xylene	+	+	?	+	(+)
m-Xylene	+	+	?	?	?
p-Xylene	+	+	?	+	?
Substituted aromatics					
Chlorobenzene	+	?	?	+	÷
1,2-Dichlorobenzene	+	?	?	?	(+)
1,4-Dichlorobenzene	+	?	?	?	(+)
Nitrobenzene	+	?	?	?	+
Polyaromatic Hydrocarbons					
Naphthalene	+	?	(+)	?	÷
Phenanthrene	+	?	?	?	÷
Indene	+	?	?	?	?
Biphenyl	+	?	?	?	?
Fluorene	+	?	?	?	?
Phenols					
Phenol	+	+	+	+	+
Cresol (o, m, p)	+	+	+	+	+
Chlorophenol	+	?	?	÷	+
2,4-Dichlorophenol	+	?	?	?	+
2,6-Dichlorophenol	+	?	?	?	+
Pentachlorophenol	?	?	?	?	+
o-Nitrophenol	+	?	?	?	+
p-Nitrophenol	+	?	?	?	+
Chlorinated aliphatics					
Tetrachloromethane	÷	+	+	+	+
1,1,1-Trichloroethane	÷	?	?	?	+
1,1-Dichloroethylene	÷	?	?	?	+
1,2-Dichloroethylene	÷	?	?	?	+
1,1,1-Trichlorethylene	÷	?	?	?	+
Tetrachloroethylene	÷	?	?	?	÷

+, Degradable; (+), possibly degradable; ÷, not degradable; ?, not tested or not conclusive; NO_3, nitrate reduction; Fe(III), iron reduction; SO_4, sulphate reduction; CH_4, methane production.

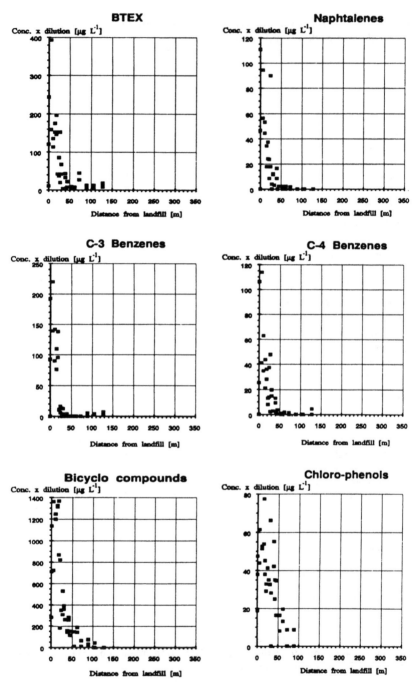

Figure 9. Concentrations of different groups of specific organic compounds in groundwater downgradient of Vejen Landfill (Denmark) as a function of distance. All concentrations have been corrected for dilution according to chloride concentration (Lyngkilde & Christensen, 1992*b*).

seen. It should be noted that for the iron reducing zone extremely little information is available.

Very little of the information behind Table 3 originates from leachate affected aquifers and very little insight as to degradation of specific organic compounds in leachate plumes is gained from Table 3. (Degradation studies directly involving leachate contaminated aquifers are Wilson *et al.* (1986), Gibson & Suflita (1986), Smolenski & Suflita (1987), Kjeldsen *et al.* (1990), and Lyngkilde *et al.* (1992).)

Experiences from detailed mapping of occurrence and distribution of specific organic compounds in leachate plumes are few (DeWalle & Chian, 1981; Reinhard & Goodman, 1984; Barker *et al.*, 1986, and Lyngkilde & Christensen, 1992*b*). Interpretation of such field observations is often hampered by insufficient characterization of redox conditions and uncertainty about dilution effects. Lyngkilde & Christensen (1992*b*) corrected specific organic compound concentrations for dilution according to chloride content, and found indication of degradation of many specific compounds in the anaerobic part of the leachate plume at the Vejen Landfill (Fig. 9). The majority of the compounds were apparently degraded under iron-reducing conditions, which is interesting in the light of the lack of basic understanding of this redox zone. Chlorinated aliphatic compounds were not present in the Vejen Landfill leachate plume, but 1,1,1-trichlorethane and trichloroethylene were found to degrade under strictly anaerobic conditions in the North Bay Landfill plume (Ontario, Canada) according to Barker *et al.* (1986).

The field studies of leachate plumes, although few and associated with much uncertainty, tend to indicate a substantial attenuation of specific organic compounds under anaerobic conditions. This may indicate substantial degradation capacity of these zones with respect to specific organic compounds and there should be more focus on these aspects in future research. However, not all compounds are necessarily degraded in the plume. Benzene, some pesticides and other compounds may be rather persistent under anaerobic conditions and may in the future prove to be the compounds defining the specific organic compound plume. In the case of the Vejen Landfill, the pesticide Mecoprop® did not degrade in the anaerobic zone.

Figure 10 (*overleaf*). Concentration contour plots of the Borden Landfill leachate plume with respect to chloride, sulphate, alkalinity (HCO_3^-), calcium, magnesium, sodium, potassium and ammonium (Nicholson *et al.*, 1983).

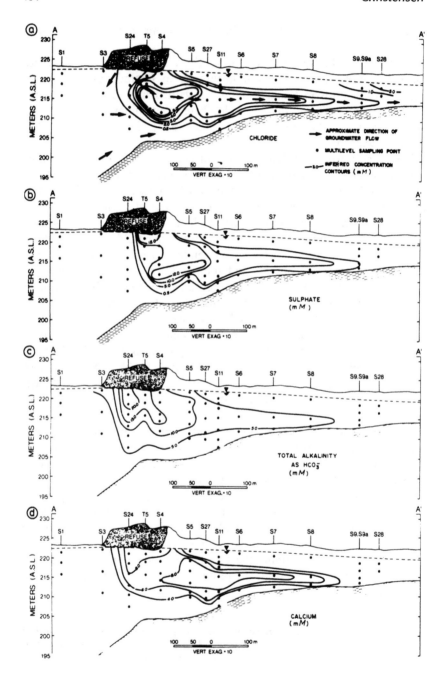

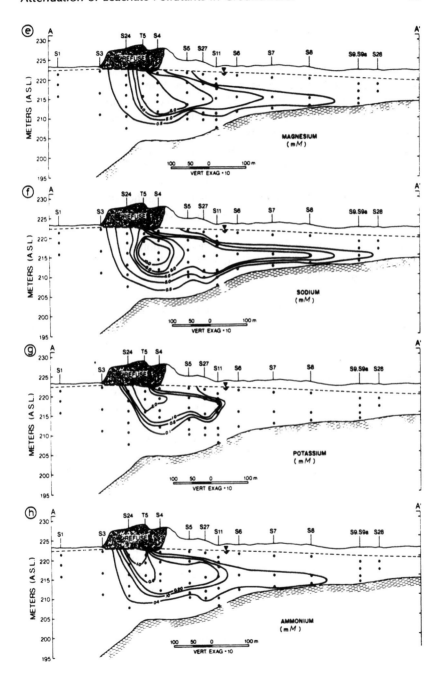

INORGANIC MACROCOMPONENTS

The term macrocomponent refers here to common inorganic constituents present at elevated concentrations (several mg litre^{-1} in leachate and/or polluted groundwater) and includes the anions chloride, hydrogen carbonate and sulphate, and the cations calcium, magnesium, potassium, sodium, ammonium, iron and manganese.

These inorganic macrocomponents do not as such constitute a severe groundwater polution problem. However, drinking water quality standards usually include most of these macrocomponents, and usually leachate concentrations exceed these standards significantly. In addition, some of these components are important in controlling redox environments and attenuation of heavy metals.

The most extensive field report of macrocomponent distribution in a leachate plume is given by Nicholson *et al.* (1983) on the Borden Landfill plume (Canada). As a field reference for the following discussions, Fig. 10 presents the Borden Landfill plume concentration contours with respect to chloride (as a reference), sulphate, total alkalinity as hydrogen carbonate, calcium, magnesium, sodium, potassium and ammonium. Later, Fig. 11 presents the concentration contours of iron and manganese.

The macrocomponents may be attenuated, besides by dilution, by redox processes, ion exchange and precipitation. In the concentrated part of the pollution plume, complexation may also occur, decreasing the attenuation by increasing component solubility and mobility. Component concentrations may also in some cases increase by dissolution of solids present in the aquifer. The most likely processes of importance for the behaviour of macrocomponents in leachate plumes are summarized in Table 4.

Anions

Chloride is not considered to undergo any chemical or physico-chemical reactions in the aquifers and as such is considered inert or conservative. For this reason, chloride was used in the previous discussion on dispersion.

Carbon dioxide pressure and *alkalinity* are usually high in leachates owing to the decomposition and dissolution processes in the landfill (see Christensen & Kjeldsen, 1989). Leachate from the methanogenic phase

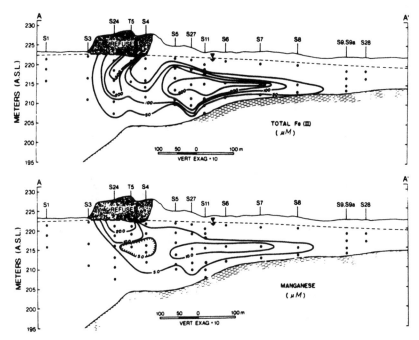

Figure 11. Concentration contour plots of the Borden Landfill leachate plume with respect to iron and manganese (Nicholson *et al.*, 1983).

of the landfill stabilization, which is by far the longest and most important phase, often has a pH around neutral. As a consequence of this the leachate plume probably shows increased alkalinity compared to the uncontaminated groundwater. If the aquifer is acidic, the leachate alkalinity will act as a buffer increasing pH in the center of the plume, and, if the aquifer is neutral and for example contains calcite, the calcite will be dissolved upon exposure to leachate carbon dioxide, resulting in increased dissolved alkalinity. In the absence of volatile fatty acids, alkalinity in the leachate plume is practically equivalent to hydrogen carbonate (HCO_3^-). The importance of hydrogen carbonate and carbonate in the plume, besides pH-buffering, is to form complexes with calcium, magnesium, sodium, iron and manganese and to form precipitates with calcium, magnesium, iron, manganese and maybe some heavy metals. The significance of these processes, of course, depends on species concentrations and pH.

Sulphate is present in the Borden Landfill leachate plume at elevated

TABLE 4. Summary of Processes Affecting Macrocomponent Behaviour in Leachate Pollution Plumes

Process	Anions			Cations						
	Cl^-	HCO_3^-/CO_3^{2-}	SO_4^{2-}	Ca^{2+}	Mg^{2+}	Na^+	K^+	NH_4^+	Fe^a	Mn^a
Dilution	+	+	+	+	+	+	+	+	+	+
Complexation[g]	$-^b$	+	+	+	+	$(+)^c$	$(+)^c$	−	+	+
Redox processes	−	−	$(+)^c$	−	−	−	−	$(+)^d$	+	+
Ion exchange	−	−	$(+)^e$	+	+	+	+	+	+	+
Precipitation/ dissolution	−	+	$(+)^f$	+	+	−	−	−	+	+

[a] Fe as Fe^{2+} or Fe^{3+}, Mn as Mn^{2+} or Mn^{4+}.
[b] Chloride does form many dissolved complexes, primarily with heavy metals, Ca^{2+} and Mg^{2+}, but usually only a small part of the total chloride concentration is complexed.
[c] Complexes of Na^+ and K^+ are usually not very important but may appear in the leachate.
[d] Ammonium may be oxidized under aerobic conditions.
[e] Anion exchange is usually of only minor importance.
[f] Usually not very likely.
[g] Complexation is not an attenuation process, since complexation results in increased solubility and mobility.

concentrations (see Fig. 10), but such high concentrations supposedly are not common in leachate plumes. Sulphate presumably will be reduced to sulphide in an active, methanogenic landfill or in the methanogenic and sulphate-reducing zone in the leachate plume. However, it cannot be excluded that landfill cells rich in demolition waste may release leachate with high sulphate concentrations, which upon mixing in the plume may result in detectable sulphate concentrations in the plume. Sulphide generated in the plume supposedly will precipitate with ferrous iron and perhaps heavy metals also present. Dissolved sulphide concentrations (S^{2-}, HS^-, H_2S) are believed to be very modest in leachate plumes (confer previous discussion on redox zones). Bisdom *et al.* (1983) identified amorphous and crystalline ferrosulphides in soil columns loaded with leachate.

Anion exchange sites on the aquifer material are few (far fewer exchange sites for anions than for cations) but may contain small amounts of sulphate. Upon exposure to leachate rich in other anions (primarily chloride and hydrogen carbonate) and maybe leading to decreasing anion exchange capacity by increasing pH and reducing oxide contents of the solids, sulphate may be released from the aquifer material. This has been observed in laboratory column studies by Kjeldsen (1986). Anion exclusion of sulphate is supposedly only important at the front of the plume.

Cations

Aquifer material usually has a low cation exchange capacity (CEC is typically of the order of 1 meq/100 g) compared to topsoils and sediments with high concentrations of clay and organic matter. However, the exchange capacity is still very significant: in unpolluted aquifers, cations associated with the exchange sites typically make up 80% of the total amount of cations per volume of aquifer, while in the central part of a leachate pollution plume the corresponding value is of the order of 40% (for calculations made on the Borden aquifer sediment, see Nicholsen *et al.*, 1983).

Several equations exist describing cation exchange, usually employing a selectivity coefficient expressing the relative affinity of the two cations in question for the exchange sites. Box 4 presents the Gaines-Thomas ion exchange equation and gives ranges of selectivity coefficients. The cation exchange equation—like other available equations (see e.g. Harmsen,

Box 4. Ion exchange according to Gaines-Thomas.

Exchange of cations may be described by several equations (Kerr, Gapon, Wanselow, Davis, Gaines-Thomas, etc.; see Harmsen, 1979). The expression by Gaines-Thomas assumes the reaction

$$Z_B \cdot \bar{A} + Z_A \cdot B^{+Z_B} \rightleftharpoons Z_A \cdot \bar{B} + Z_B \cdot A^{+Z_A}$$

and the Law of Mass Action

$$\frac{(\bar{B})^{Z_A} \cdot (A^{+Z_A})^{Z_B}}{(\bar{A})^{Z_B} \cdot (B^{+Z_B})^{Z_A}} = K$$

modified to

$$\left[\frac{\bar{N}_B}{m_B}\right]^{Z_A} \cdot \left[\frac{m_A}{\bar{N}_A}\right]^{Z_B} = K_{B/A}$$

where:

A and B = cations
Z = valency
bar = indicates 'ion exchange sites'
( ) = activities
K = equilibrium constant
$\bar{N}$ = equivalent fraction of CEC: cation exchange capacity
m = molar activity in solution
$K_{B/A}$ = selectivity coefficient

Usually Ca^{2+} is used as a reference cation.

Typical order of magnitude for selectivity coefficients:

$\overline{Ca} + Na^+$ $K_{Na/Ca}$, 0·1–0·4

$\overline{Ca} + K^+$ $K_{K/Ca}$, 5–50

$\overline{Ca} + NH_4^+$ $K_{NH_4/Ca}$, 2–50

$\overline{Ca} + Mg^{2+}$ $K_{Mg/Ca}$, 0·5–2

Note: Very few data exist for aquifers and the above listed numbers should be used with discretion.

1979)—shows that the ratio of cations on the exchange sites depends not only on the selectivity coefficients but also on the actual ratio of the cations in the solute. It is also noted that the exchange equations only employ the relative ratio of the cations. Since the number of exchange sites available at constant pH is independent of the ionic strength of the solution, the apparent retardation of a cation owing to ion exchange will be larger in the dilute part of the plume than in the concentrated part of the plume.

In a leachate plume several cations will be present at the same time, having different affinities for the exchange sites and different concentrations. This creates a complex system of travelling fronts and peaks in the plume as the plume composition changes owing to ion exchange. This complexity makes it impossible to illustrate in a simple way the overall migration retardation by ion exchange. Figure 12 shows effluent breakthrough curves for several cations in laboratory soil columns exposed to leachate (Kjeldsen & Christensen, 1984). Although relative migration velocities can be estimated from soil column studies and used for general ranking of the cation migration, it should be kept in mind that such estimates are valid only for that specific combination of leachate and soil/aquifer material.

Cation exchange, as the name implies, is an exchange of cations, implying that when a cation associates with an exchange site another cation is being expelled. In many aquifers, the cation exchange sites are saturated with calcium, magnesium and, at low pH, with protons. In saline soils, sodium will dominate. Upon exposure to leachate, having a different relative cationic composition than the natural groundwater, these saturating cations will be expelled and move with the leachate front in concentrations in excess of the leachate concentrations. This is often referred to as the 'hardness halo' and has often been reported in the literature (e.g. Anderson & Dornbush, 1967; Griffin *et al.*, 1976; Campbell *et al.*, 1983; Nicholson *et al.*, 1983).

Potassium, K^+, has a high affinity for ion exchange and will typically be much retarded in the leachate plume. Absorption into crystal lattices of, e.g. illite may constitute an additional attenuation mechanism.

Ammonium, NH_4^+, also exhibits a high affinity for ion exchange and will be retarded in the plume. Ammonium may also partly become incorporated into crystal lattices of clay minerals. Dissociation of ammonium (NH_4^+) to free ammonia ($NH_3 + H^+$) is not very likely in aquifers ($pK_a = 9\cdot3$). If ammonium is present in the plume when it reaches aerobic zones, oxidation may occur resulting in elevated nitrate concentrations. However, it is assumed that in most cases organic compounds will move faster than ammonium and deplete oxygen before the arrival of ammonium.

Sodium, Na^+, is considered to have the least affinity of the relevant cations for ion exchange and will be only slightly retarded in the plume.

Calcium, Ca^{2+}, has a high affinity for exchange sites, but since calcium

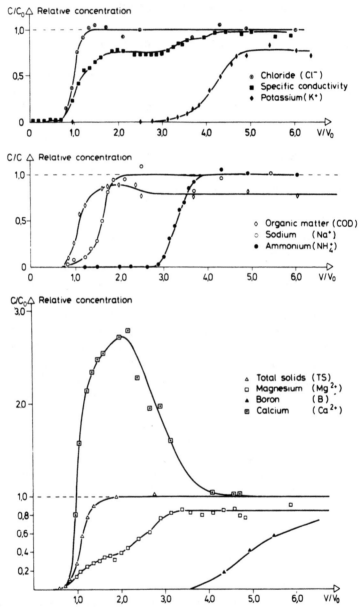

Figure 12. Solute breakthrough curves illustrating cation migration in a sandy soil column exposed to landfill leachate (Kjeldsen & Christensen, 1984). V/V_o expresses the number of pore volumes of leachate that has passed the column.

often is the dominating base saturating ion, it will often move with the leachate front. Attenuation may take place behind the front, depending on the selectivity coefficients and plume composition. Calcium may also be heavily involved in complex formation (e.g. $CaHCO_3^+$) and in dissolution/precipitation reactions e.g. involving calcite ($CaCO_3$) and maybe dolomite ($CaMg(CO_3)_2$). The precipitation and dissolution processes are closely linked to the dissolved carbonate compounds. It should be mentioned that supersaturation with respect to calcite has been reported (see Nicholson *et al.*, 1983).

Magnesium, Mg^{2+}, resembles calcium in many ways, although its affinity for ion exchange, occurrence as base saturation ion and likelihood for forming precipitates are slightly less than for calcium.

Proton, H^+, expressed through pH, is involved in many reactions, but mentioned here because of its high affinity for ion exchange sites at low to moderate pH. This is often an overlooked aspect and very little data is available describing proton exchange on to aquifer material, currently making prediction of cation transport in leachate plumes involving pH changes an impossible task.

Iron and Manganese

Iron and manganese are treated separately because they, in contrast to the other cations, are subject to redox processes (see Box 1).

In aerobic aquifers, iron and manganese are associated with the sediment as solid oxyhydroxy compounds, and practically absent from the water phase. Iron is usually present in much higher concentrations than manganese.

Iron concentrations in the leachate are high during the acid phase, but moderate in the methanogenic phase (Ehrig, 1983). Iron solubility is supposedly controlled by sulphides or carbonates. Partial complexation by organic matter cannot be excluded. Entering the aquifer, iron will be subject to ion exchange as well. This may substantially retard the migration of iron as observed by Kjeldsen & Christensen (1984) in laboratory columns loaded with leachate. If sulphate reduction takes place, iron may precipitate as sulphide (see Bisdom *et al.*, 1983). In the following zone of the leachate plume, the presence of degradable organic matter in the leachate will result in reductive dissolution of iron compounds from the aquifer material. On the sediment, iron in its oxidized form will be present as various amorphous and crystalline compounds of varying vulnerability to reduction and dissolution. Kjeldsen & Christensen (1984) observed release of iron from laboratory soil columns exposed to leachate resulting in substantially elevated iron concentrations in the effluent, but determination of iron profiles in the columns before and after exposure did not reveal any detectable reduction in soil iron. The availability of iron oxyhydroxides and the degradability of the dissolved organic matter will govern the characteristics of the iron-reducing zone (see Fig. 5). The released ferrous iron will also in this zone be subject to ion exchange, but may be solubility controlled by ferrous carbonate (siderite) as discussed by Nicholson *et al.* (1983). If the reduced iron migrates to the aerobic part of the plume, oxidation and precipitation as ferric oxyhydroxides is likely.

Manganese behaves similarly to iron, but is less significant because of lower concentrations. From a thermodynamic point of view, manganese reduction should occur prior to iron reduction. This was also observed by Kjeldsen & Christensen (1984) who found a complete depletion of soil manganese owing to reductive solubilization by leachate.

The complex pattern of processes governing attenuation as hypothesized above is only partly supported by experimental data and field observations. Kinetic aspects are supposedly involved and much still remains to be learned before the attenuation of iron and manganese in leachate plumes is understood.

HEAVY METALS

Heavy metals do not usually constitute a frequent groundwater pollution problem at solid waste landfills, partly because landfill leachates usually contain only modest concentrations of heavy metals (see Ehrig, 1983) and partly because the heavy metals are subject to strong attenuation by sorption and precipitation in the plume. A survey in Germany (Arneth *et al.*, 1989), including 92 abandoned waste disposal sites, revealed that for cadmium (Cd) and lead (Pb), 78% and 85% respectively of downstream groundwater samples were below detection limits (Cd, $1 \mu g$ litre^{-1}; Pb, $10 \mu g$ litre^{-1}) and only 3% of the samples exceeded the drinking water standards (Cd, $5 \mu g$ litre^{-1}; Pb, $50 \mu g$ litre^{-1}).

Heavy metals considered here are cadmium (Cd), zinc (Zn), nickel (Ni), copper (Cu), lead (Pb) and chromium (Cr). Metals like mercury (Hg) and cobalt (Co) are rare in leachates and, together with the metalloid arsenic (As), not dealt with in this context. Iron (Fe) and manganese (Mn) are usually not considered heavy metals and are dealt with in the section on inorganic macrocomponents.

Heavy metal concentrations in leachates usually refer to total dissolved metal, determined, e.g. on an acidified sample after removal of solids. However, different metal species are supposedly present in many landfill leachates owing to complexation with organic and inorganic ligands. Lun & Christensen (1989) determined cadmium species in landfill leachate and found that free divalent Cd^{2+} only made up a few per cent of the total cadmium content. Most of the complexed fraction was characterized as labile complexes that were easily redistributed. However, a small fraction (5–15%) was characterized as stable complexes, here defined by lack of ability to exchange with a cation exchange resin. The stable complexes were considered to be organic. Knox & Jones (1979) showed that both low molecular weight compounds (<500) comparable to simple carboxylic acids and high molecular weight compounds (>10 000) contributed significantly to complexation. In the case of cadmium, chloride is also an important ligand resulting in substantial fractions of monochlorocomplexes and maybe dichlorocomplexes. The presence of metal complexes in leachate complicates the discussion of attenuation mechanisms in aquifers, because the species may behave differently and because most of the available information on attenuation mechanism only concerns free divalent metal ions. However, the presence of metal

TABLE 5. Summary of Processes Affecting Heavy Metal Behaviour in Leachate Pollution Plumes

Process	Cd	Zn	Ni	Cu	Pb	Cr
Dilution	+	+	+	+	+	+
Complexation[a]	+	+	+	+	+	+
Redox processes	−	−	−	−	−	−[b]
Sorption	+	+	+	+	+	+
Precipitation						
Sulphides	+	+	+	+	+	−
Carbonates	+	+	−	−	+	−
Other	−	−	−	+	+	+

[a] Complexation is not an attenuation process, since complexation results in increased solubility and mobility.
[b] Cr as Cr(III). Cr(VI) may appear as chemical waste, but presumably is rapidly reduced to Cr(III) in the landfill.

complexes supposedly increases metal solubility and mobility and cannot be neglected.

Heavy metals in leachate plumes will be attenuated, besides by dilution, primarily by sorption and precipitation. Table 5 summarizes the processes involved in heavy metal attenuation in aquifers.

Sorption

The term 'sorption' covers all surface related reactions such as adsorption, absorption, surface complexation, surface precipitation and ion exchange. Divalent metal cations have a high affinity for sorption on to negatively charged sites associated with clay minerals, organic matter, oxides of iron, manganese, aluminium and silicon, and calcite ($CaCO_3$). The equilibrium distribution between sorbed and dissolved metal is usually expressed by the distribution coefficient, K_d. As for the specific organic compounds subject to sorption, the relative transport velocity can be predicted by determining the distribution coefficient. It should be emphasized that such predictions are only valid if the governing process is sorption. If precipitation has also taken place in the experiments conducted for estimating K_d, application of the relative transport equation (see Box 2) is not warranted.

K_d-values have been determined in the laboratory for many divalent free metals in the order of ten to a few thousands, depending on metal, pH, and soil material (see Anderson & Christensen, 1988). A typical value of 100 litres kg^{-1} for cadmium (Cd^{2+}) at pH 6 for a coarse soil or aquifer material yields an estimated transport velocity of 0·2% of the water transport velocity, which is extremely slow. However, K_d-values determined in leachates (e.g. Christensen, 1985; Kjeldsen, 1986), are typically one or two orders lower owing to complexation, metal competition and ionic strength effects on activity coefficients. A K_d value of 5 litres kg^{-1}, however, still results in transport velocities of only 4% of the water flow velocity.

Precipitation

The heavy metals in leachate plumes may be solubility governed by sulphides and carbonates. For some metals phosphates and oxides may also limit the solubility.

In the sulphate reducing zone, even small concentrations of sulphides will precipitate all heavy metals, except Cr, which does not form insoluble sulphides. The solubility of heavy metal sulphides is very low and even reduced pH or strong complexation cannot keep the metals in solution. Reduced iron will also precipitate as sulphide and heavy metals may be occluded in this process (see Bisdom *et al.*, 1983).

Carbonates may also control the solubility of Cd, Zn and Pb outside the sulphate reducing zone, although carbonates are much more soluble, allowing metal concentrations to exceed drinking water standards several-fold. Lead may also be limited by chloropyromorphite precipitation ($Pb_5(PO_4)_3Cl$) according to Christensen & Nielsen (1987).

Precipitation reactions may be relatively slow, maybe requiring months to establish equilibrium. However, in the context of groundwater this is not critical although the kinetics of the reaction may lead to stretching of contaminant plumes. Predictions of solute equilibrium concentrations are very uncertain because of potential complexation by ill-defined organic substances, because of possible supersaturation, and because of uncertain solubility products (e.g. for carbonates). However, for the sulphate reducing zone, there is probably little need for prediction of heavy metal concentrations at the level of analytical detection limits.

Experimental Evidence

In a leachate plume the behaviour of heavy metals is simultaneously controlled by sorption, maybe precipitation, and complexation; proper evaluations of metal migration must account for this complex system and the actual metal concentrations.

Several reported investigations on metal migration in leachate loaded soils have employed extremely high, and hardly relevant, concentrations of metals by spiking the leachate (Griffin *et al.*, 1977; Alesii *et al.*, 1980; Fuller *et al.*, 1980; Loch *et al.*, 1981; Campbell *et al.*, 1983). However, in most cases substantial attenuation was observed, yielding apparent transport velocities of 10–80% of the water transport velocity.

Experiments with actual (unspiked) metal leachate concentrations have shown very restricted migration of heavy metals; Kjeldsen & Christensen (1984) and Kjeldsen (1986) found apparent transport velocities, based on determined metal profiles in soil columns after leachate exposure, of the order of 0·5–2% for cadmium and 0·4–3% for zinc. It is important to stress that the numbers are apparent transport velocities and that, owing to the likely significance of precipitation in the columns, they cannot be extrapolated.

Although most of the metals were retained in the upper few centimetres of the soil columns, Kjeldsen & Christensen (1984) and Kjeldsen (1986) did see cadmium and zinc in the column effluent, indicating that several species were present and subject to different degrees of attenuation. The significance of organic complexes in terms of increasing metal mobility has been demonstrated by Hoeks et al. (1979, 1984), Loch et al. (1981) and Christensen (1989) in laboratory columns. Christensen (1989) found that stable cadmium complexes present in leachate were not attenuated in soil columns. However, the long-term stability of such complexes, with respect to exchange with iron and calcium in the plume and degradation of the ligand, is not known.

The most contaminated wells at the North Bay Landfill (Ontario, Canada) showed low concentrations of all heavy metals: Cd, 2–8 μg litre^{-1}; Zn, 3–11 μg litre^{-1}; Ni, 10–80 μg litre^{-1}; Pb, 16–67 μg litre^{-1}, and Cr, 33–85 μg litre^{-1} (Barker et al., 1986). The complexation capacity of the organic matter in the samples, determined by Cu-titration, exceeded the actual concentration of metals by orders of magnitude. Although a low input of heavy metals could be the cause of the low concentrations in the plume, attenuation by sorption and precipitation was considered significant.

In the Vejen Landfill leachate plume (Denmark) very low concentrations of heavy metals were identified: Cd, <0·2 μg litre^{-1}; Zn, <100–280 μg litre^{-1}; Ni, <5–7 μg litre^{-1}; Pb, <0·5–1·5 μg litre^{-1}, and Cu, <10 μg litre^{-1} (Andersen et al., 1991). Analysis of leachate revealed concentrations that were only slightly above what was found in the plume. All the abovementioned concentrations in the Vejen Landfill plume were below drinking water standards.

In both the North Bay Landfill and the Vejen Landfill leachate plumes, sulphate reducing environments have been identified, supposedly providing significant attenuation of heavy metals by sulphide precipitation. In the absence of sulphide precipitation, sorption also is a significant attenuation mechanism although strong organic complexes may lead to increased migration velocities. It appears that the strong attenuation of heavy metals in leachate plumes will restrict heavy metals to the most concentrated part of the plume near to the landfill.

CONCLUSION

Leachate migrating into the groundwater will be subject to attenuation involving, in addition to dilution, many different processes: redox processes, biodegradation, sorption, complexation, precipitation/dissolution and ion exchange. It is only recently (the last 10–20 years) that we are starting to understand these processes in aquifers, and much still remains to be learned before we shall adequately understand the complex environments developing in leachate plumes. Aquifers, although often characterized by coarse sandy material, apparently do provide some attenuation of many pollutants found in landfill leachate, but we still do not understand the governing factors and actual prediction of plume migration, except for conservative pollutants, as needed for proper risk assessment, and design of safety zones and control monitoring programs still lies in the future.

REFERENCES

Abdul, A. S., Gibson, T. L. & Rai, D. N. (1987). Statistical correlations for predicting the partition coefficient for nonpolar organic contaminants between aquifer organic carbon and water. *Hazardous Waste & Hazardous Materials*, **4**, 211–22.

Alesii, B. A., Fuller, W. H. & Boyle, M. V. (1980). Effect of leachate flow rate on metal migration through soil. *Journal of Environmental Quality*, **9**(1), 119–26.

Andersen, T. V., Holm, P. E. & Christensen, T. H. (1991). Heavy metals in a municipal landfill leachate pollution plume. In *Heavy Metals in the Environment, International Conference, Edinburgh, 16–20 September 1991*, vol. 2. CEP Consultants, Edinburgh, UK, pp. 252–5.

Anderson, J. R. & Dornbush, J. N. (1967). Influence of sanitary landfill on ground water quality. *American Water Works Association Journal*, **59**, 457–70.

Anderson, P. R. & Christensen, T. H. (1988). Distribution coefficients of Cd, Co, Ni, and Zn in soils. *Journal of Soil Science*, **39**, 15–22.

Arneth, J.-D., Milde, G., Kerndorff, H. & Schleyer, R. (1989). Waste deposit influences on groundwater quality as a tool for waste type and site selection for final storage quality. In *The Landfill*, ed. P. Baccini, Lecture Notes in Earth Sciences, Vol. 20. Springer Verlag, Berlin, Germany, pp. 399–424.

Barker, J. F., Tessmann, J. S., Plotz, P. E. & Reinhard, M. (1986). The organic geochemistry of a sanitary landfill leachate plume. *Journal of Contaminant Hydrology*, **1**, 171–89.

480   *Christensen*

Beeman, R. E. & Suflita, J. M. (1987). Microbial ecology of a shallow unconfined ground water aquifer polluted by municipal landfill leachate. *Microbial Ecology*, **14**, 39–54.

Bisdom, E. B. A., Boekestein, A., Curmi, P., Legas, P., Letsch, A. C., Loch, J. P. G., Nauta, R. & Wells, C. B. (1983). Submicroscopy and chemistry of heavy metal contaminated precipitates from column experiments simulating conditions in a soil beneath a landfill. *Geoderma*, **30**, 1–20.

Bjerg, P. L., Christensen, T. H. & Ammentorp, H. C. (1991). Field experiment on multi component ion exchange in a sandy aquifer. In *Transport and Mass Exchange Process in Sand and Gravel Aquifers: Field and Modelling Studies*, Proceedings of the International Conference, Ottawa, Canada, 1–4 October 1990, vol. 2. Atomic Energy of Canada, Chalk River, Ontario, Canada, pp. 841–9.

Boucher, F. R. & Lee, G. F. (1972). Adsorption of lindane and dieldrin pesticides on unconsolidated aquifer materials. *Environmental Science & Technology*, **6**, 538–43.

Brown, K. W. & Donnelly, K. C. (1988). An estimation of the risk associated with the organic constituents of hazardous and municipal waste landfill leachate. *Hazardous Waste & Hazardous Materials*, 5(1), 1–30.

Brusseau, M. L. & Rao, P. S. C. (1989). Sorption nonideality during organic contaminant transport in porous media. *CRC Critical Review of Environmental Control*, **19**(1), 33–99.

Brusseau, M. L. & Reid, M. E. (1991). Nonequilibrium sorption of organic chemicals by low organic-carbon aquifer materials. *Chemosphere*, **22**, 341–50.

Brusseau, M. L., Jessup, R. E. & Rao, P. S. C. (1991*a*). Nonequilibrium sorption of organic chemicals: elucidation of rate-limiting processes. *Environmental Science & Technology*, **25**, 134–42.

Brusseau, M. L., Larsen, T. & Christensen, T. H. (1991*b*). Rate-limited sorption and nonequilibrium transport of organic chemicals in low organic carbon aquifer materials. *Water Resources Research*, **27**, 1127–45.

Campbell, D. J. V., Parker, A., Rees, J. F. & Ross, C. A. M. (1983). Attenuation of potential pollutants in landfill leachate by lower greensand. *Waste Management and Research*, **1**(1), 31–52.

Champ, D. R., Gulens, J. & Jackson, R. E. (1979). Oxidation–reduction sequences in ground water flow systems. *Canadian Journal of Earth Science*, **16**, 12–23.

Chian, E. S. K. & DeWalle, F. B. (1977). Characterization of soluble organic matter in leachate. *Environmental Science & Technology*, **11**, 158–63.

Christensen, T. H. (1985). Cadmium soil sorption at low concentrations. IV: Effect of waste leachates on distribution coefficients. *Water, Air, & Soil Pollution*, **26**, 265–74.

Christensen, T. H. (1989). Cadmium soil sorption at low concentrations. VII: Effect of stable solid waste leachates complexes. *Water, Air, & Soil Pollution*, **44**, 43–56.

Christensen, T. H. & Kjeldsen, P. (1989). Basic biochemical processes in landfills. In *Sanitary Landfilling: Process, Technology and Environmental Impact*, ed. T. H. Christensen, R. Cossu and R. Stegmann. Academic Press, London, UK, pp. 29–49.

Christensen, T. H. & Nielsen, B. G. (1987). Retardation of lead in soils. In *Heavy Metals in the Environment, International Conference, New Orleans, September 1987*, vol. 1. CEP Consultants, Edinburgh, UK, pp. 319–21.

Christensen, T. H., Kjeldsen, P., Christensen, S., Hjelmar, O., Jansen, J. la Cour, Kirkegaard, C., Madsen, B., Olsen, N., Refsgaard, J. C. & Toudal, J. K. (1985). *Ground Water Monitoring at Sanitary Landfills* (in Danish). Polyteknisk Forlag, Lyngby, Denmark.

Curtis, G. P., Roberts, P. V. & Reinhard, M. (1986). A natural gradient experiment on solute transport in a sand aquifer. 4. Sorption of organic solutes and its influence on mobility. *Water Resources Research*, **22**, 2059–67.

DeWalle, F. B. & Chian, E. S. K. (1981). Detection of trace organics in well water near a solid waste landfill. *American Water Works Association Journal*, **73**, 206–11.

Dorgarten, H. W. (1989). Finite element simulation of immiscible and slightly soluble pollutants in soil and groundwater. In *Contaminant Transport in Groundwater*, Proceedings of the International Symposium, Stuttgart, Germany, 4–6 April, ed. H. E. Kobus, and W. Kinzelbach. A. A. Balkema, Rotterdam, The Netherlands, pp. 389–96.

Duijvenbooden, W. van & Kooper, W. F. (1981). Effects on groundwater flow and groundwater quality of a waste disposal site in Noordwijk, The Netherlands. In *Quality of Groundwater*. Proceedings of an international symposium, Noordwijkerhout, The Netherlands, March, Studies in Environmental Science Vol. 17. Elsevier, Amsterdam, The Netherlands, pp. 253–60.

Ehrig, H.-J. (1983). Quality and quantity of sanitary landfill leachate. *Waste Management & Research*, **1**, 53–68.

Exler, H. J. (1972). Ausbreitung und Reichweite von Grundwasserverunreinigungen im Unterstrom einer Mülldeponie. *GWF Wassser Abwasser*, **113**, 101–12.

Freeze, R. A. & Cherry, J. A. (1979). *Groundwater*. Prentice-Hall, Englewood Cliffs, NJ.

Freyberg, D. L. (1986). A natural gradient experiment on solute transport in sand aquifer. 2. Spatial moments and the advections and dispersion of non-reactive tracers. *Water Resources Research*, **22**(13), 2031–46.

Fuller, W. H., Amoozegar-Fard, A., Niebla, E. E. & Boyle, M. (1980). Influence of leachate quality on soil attenuation of metals. In *Disposal of Hazardous Waste*, Proceedings of the 6th Annual Research Symposium, Chicago, IL. US Environmental Protection Agency, Cincinnati, OH.

Garabedian, S. P., LeBlanc, D. R., Gelhar, L. W. & Celia, M. A. (1991). Large-scale natural gradient tracer test in sand and gravel, Cape Cod, Massachusetts. 2. Analysis of spatial moments for nonreactive tracer. *Water Resources Research*, **27**(5), 911–24.

Ghiorse, W. C. & Wilson, J. T. (1988). Microbial ecology of the terrestrial subsurface. *Advances in Applied Microbiology*, **33**, 107–72.

Gibson, S. A. & Suflita, J. M. (1986). Extrapolation of biodegradation results to groundwater aquifers: Reductive dehalogenation of aromatic compounds. *Applied & Environmental Microbiology*, **52**, 681–8.

Golwer, A., Mattheß, G. & Schneider, W. (1969). Selbstreinigungsvorgänge im aeroben und anaeroben Grundwasserberich. *Vom Wasser*, **36**, 65–92.

Griffin, R. A., Cartwright, K., Shrimp, N. F., Steele, J. D., Ruch, R. R., White, W. A., Hughes, G. M. & Gilkeson, R. H. (1976). *Attenuation of Pollutants in Municipal Landfill Leachate by Clay Minerals. Part 1—Column Leaching and Field Verification*, Environmental Geology Notes No. 78. Illinois State Geological Survey, Urbana, IL.

Griffin, R. A., Frost, R. R., Au, A. K., Robinson, G. D. & Shimp, N. F. (1977). *Attenuation of Pollutants in Municipal Landfill Leachate by Clay Minerals. Part 2—Heavy Metal Adsorption*, Environmental Geology Notes No. 79. Illinois State Geological Survey, Urbana, IL.

Harmsen, K. (1979). Theories of cation adsorption by soil constituents: Discrete-site models. In *Soil Chemistry, B. Physico-chemical Models*, ed. G. H. Bolt, Developments in Soil Science, 5B. Elsevier, Amsterdam, The Netherlands, pp. 77–139.

Hoeks, J., Beker, D. & Borst, R. J. (1979). Soil column experiments with leachate from a waste tip. II. Behaviour of leachate components in soil and groundwater. Instituut voor Cultuurtechniek en Waterhuishouding, Wageningen. The Netherlands.

Hoeks, J., Hoekstra, H. & Ryhiner, A. H. (1984). Soil column experiments with leachate from a waste tip. III. Behaviour of non-acidic, stabilized leachate in soil (in Dutch). Instituut vor Cultuurtechniek en Waterhuishouding, Wageningen, The Netherlands.

Jensen, K. H., Bitsch, K. & Bjerg, P. L. (1992). Large-scale dispersion experiments in a sandy aquifer in Denmark: Observed tracer movements and numerical analysis. Submitted.

Karickhoff, S. W. (1984). Organic pollutant sorption in aquatic system. *Journal of Hydraulic Engineering*, **110**, 707–35.

Kimmel, G. E. & Braids, O. C. (1974). Leachate plumes in a highly permeable aquifer. *Ground Water*, **12**, 388–92.

Kjeldsen, P. (1986). Attenuation of landfill leachate in soil and aquifer material. PhD thesis, Department of Environmental Engineering, Technical University of Denmark, Lyngby, Denmark.

Kjeldsen, P. & Christensen, T. H. (1984). Soil attenuation of acid phase landfill leachate. *Waste Management & Research*, **2**, 247–63.

Kjeldsen, P., Kjølholt, J., Schultz, B., Christensen, T. H. & Tjell, J. C. (1990). Sorption and degradation of chlorophenols, nitrophenols and organophosphorus pesticides in the subsoil under landfills—laboratory studies. *Journal of Contaminant Hydrology*, **6**, 165–84.

Knox, K. & Jones, P. H. (1979). Complexation characteristics of sanitary landfill leachates. *Water Research*, **13**, 839–46.

Larsen, T., Kjeldsen, P. & Christensen, T. H. (1992a). Sorption of hydrophobic hydrocarbons on three aquifer materials in a flow through system. *Chemosphere*, **24**, 439–51.

Larsen, T., Kjeldsen, P. & Christensen, T. H. (1992b). Correlation of benzene, 1,1,1-trichloroethane, and naphthalene distribution coefficients to the characteristics of aquifer materials with low organic carbon content. *Chemosphere*, in press.

Larsen, T., Christensen, T. H., Pfeffer, F. M. & Enfield, C. G. (1992c). Landfill leachate effects on sorption of organic micropollutants onto aquifer materials. *Journal of Contaminant Hydrology*, **9**, 307–24.

Lee, L. S., Rao, P. S. C., Brusseau, M. L. & Ogwada, R. A. (1988). Nonequilibrium sorption of organic contaminants during flow through columns of aquifer materials. *Environmental Toxicology and Chemistry*, **7**, 779–93.

Lesage, S., Jackson, R. E., Priddle, M. W. & Riemann, P. G. (1990). Occurrence and fate of organic solvent residues in anoxic groundwater at the Gloucester Landfill, Canada. *Environmental Science & Technology*, **24**, 559–66.

Loch, J. P. G., Lagas, P. & Haring, B. J. A. M. (1981). Behaviour of heavy metals in soil beneath a landfill; Results of model experiments. *The Science of the Total Environment*, **21**, 203–13.

Lun, X. Z. & Christensen, T. H. (1989). Cadmium complexation by solid waste leachates. *Water Research*, **23**, 81–4.

Lyngkilde, J. & Christensen, T. H. (1992a). Redox zones of a landfill leachate pollution plume (Vejen, Denmark). *Journal of Contaminant Hydrology*, in press.

Lyngkilde, J. & Christensen, T. H. (1992b). Fate of organic contaminants in the redox zones of a landfill leachate pollution plume (Vejen, Denmark). *Journal of Contaminant Hydrology*, in press.

Lyngkilde, J., Christensen, T. H., Gillham, R., Larsen, T., Kjeldsen, P., Skov, B., Foverskov, A. & O'Hannesin, S. (1992). Degradation of specific organic compounds in leachate polluted groundwater. This book, Chapter 4.3.

MacFarlane, D. S., Cherry, J. A., Gillham, R. W. & Sudicky, E. A. (1983). Migration of contaminants in groundwater at a landfill: A case study. 1. Groundwater flow and plume delineation. *Journal of Hydrology*, **63**, 1–29.

Nicholson, R. V., Cherry, J. A. & Reardon, E. J. (1983). Migration of contaminants in groundwater at a landfill: A case study. 6. Hydrogeochemistry. *Journal of Hydrology*, **63**, 131–76.

Reinhard, M. & Goodman, N. L. (1984). Occurrence and distribution of organic chemicals in two landfill leachate plumes. *Environmental Science & Technology*, **18**, 953–61.

Rowe, R. K. (1992). Diffusive transport of pollutants through clay liners. In *Landfilling of Waste: Lining and Leachate Collection*, ed. T. H. Christensen, R. Cossu, & R. Stegmann. Elsevier, London, UK, in press.

Schwarzenbach, R. P. & Westall, J. (1981). Transport of nonpolar organic compounds from surface water to groundwater. Laboratory sorption studies. *Environmental Science & Technology*, **15**, 1360–7.

Smolenski, W. J. & Suflita, J. M. (1987). Biodegradation of cresol isomers in anoxic aquifers. *Applied and Environmental Microbiology*, **53**, 710–16.

Soyupak, S. (1979). Modifications to sanitary landfill leachate organic matter migration through soil. PhD thesis, University of Waterloo, Ontario, Canada.

Weis, M., Abbt-Barun, G. & Frimmel, F. H. (1989). Humic-like substances from landfill leachates—characterization and comparison with terrestrial and aquatic humic substances. *The Science of the Total Environment*, **81/82**, 343–52.

Wilson, B. H., Smith, G. B. & Rees, J. F. (1986). Biotransformations of selected alkylbenzenes and halogenated aliphatic hydrocarbons in methanogenic aquifer material: A microcosm study. *Environmental Science & Technology*, **20**, 997–1002.

4.3 Degradation of Specific Organic Compounds in Leachate-Polluted Groundwater

JOHN LYNGKILDE,[a] THOMAS H. CHRISTENSEN,[a] ROBERT GILLHAM,[b] THOMAS LARSEN,[a] PETER KJELDSEN,[a] BENT SKOV,[a] ANJA FOVERSKOV[a] & STEPHANIE O'HANNESIN[b]

[a]Department of Environmental Engineering, Technical University of Denmark, Building 115, DK-2800 Lyngby, Denmark
[b]Waterloo Centre for Groundwater Research, University of Waterloo, Waterloo, Ontario, Canada N2L 3G1

INTRODUCTION

Based on typical concentrations in landfill leachates and the extremely low concentrations accepted for meeting drinking water quality standards, specific organic compounds seem to constitute the major risk for pollution of groundwater by landfill leachate.

The attenuation mechanisms in aquifers for specific organic compounds are, besides dilution, primarily sorption and degradation. The sorption process retards the migration of the specific organics, but in aquifers, often containing very little solid organic carbon being the major sorbent, the retardation is only of significant, long-term importance for very hydrophobic compounds (e.g. having log $K_{ow} > 5$). However, many of the specific compounds in landfills have low log K_{ow}-values and hence constitute a migration risk unless they are degraded in the aquifer environment. This makes degradation a key issue in evaluation of the groundwater pollution risk at landfills, in determining appropriate remedial action schemes if a leachate release has occurred, and in consideration of organic compounds as potential groundwater quality control monitoring parameters.

Degradation of xenobiotic organic compounds has gained much research focus during the last decade and many compounds have proved to be degradable under optimal conditions, primarily being governed by

TABLE 1. Evaluation of Compound Degradation at 10°C in Batches Simulating Aerobic, Anoxic and Anaerobic Environments with a Primary Substrate at 50 mg COD litre⁻¹ (anaerobic: 200 mg COD litre⁻¹) Originating from Methane Phase Leachate

Compound	Aerobic/oxygen				Anoxic/nitrate				Anaerobic/methane generating			
	Excellent	Good	Medium	No	Excellent	Good	Medium	No	Excellent	Good	Medium	No
Benzene		×						×				×
Toluene		×						×				×
o-Xylene		×						×				×
1,4-Dichlorobenzene			×					×				×
1,2-Dichlorobenzene			×					×				×
Indene	×										(×)	
Nitrobenzene			×				×			×		
Naphthalene	×							×				×
Biphenyl	×							×				×
Fluorene	×							×				×
Dibenzothiophene	×							×				×
Phenanthrene	×					×						×
Phenol	×							×				×
o-Cresol	×						×					×
2,4-Dichlorophenol		×						×				×
2,6-Dichlorophenol			×					×				×
4,6-o-Dichlorocresol			(×)					×				×
o-Nitrophenol		×						×		×		
p-Nitrophenol		×					×				×	
1,1,1,-Trichloroethane				×				×				×
Tetrachloromethane				×	×				×			
Trichloroethylene				×				×				*
Tetrachloroethylene				×				×				×

( ): Abiotic removal. Comparable removal in sterile batch.

* Data missing.

redox conditions, nutrient availability and, of course, time for adaptation and actual metabolic activity. However, relatively little information is available on the degradation of specific organics in a groundwater environment affected by landfill leachate. This environment supposedly contains several redox environments ranging from strict methanogenic conditions just outside the landfill to aerobic conditions in the far outskirts of the plume (cf. Christensen, 1992). The environment is also characterized by the presence of non-specific organic matter leached from the landfill together with the relatively lower concentrations of specific compounds. Typical concentrations range for a municipal landfill receiving only minor quantities of industrial waste for the non-specific compounds from 500 to 3000 mg COD litre^{-1} and for each of the specific compounds from 10 to 1000 μg litre^{-1}. Occasionally, higher concentrations may appear.

Table 1 summarizes the preliminary results on the degradation of several specific compounds in the presence of diluted methane phase leachate (corresponding to 50–200 mg COD litre^{-1}) at aerobic, anoxic (denitrifying) and anaerobic (methane generating) conditions. The experiments were conducted in completely stirred batch reactors as described later (see also Lyngkilde *et al.*, 1988). The results presented in Table 1 indicate that most of the compounds potentially are degradable. However, the results are obtained in the laboratory under very manipulated conditions, and it might be questioned whether these results apply to field conditions.

This chapter describes experiences comparing the degradation of benzene, toluene and xylene (BTX) in laboratory batch reactors and in in-situ testers in a landfill leachate-affected aquifer. The comparison involves both aerobic and anaerobic redox conditions and the BTX have been selected as model compounds because they are very common, mobile constituents of landfill leachate and they show very distinctly different behaviour in the aerobic and the anaerobic environment.

MATERIALS AND METHODS

Batch Experiments

The batch experiments were conducted in 10-litre stirred glass containers equipped with supply and sampling gates, see Fig. 1. Diluted methane phase leachate was supplied to the glass containers together with

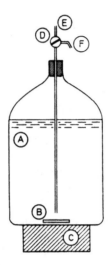

Figure 1. Completely stirred batch reactor. A, glass container for diluted leachate; B, rod for stirring; C, magnetic stirrer; D, three-way valve; E, input of oxygen (air) or nitrogen for sampling; F, sampling gate.

a nutrient/mineral solution and an inoculum of groundwater from the polluted aquifer. The aerobic environment was controlled by introduction of atmospheric air, and measurement of oxygen in the water phase, while the anaerobic environment was controlled by addition of sulphide and measurement of methane generation. The BTX were added to the reactor at time zero in concentrations of the order of 100 μg litre^{-1} each. Samples were obtained at varying time intervals by introducing a slight pressure in the batches by addition of air (aerobic) or nitrogen (anaerobic) and releasing the pressure through the sampling gate. The samples were extracted immediately after sampling.

Sterile controls (2000 mg litre^{-1} sodium azide) were run in parallel.

In-situ Testers

The concepts and techniques of the in-situ tester have been developed by Gillham *et al.* (1990*a,b*). The idea is to establish a batch-like volume of pollutants in a semi-confined sub-compartment of the aquifer imposing a minimum of disturbance on the system.

The tester is shown in Fig. 2. All equipment except the tubing is made of stainless steel, unless not contacting the fluid studied. The tubing is

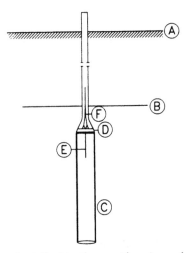

Figure 2. In-situ tester installed in the aquifer. A, surface; B, groundwater table; C, test chamber; D, screen; E; spike; F, Teflon tubing.

made of Teflon. The tester has the shape of a semi-open column and contains approximately 2 litres of the aquifer, corresponding to a pore volume of 0·6–0·7 litre. The top of the tester is supplied with a screen to which fluids can be pumped. The centred spike inside the tester also contains a screen allowing for injection and sampling.

Installation of in-situ tester.   A 10-cm boring was made by a hand auger and a bucket auger. At the desired depth the tester was placed in a well casing, and hammered down 65 cm to fill the tester with aquifer material and groundwater. Afterwards the well casing was withdrawn. Approximately 5 litres was pumped from the screen and 1 litre from the centre spike by means of a peristaltic pump in order to develop the tester.

Two testers were installed at the Vejen Landfill in Jutland, Denmark, one in the strict anaerobic zone approximately 65 m directly downstream from the major leakage area of the landfill and the other in the aerobic zone approximately 140 m from the landfill located in the outskirts of the plume. The typical water quality of the two zones is presented in Table 2. The landfill was established 25 years ago and the leakage of the leachate into the surrounding shallow, well aerated aquifer has taken place for more than 15 years.

TABLE 2. Typical Groundwater Quality of the Anaerobic and the Aerobic Zones of the Leachate-Affected Aquifer (all units mg litre^{-1} except pH.)

Parameter	Anaerobic	Aerobic
pH	5·8	5·2
Oxygen	0	7·9
Nitrate	0	122
Sulphate	92	24
Iron	51	0·9
Manganese	2·8	0·3
Ammonia	1·3	0·2
COD	24	3
Chloride	420	30

Experimental procedures. An empty 4-litre Teflon bag previously flushed with nitrogen was connected to the tester and filled with groundwater withdrawn from the inside of the installed tester by a peristaltic pump. A 100-ml spike of BTX and tritiated water (to act as a tracer for control of dilution in the test period) was added to the bag and mixed with the groundwater. The spiked groundwater was re-entered into the tester. A volume of 4 litres corresponding to 5–6 pore volumes assured that all dissolved organic compounds with retardation factors less than five will be evenly distributed within the tester volume. The BTX has retardation factors in the investigated aquifer of the order of 1·01–1·10 (Larsen *et al.*, 1989). The injected BTX is assumed to equilibrate with the aquifer material within 5 h (see Brusseau *et al.* (1989) on the kinetics of the sorption process for the studied aquifer).

Samples were obtained from the tester at various time intervals by a syringe sampling device as shown on Fig. 3. The initial 20 ml were discarded, and the following 10 ml sealed in the glass vial, conserved by sodium azide and kept cool during transport to the laboratory. Ten ml were used for extraction and 1 ml for scintillation counting. The maximum number of samples to be taken from the tester is 7–8 before dilution with groundwater from outside the tester appears.

Chemical analysis. Analytical determination of BTX is performed by extraction of the water samples with pentane in a ratio of 100:1. Isopropylbenzene is used as internal standard. Three μl of the pentane phase are injected on 30 m J&W DB-5; film thickness 1·5 μm: 0·53 μm

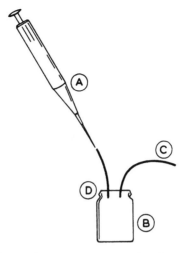

Figure 3. Sampling device for in-situ testers. A, 100-ml plastic syringe; B, 10-ml Hypovial; C, tubing connected to the tester unit; D, stainless steel capsule.

megabore column in a Carlo Erba Mega 500 gas chromatograph equipped with a FID detector.

In the batch experiment, ATP-determination was made as a measure of microbial activity according to the luceferin–luceferase method (Leach, 1981).

RESULTS AND DISCUSSION

The results of the experiments are shown in Fig. 4 for the anaerobic environment and in Fig. 5 for the aerobic environment in terms of dissolved concentrations of BTX (ppb equals μg litre^{-1}) as a function of time (days). The tritium data for the testers showed constant values during the experimental period indicating that any compound losses were not due to dilution by unspiked water entering the in-situ testers.

Figure 4 shows that in the anaerobic environment no degradation of BTX is seen in either batch or in-situ testers, since the concentrations ıre relatively constant over time. The observed fluctuation is by no means unusual for experiments working in this low concentration range.

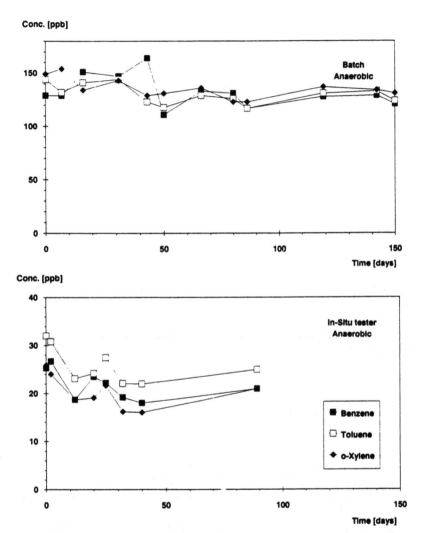

Figure 4. Concentrations of BTX as a function of time in anaerobic batch reactor and in-situ tester, respectively.

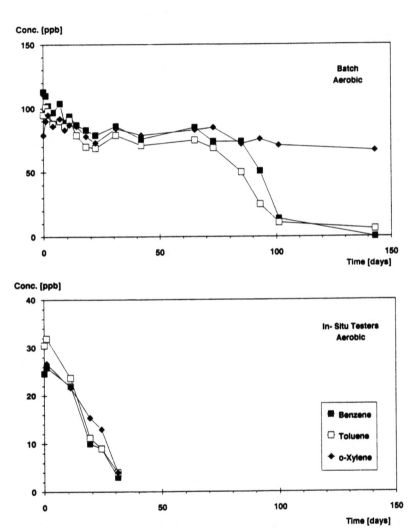

Figure 5. Concentrations of BTX as a function of time in aerobic batch reactor and in-situ tester, respectively.

The lack of BTX degradation observed in the anaerobic batch was not due to lack of microbial activity, since both measured ATP-levels and observed degradation of chlorinated aliphatics in the same batches indicate microbial activity.

Figure 5 shows that benzene and toluene are degraded in aerobic environments simulated both by the laboratory batch and by the in-situ tester. In the batch, a lag phase of the order of three months was observed, but when the degradation of benzene and toluene initiated, solute concentrations decreased from approximately $90 \, \mu g \, litre^{-1}$ to $10 \, \mu g \, litre^{-1}$ within 30 days. In the tester no lag-phase was observed and within 30 days the concentrations decreased from approximately $30 \, \mu g \, litre^{-1}$ to nearly zero.

In the aerobic environment, *o*-xylene apparently does not behave identically in the batch system and in the in-situ tester. In the in-situ tester *o*-xylene is degraded rapidly without the presence of the lag phase, while no degradation seems to take place in the batch within the reported 150-day period. In other completed aerobic batches, *o*-xylene has been degraded after much shorter lag-phases. In some cases, the degradation has ceased at a concentration level of approximately $10 \, \mu g \, litre^{-1}$. After a new lag-phase at this concentration level, complete degradation of *o*-xylene has been observed. This phenomenon has also been observed by Jørgensen & Aamand (1987) and is supposedly related to some co-metabolic effects. Although not conclusive, the observations on *o*-xylene may indicate that the in-situ tester with its large volume of aquifer material constitutes a more diverse seeding than the batch reactor or a less stressed system demanding no adaptation to degrade BTX.

The results from the two experimental methods show that the degradation of BTX depends on the redox environment: no degradation was observed in the anaerobic environment, while a relatively rapid degradation was observed in the aerobic environment. Currently it cannot be concluded whether the observed lag phases in batch experiments are a result of no previous exposure of the microbial consortium to the specific BTX compounds or a result of a general initial stress on the active microbial consortium applied to the batch. Analogously, it cannot be concluded whether the lack of lag-phases in the in-situ experiments is due to a previous exposure to diluted leachate containing these compounds or due to the limited stress imposed on the in-situ microbial consortium. Further research must be conducted to obtain general conclusions.

CONCLUSIONS

Investigation of BTX degradability in landfill leachate-affected environments by completely stirred batch reactors and by in-situ testers installed in an actually polluted aquifer gave overall comparable results: BTX was degraded in aerobic environments but not in anaerobic environments. However, lag phases of at least three months were observed in the batch reactors while very rapid degradation was observed in the aquifer.

The results of this investigation indicate that simple batch experiments may be valuable in evaluation of the biodegradability of specific compounds in landfill leachate-polluted groundwater. Referring to Table 1, this means that the degradation potential is significantly increased when through dilution and re-aeration the plume becomes aerobic.

REFERENCES

Brusseau, M., Larsen, T. & Christensen, T. H. (1991). Rate-limited sorption and nonequilibrium transport of organic chemicals in low organic carbon aquifer materials. *Water Resources Research*, **27**, 1137–45.

Christensen, T. H. (1992). Attenuation of leachate pollutants in groundwater. This book, Chapter 4.2.

Gillham, R. W., Robin, M. J. L. & Ptacak, C. J. (1990a). A device for in situ determination of geochemical transport parameters, 1. Retardation. *Ground Water*, **28**(5), 666–72.

Gillham, R. W., Starr, R. C. & Miller, D. J. (1990b). A device for in situ determination of geochemical transport parameters 2. Biochemical reactions. *Ground Water*, **28**(6), 858–62.

Jørgensen, C. & Aamand, J. (1987). Co-metabolic transformation of *o*-xylene in groundwater. Presented at 8th International Symposium on Environmental Biogeochemistry, 14–18 Sept., Nancy, France.

Larsen, T., Kjeldsen, P., Christensen, T. H., Skov, B. & Refstrup, M. (1989). Sorption of specific organics in low concentrations on aquifer materials of low organic carbon content: Laboratory experiments. In: *Contaminant Transport in Groundwater*, ed. H. E. Kobus & W. Kinzelbach. A. A. Balkma, Rotterdam, pp. 133–40.

Leach, F. R. (1981). ATP determination of Firefly luciferase. *J. Applied Biochemistry*, **3**, 473–517.

Lyngkilde, J., Tjell, J. C. & Foverskov, A. (1988). The degradation of specific organic compounds with landfill leachate as a primary substrate. In *Contaminated Soil '88*, ed. K. Wolf, W. J. van den Brink & F. J. Calon. Kluwer Academic Publishers, Hingham, MA, pp. 91–100.

4.4 Groundwater Control Monitoring at Sanitary Landfills

THOMAS H. CHRISTENSEN, PETER KJELDSEN

Department of Environmental Engineering, Technical University of Denmark, Building 115, DK-2800 Lyngby, Denmark

&

JES LA COUR JANSEN

Water Quality Institute, Agern Allé 11, DK-2970 Hørsholm, Denmark

INTRODUCTION

At most of the modern sanitary landfills extensive and costly measures are taken in order to prevent pollution of groundwater. To document the performance of these measures groundwater control monitoring programmes are often established. These programmes serve both the landfill company and the public. In fact such programmes are often demanded by the authorities when approving or licensing a landfill.

Groundwater quality has been measured in many countries for several years in connection with water supplies based on groundwater abstraction. However, it should be realized that control monitoring of the groundwater quality in the vicinity of a landfill is a somewhat different task.

First of all, it must be realized that the monitoring programme is aimed at detecting the unexpected, i.e. a release of leachate into the aquifer in spite of the substantial measures taken to avoid this (top covers, liner systems, drainage systems). This means that a monitoring programme should, with a high probability, detect the unlikely event. In addition to this the unlikely event is very uncertain: it is not known where the leakage point will appear, the exact migration pathway of the leachate plume and, obviously, when the release of leachate into the aquifer might take place. This indicates that a simple, reliable monitoring system is not easy to establish.

Secondly, it must be realized that although the landfill leachate composition changes over the years (Ehrig, 1989) elevated concentrations can be expected for many of the parameters for a couple of hundred years (Belevi & Baccini, 1992). This means that, although the annual costs of the monitoring programme may be low when compared with the operational costs of the landfilling, the groundwater monitoring cost is going to continue long after the closure of the landfilling. This indicates that the annual costs of the monitoring must be evaluated thoroughly when designing the programme and when specifying its degree of reliability.

Thirdly, it must be realized that the purpose of the monitoring programme is not to detect an unacceptable decrease in groundwater quality, but to provide an early warning—if the unexpected pollution of the groundwater should happen—allowing sufficient time to establish remedial actions before significant water supply wells or surface recipients such as rivers and lakes are damaged. This means that the monitoring programme may make use of indicators which as such do not necessarily characterize the quality of the groundwater, but indicate that something may have changed. Then further action can be taken to investigate the character and extent of the change. If the alarm is due to a substantial release of leachate into the aquifer, a first phase remedial action plan specifying responsibilities and action steps to be taken should be enforced immediately in order to limit the damage. Such a first phase remedial action plan should be developed in connection with the design of the groundwater monitoring programme, because limited possibilities of remedy with respect to time may demand very early alarms by the monitoring programme.

The design of a groundwater monitoring programme will depend on local conditions such as hydrogeology, sensitivity and importance of groundwater resources, remedial action possibilities and the like. General guidelines are therefore difficult to provide, but some features such as selection of monitoring principles, programme phases, location of wells, selection of indicator parameters and control charts for the collected data contain general elements and are discussed in the following sections based on investigations and experiments of the last five to ten years in Denmark.

MONITORING PRINCIPLES

Indirect Methods Versus Groundwater Sampling

New approaches are currently emerging on monitoring landfill confinement and liner integrity by geoelectrical principles (e.g. Gervasoni & Piepoli, 1989). Geoelectrical resistivity measurements in the strata outside the landfill may also potentially be used for determining leachate pollution of groundwater. So far, this technique has been a successful tool in some cases in delineating contaminant plumes. Fibre optic sensors may also one day prove a useful tool in groundwater control monitoring programmes. However, as a routine technique administrated at the local level at a large number of landfills, only sampling of groundwater wells followed by chemical analysis in the laboratory appears currently to be recommendable. All the elements of a control programme based on groundwater samples are proven technology and only a rational and site-specific combination of these elements is needed to design a monitoring programme.

References for Comparison

Basing the control programme on groundwater samples means that the collected data must be compared to a reference to identify whether the data deviate from the normal situation and hence indicate that the groundwater may be affected by the landfill.

The use of upstream wells sampled in connection with the control monitoring of the downstream wells has often been proposed. However, a close examination in terms of average values and standard deviations of selected parameters of uncontaminated control wells at seven Danish landfills has shown that in many cases even closely situated wells may show significant differences with respect to average and standard deviation. A control procedure based on comparing the water quality in an upstream and a downstream well may then detect only very large releases of leachate and then be of limited value. Figure 1 summarizes the data for the seven landfills (A–G) for specific conductivity, chloride, chemical oxygen demand (COD) and ammonia. The data presented cover between 8 and 27 samples for each well over a period of 4–6 years. As a consequence of this data analysis, comparison of monitoring well

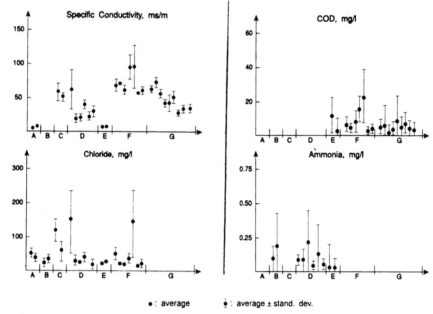

• : average   ⬩ : average ± stand. dev.

Figure 1. Summary of averages and standard deviations of unpolluted control wells at seven landfills in Denmark (A–G) based on 8–27 samples covering 4–6 years. From Christensen *et al.* (1985).

water quality with previous data from the same well must be recommended as the basic monitoring principle. However, this does not abandon all monitoring of upstream wells. Use of one or two upstream wells is recommended as a general check on any trends or variation in groundwater quality caused by activities other than the landfill in question.

Variations and Statistical Evaluation

Data on the water quality of an individual monitoring well will over a period of time show variation as indicated in Fig. 1. The unpolluted background, the reference, is thus characterized by a level and a variation typically expressed through the average and the standard deviation. To determine which statistical distribution describes the groundwater quality of shallow monitoring wells a substantial amount of data (more than 30) is required. Normal (Gaussian) distributions and normal distributions of the logarithm of the observations are usually proposed. The analysis of the data summarized in Fig. 1 did not indicate any advantage in applying the log-normal distribution, thus a normal distribution was used for the further development of the monitoring programme.

With long time series of data available from each well more advanced statistical tests for the determination of statistical distribution and for test of trends in averages, variations and cyclic fluctuations can be performed. However, such data are never available in practice when a monitoring programme is to be designed and, in practice, sufficient time is not available to establish such a background data set. Frequent sampling can of course be enforced but the sampling must cover a long period of time to be representative for the variations that need to be determined. For a shallow groundwater well, this period is supposedly more than two years. So the designer of the monitoring programme is faced with the problem of determining when a sample shows characteristics that deviate from a very uncertain reference level. This calls for application of simple statistical control charts making the most of the available data and for a phasing of the monitoring programme over time allowing for adjustment of control rules as the database is improved.

Since the monitoring programme aims at an early warning for relatively small leachate releases into groundwater it must be accepted that the programme may signal false alarms, i.e. taking a slightly increased value as an indication of leachate pollution although the high value was due only to statistical variation. If the risk of false alarms must be kept low only very significant leachate releases will be observed at a late time, which is not what is wanted. However, starting extensive investigations of the leachate release based on a false alarm is not very attractive and a system must be established that can identify any false alarms which inevitably will appear one day.

PHASES OF A MONITORING PROGRAMME

Phasing of a monitoring programme seems appropriate both with respect to time and the alarm situation. This is discussed below.

Time Phasing

Immediately after the decision to establish the landfill and the attainment of basic hydrogeological data for the area, the future groundwater control monitoring wells should be installed to allow for time to determine the reference quality. During the landfill construction period or at least for a one-year-long period, 10–12 samplings are made in all control wells to determine their individual average and standard deviation. On this basis the first set of control charts is developed, as discussed later. Thereafter, the monitoring programme enters a routine phase prescribing 2–6 samplings per year depending on local conditions. After sampling by routine for 2–5 years, and assuming that no unexpected results have appeared, the control charts are revised based on the total database now containing approximately 25 samples. With this amount of data available, fair estimates of average and standard deviations can be obtained and the uncertainty of the control system diminished significantly. After another 10 years it may be beneficial to go back and re-evaluate the control programme, in particular checking the assumptions as to frequency distribution and slowly developing trends. A time phasing of a monitoring programme is illustrated in Fig. 2.

Action Phasing

Since extensive measures are taken to limit the risk of groundwater pollution at the landfill, the groundwater control monitoring programme usually will be in its routine phase, where data are collected as specified and plotted on control charts, indicating that the situation is in control. However, once in a while, and maybe due to a release of leachate, one of

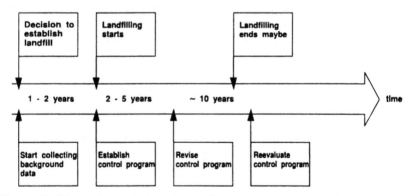

Figure 2. Illustration of the time phases of a groundwater control monitoring programme; the time estimates presented are given for illustration purposes only.

the control parameters will signal an alarm. In this situation it is important to have precise instructions on what to do. Since it cannot be excluded that the alarm is due to an error (erroneous alarm) or simply is a statistical phenomenon (false alarm) it is recommended to immediately conduct a supplementary control sampling of all wells enforcing strict quality control on all steps in the monitoring. This phase is called the check phase. If the supplementary data set does not release an alarm, the situation is not likely to be out of control. However, before returning to the normal routine phase it may be worthwhile to consider whether the previous data set was affected by errors in terms of changes in sampling or analytical procedures. If modification of procedures was the cause, which is not unlikely considering the length of time involved, this should be revised and additional precautions taken.

On the other hand, if the new data set supports the previous alarm-releasing data set, the situation is out of control as defined by the

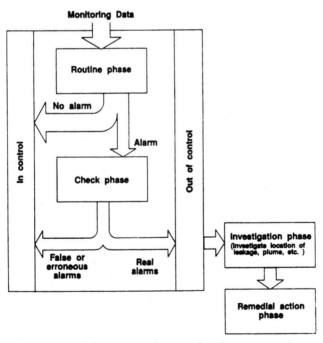

Figure 3. Illustration of the action phases related to a groundwater control monitoring programme.

control monitoring system and the first steps of an investigation phase must be taken to identify the location of the leakage, the extent of the leachate plume and its major pollutants. The elements of such an investigation phase are highly dependent on local conditions and are outside the scope of this discussion. If the contamination is severe, this investigation may lead to remedial actions.

The action phases of a groundwater control monitoring programme are illustrated in Fig. 3.

MONITORING WELLS

The number and location of monitoring wells must be determined as a compromise between requested level of probability for detecting a leachate leakage into the aquifer and the cost of installing the wells and operating the control monitoring programme in the future.

The monitoring wells should, of course, be located so that they possess the largest possibility of detecting the released leachate. However, this is no simple task due to the heterogeneity of the surrounding geological strata and the uncertainty about the locations and extent of the leakage. Local variations in hydraulic conductivity over the vertical profile, the potential density effect associated with the leachate, and maybe seasonal variations in hydraulic potential pattern, make it nearly impossible to predict the migration pathway of a leachate leakage over distances of 30–200 m which supposedly is the distance most likely to be considered for monitoring purposes.

The transverse dispersion, the spreading of the plume perpendicular to the flow direction, may increase the width of the plume and hence potentially increase the detection probability of the wells that are located further away from the landfill. However, very few experimental field data exist on the transverse dispersivity and an overestimation leads to a spacing of monitoring wells that, in practice, allows a narrow leachate plume to escape in between two wells located close to the fill. On the other hand an underestimation, i.e. a much larger dilution takes place than expected, may have the consequence that monitoring wells located far away from the landfill can detect only very large releases of leachate. It should be realized that the wider the plume becomes, the more diluted the samples will become and the more groundwater will be polluted before the monitoring well can detect the presence of the leachate.

The two factors to be determined in respect of the localization of monitoring wells are the downstream distance from the landfill and the spacing between the wells. In some cases, legal constraints and access to neighbouring land force the monitoring wells to be very close to the landfill and increase the number of monitoring wells in order to maintain a fair probability of detecting a leakage.

Theoretically the probability of detecting a leachate release may be calculated for specific conditions of hydrogeology, leachate composition, leachate release, etc., assuming homogeneous isotropic geological strata. But as previously mentioned such conditions hardly ever exist and information on spatially distributed parameters at the small scale in question here is an impractical task. An extreme case is karst terrains where fracture transport dominates and in practice is impossible to predict, unless site-specific tracer experiments are conducted. However, it is important to realize that the location of the monitoring wells governs the probability of detecting a leachate release and that the requirement that even a small leachate release should be detected with a high probability significantly increases the number of monitoring wells to be installed. In each case these aspects must be evaluated as closely as possible. As an indication, it is mentioned that evaluation of these aspects in Denmark (sandy alluvial aquifers, sandy moraines) has indicated that the monitoring wells should be located at distances of 50–100 m from the border of the landfill and that the spacing should not exceed half of that distance.

The wells are made by various techniques, and unless they perforate aquitards, screens covering the full depth in question are usually preferred. If leachate contamination is observed, the actual location of the plume in the vertical profile must be made to improve prospective remedial actions.

SELECTION OF MONITORING PARAMETERS

Selection of proper monitoring parameters is a key aspect in establishing an efficient groundwater control monitoring programme. The parameters to be chosen are those revealing a significantly increased value—as compared to the background—as early as possible after a leachate release. This could lead to the use of xenobiotic compounds—not naturally present in the groundwater—as monitoring parameters. Those to be considered are isotopes purposely added to the landfill or specific organic compounds disposed of at the landfill. But adding isotopes—to be used as leakage indicators—is usually not accepted by the health authority and most of the mobile specific compounds may be subject to degradation and hence lose their significance as indicators.

The use of inorganic ions as indicators seems to be recommendable because of their stable nature and the lack of problems with respect to sampling and analysis.

Monitoring Well Response to Pollution

Three types of theoretical response curves for parameters determined in water samples from a monitoring well subject to leachate contamination are presented in Fig. 4. The concentration measured is normalized by the average concentration of the uncontaminated samples (y-axis) while the time scale is normalized by the distance to the well divided by the average water pore velocity (x-axis). The latter means that the leachate water, on average, reaches the monitoring well at time 1. Example (a) represents a conservative parameter, the background concentration of which is subject to significant uncertainty. The increase in the parameter follows an S-shaped curve due to the longidutinal dispersion. The maximum value reached ($C_{max}/C_{background}$) depends on the concentration in the leachate, the concentration in the uncontaminated groundwater and the dilution caused by transverse dispersion (horizontal and vertical).

If the leachate concentration is subject to a long-term decreasing trend, the response curve will also show a decreasing trend after the breakthrough. Due to the considerable uncertainty of the background value a significantly affected sample will not be detected until a major part of the breakthrough has occurred. In Example (b) the uncertainty of the background value is less, but the parameter is delayed relative to the water because of sorption onto the aquifer material and a significantly affected sample is detected at a late stage. In Example (c) the parameter is also subject to sorption processes, but the contrast between concentrations in leachate and unaffected groundwater is so large that the first minute part of the breakthrough results in a significantly affected sample even before the leachate water on average reaches the monitoring well.

The most efficient indicator parameter is the one that detects the earliest significantly affected sample. The time for the first significantly affected sample can be calculated for each considered parameter by

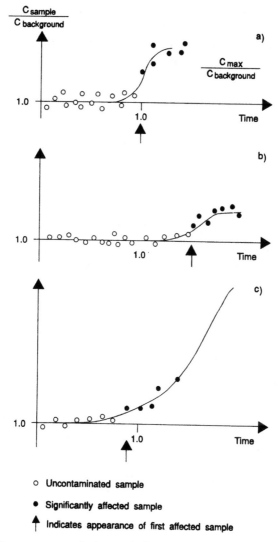

Figure 4. Three types of theoretical response curves for parameters determined in water samples from a monitoring well subject to leachate contamination.

combining a simple statistical t-test with the solution of the one-dimensional solute transport equation (convective–dispersive–adsorptive). Christensen & Kjeldsen (1984) showed that the mathematical expression can be approximated by

$$2CV = TD \frac{C_L - C_{bg}}{C_{bg}} \left[0.5 \, \mathrm{erfc} \left(\left(\frac{P}{4T_a(1+R)} \right)^{0.5} . (1 + R - T_a) \right) \right.$$

$$\left. + 0.5 \exp(P) . \mathrm{erfc} \left(\left(\frac{P}{4T_a(1+R)} \right)^{0.5} . (1 + R + T_a) \right) \right]$$

where

   CV = Coefficient of variation for the background
          water quality (stand. dev./average)
   C_L = Concentration in leachate
   C_{bg} = Background concentration in groundwater
   TD = Transverse dilution (dimensionless)
   R = Distribution ratio (amount of compound adsorbed divided by
          amount of compound in solution per volume of aquifer)
          (dimensionless)
   P = Pecklet's Number = $L\alpha^{-1}$ (dimensionless)
   L = Distance from pollution source to monitoring well (m)
   α = Dispersivity (m)
   T_a = Dimensionless time for alarms (actual time of sample multiplied
          by water pore velocity and divided by L). T_a is the parameter
          for which to solve.

The equation shows that the smaller the uncertainty of the background level (CV), the larger the contrast between leachate and groundwater concentrations ($C_L - C_{bg}$), and the less retarded (R) the compound is, the better the parameter will be as an indicator of leachate polluting the groundwater.

The problem in the application of the equation is the lack of site-specific parameters, in particular the distribution ratios (R) and the leachate concentrations to expect (C_e). The traverse dilution is also associated with significant uncertainty, but this number is a physical parameter being identical for all the parameters of the same well and does, as such, not influence their mutual ranging. However, a high value of *TD* may result in unacceptably long times for an alarm. In such cases the leaking and the monitoring wells must be reconsidered.

The equation is able to handle those parameters that are being retarded due to sorption and ion exchange, but does not reveal that during ion exchange the cationic basic ions occupying the exchange sites are being expelled and will show increased solute concentrations as the leachate cations are entering the cation exchange process (Kjeldsen & Christensen, 1984; Christensen *et al.*, 1989). Thus the basic ions—most often calcium in northern European soils and aquifers—also appear to be a proper indicator parameter although not determined by the model.

The validity of the model for parameter selection is *per se* difficult to prove but application of the model to two actual cases of leachate release (a landfill and an incinerator slag and flyash disposal site) support the model (Christensen & Kjeldsen, 1984; Kjeldsen *et al.*, 1984). Table 1 shows the predicted ranking and the observed ranking of the indicator parameters. For this landfill chloride, ammonia, sodium and manganese proved to be the most efficient indicator parameters. It should be noted that in the actual case (four samplings per year), about two years elapsed between when the first alarm appeared for the parameters ranked A and for the parameters ranked C (see control charts section). In this context two years is a long time, if the spread of the leachate is to be minimized.

It should be emphasized that 3-5 indicator parameters should be selected for the monitoring programme since the uncertainty is diminished when more than one parameter is determined and the selection of more parameters affects the cost only slightly. The selected parameters will be site-specific but experience indicates that chloride, sodium, chemical oxygen demand, specific conductivity, ammonia and calcium are relatively often among the selected parameters.

CONTROL CHARTS

While sampling frequency, monitoring parameters to be used, and the like are obvious items to specify for a monitoring programme, the methods for evaluation of the control data obtained are often not specified. However, to obtain maximum benefit from a monitoring programme, the monitoring data must be evaluated immediately after receipt. After each sampling (when the data are available) it must be determined whether the programme is 'in control' of 'out of control'

TABLE 1. Predicted and Observed Efficiency Ranking of Monitoring Parameters at an Actual Landfill Polluting the Groundwater (after Christensen & Kjeldsen, 1984)

Parameter	Predicted efficiency[a]						Observed efficiency[a] (based on actual observations)
	Leachate (C_L)	Ground water (C_{bg})	(CV)	Distribution ratio (R)	T_{alarm}	Ranking	
Specific conductivity	300 mS/m[b]	20 mS/m	0·24	0·19	0·49	B	d
Total dissolved solids	2 100 mg/litre[b]	460 mg/litre	0·47	0·23	1·32	C	C
COD	2 160 mg/litre	45 mg/litre	0·85	0·20	0·50	B	B
Chloride	420 mg/litre	22 mg/litre	0·32	0·00	0·42	A	A
Sulphate	13 mg/litre[b]	50 mg/litre	0·35	0·00[b]	—	c	D
Ammonia	210 mg/litre	0·08 mg/litre	0·88	1·22	0·44	A	A
Sodium	210 mg/litre	10 mg/litre	0·26	0·25	0·49	B	A
Potassium	40 mg/litre	4 mg/litre	0·26	0·15	1·12	C	C
Calcium	120 mg/litre[b]	13 mg/litre	0·28	0·15	0·60	B	B
Magnesium	42 mg/litre	5 mg/litre	0·37	0·79	1·08	C	C
Manganese	12 mg/litre	0·15 mg/litre	0·53	0·00	0·33	A	A
Iron	20 mg/litre[b]	9 mg/litre	1·24	8·10	—	c	d

[a] By the lettering A (excellent), B (good), C (bad) and D (never).
[b] Estimated value.
[c] Is never increased significantly.
[d] Not measured.

(Fig. 3). This has led to the selection of simple, easy to perform, statistical tests based on control charts. The procedure proposed in Denmark (Christensen *et al.*, 1985) demands that for each monitoring parameter of each monitoring well the background water quality is characterized by an average (X) and a standard deviation (s). Based on this a plot is developed showing concentrations as a function of time and the determined control limits of $X + s$, $X + 2s$ and $X + 3s$. The corresponding control criteria are that one observation above $X + 3s$, two consecutive observations above $X + 2s$ or five consecutive observations above $X + s$ constitute an alarm.

The routine evaluation of a newly received data set consists of plotting one new observation on all the control charts (number of monitoring parameters multiplied by number of monitoring wells) and observing if any of the charts violate the control criterion.

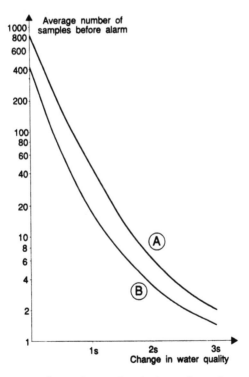

Figure 5. Average number of samples before alarm based on control charts, with (A) one upper control limit $(X + 3s)$ and with (B) three control limits.

The selected control chart is able to display a sudden significant change in water quality (one observation exceeding the upper control limit) and a slowly developing, moderate change in water quality (five consecutive observations exceeding the lower control limit). Assuming that the quality parameters are normally distributed, the average number of observations needed to obtain an alarm for a proven change in water quality level can be estimated. Figure 5 shows the average number of observations needed to obtain an alarm for a change in water quality expressed in numbers of the standard deviations. The upper curve of Fig. 5 is based on a control chart having only the upper criterion (one observation above $X + 3s$) while the lower curve is based on the introduced criterion involving three control limits. It is evident that the latter criterion is much more sensitive, but also the number of 'false alarms' (corresponding to a quality change of zero) will increase. The latter aspect emphasizes the importance of the previously introduced 'check phase' for the identification of false or erroneous alarms.

Figure 6 shows an example of a control chart for specific conductivity in a monitoring well at a newly established landfill. In this case, 21

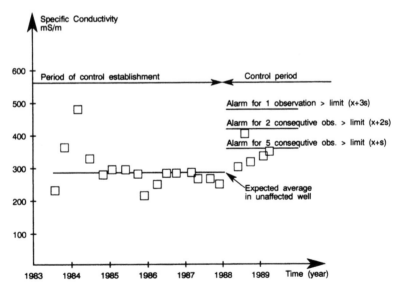

Figure 6. An example of a control chart for a specific well for specific conductivity.

observations were available when the control chart was developed. Of course, these few observations exhibiting significant variation makes in itself an uncertain control chart. It must be recommended that the control chart is re-evaluated after another 12 observations or so.

CONCLUSION

Establishment and operation of groundwater quality monitoring programmes at landfills must be properly engineered to obtain a cost-effective system that is able to yield an early alarm when a leachate release takes place. A monitoring programme must be designed paying proper attention to local conditions and long-term costs involved. In this chapter are discussed some of the major aspects involved: control principles, phases of a monitoring programme, monitoring wells, monitoring parameters and the evaluation of monitoring data.

Experience on an actual monitoring programme, in particular when a leachate release is detected, must be shared in order to improve the control monitoring at landfills.

REFERENCES

Belevi, H. & Baccini, P. (1992). Long-term leachate emissions from municipal solid waste landfills. This book (Chapter 4.1).

Christensen, T. H. & Kjeldsen, P. (1984). A rationale for selecting chemical parameters for control monitoring of the ground water quality at sanitary landfills. Proceedings, ISWA 1984. International Solid Wastes and Public Cleansing Association Congress, Philadelphia, PA, 15–20 September, 17 pp.

Christensen, T. H., Kjeldsen, P., Christensen, S., Hjelmar, O., Jansen, J. la Cour, Kirkegaard, C., Madsen, B. & Olsen, N. (1985). *Groundwater Control Monitoring at Sanitary Landfills*. Polyteknisk Forlag, Lyngby, Denmark (in Danish).

Christensen, T. H., Kjeldsen, P., Lyngkilde, J. & Tjell, J. C. (1989). Behaviour of leachate pollutants in groundwater. In *Sanitary Landfilling: Process, Technology and Environmental Impacts*, ed. T. H. Christensen, R. Cossu & R. Stegmann. Academic Press, London, Chapter 6.7.

Ehrig, H. J. (1989). Leachate Quality. In *Sanitary Landfilling: Process, Technology and Environmental Impacts*, ed. T. H. Christensen, R. Cossu & R. Stegmann. Academic Press, London, Chapter 4.2.

Gervasoni, S. & Piepoli, A. (1989). Monitoring systems of landfills. Presented at 2nd International Sanitary Landfill Symposium, Porte Conte, Italy, Oct. 1989.

Kjeldsen, P. & Christensen, T. H. (1984). Soil attenuation of acid landfill leachate. *Waste Management and Research*, **2**, 247–63.

Kjeldsen, P., Christensen, T. H. & Hjelmar, O. (1984). Selection of parameters for groundwater quality monitoring at waste incinerator residue disposal sites. *Environmental Technology Letters*, **5**, 333–44.

Index